Fadi Dohnal

Damping of Mechanical Vibrations by Parametric Excitation

Fadi Dohnal

Damping of Mechanical Vibrations by Parametric Excitation

Parametric Resonance and Anti-resonance

Südwestdeutscher Verlag für Hochschulschriften

Impressum/Imprint (nur für Deutschland/ only for Germany)
Bibliografische Information der Deutschen Nationalbibliothek: Die Deutsche Nationalbibliothek verzeichnet diese Publikation in der Deutschen Nationalbibliografie; detaillierte bibliografische Daten sind im Internet über http://dnb.d-nb.de abrufbar.
Alle in diesem Buch genannten Marken und Produktnamen unterliegen warenzeichen-, marken- oder patentrechtlichem Schutz bzw. sind Warenzeichen oder eingetragene Warenzeichen der jeweiligen Inhaber. Die Wiedergabe von Marken, Produktnamen, Gebrauchsnamen, Handelsnamen, Warenbezeichnungen u.s.w. in diesem Werk berechtigt auch ohne besondere Kennzeichnung nicht zu der Annahme, dass solche Namen im Sinne der Warenzeichen- und Markenschutzgesetzgebung als frei zu betrachten wären und daher von jedermann benutzt werden dürften.

Verlag: Südwestdeutscher Verlag für Hochschulschriften Aktiengesellschaft & Co. KG
Dudweiler Landstr. 99, 66123 Saarbrücken, Deutschland
Telefon +49 681 37 20 271-1, Telefax +49 681 37 20 271-0, Email: info@svh-verlag.de
Zugl.: Wien, TU, Diss., 2005

Herstellung in Deutschland:
Schaltungsdienst Lange o.H.G., Berlin
Books on Demand GmbH, Norderstedt
Reha GmbH, Saarbrücken
Amazon Distribution GmbH, Leipzig
ISBN: 978-3-8381-0343-3

Imprint (only for USA, GB)
Bibliographic information published by the Deutsche Nationalbibliothek: The Deutsche Nationalbibliothek lists this publication in the Deutsche Nationalbibliografie; detailed bibliographic data are available in the Internet at http://dnb.d-nb.de.
Any brand names and product names mentioned in this book are subject to trademark, brand or patent protection and are trademarks or registered trademarks of their respective holders. The use of brand names, product names, common names, trade names, product descriptions etc. even without a particular marking in this works is in no way to be construed to mean that such names may be regarded as unrestricted in respect of trademark and brand protection legislation and could thus be used by anyone.

Publisher:
Südwestdeutscher Verlag für Hochschulschriften Aktiengesellschaft & Co. KG
Dudweiler Landstr. 99, 66123 Saarbrücken, Germany
Phone +49 681 37 20 271-1, Fax +49 681 37 20 271-0, Email: info@svh-verlag.de

Copyright © 2009 by the author and Südwestdeutscher Verlag für Hochschulschriften Aktiengesellschaft & Co. KG and licensors
All rights reserved. Saarbrücken 2009

Printed in the U.S.A.
Printed in the U.K. by (see last page)
ISBN: 978-3-8381-0343-3

Abstract

The goal of this work is to investigate the method of damping vibrations by means of parametric excitation in mechanical systems with and without self-excitation. From previous studies it is known that a time-periodic stiffness coefficient may introduce additionally damping into a self-excited system if the parametric excitation frequency is near to a parametric combination frequency of the undamped system – a so-called parametric anti-resonance frequency. These works represent the starting point of this work. This investigation extends the proposed method to different types of time-periodic coefficients and develops a deeper insight into the method of vibration suppression.

This work examines mechanical systems with the most general linear equations of motion with simultaneously varying time-periodic stiffness, damping and inertia coefficients. To achieve damping by parametric excitation, at least two degrees of freedom are necessary. Such systems may be lumped mass systems or may represent two modes of a continuous structure. Based on a perturbation of first order, a thorough stability analysis for the proposed method of damping by parametric excitation is carried out for generic systems with two degrees of freedom. The analytically derived formulae represent a reliable tool to describe the additional stability domain created by parametric excitation near an anti-resonance frequency as well as the instability domain caused by parametric excitation near resonance frequencies. This study can be used as a comprehensive guide for designing a device for vibration suppression by parametric excitation.

Summarizing the results, it is shown that time-periodic stiffness coefficients cannot create additional damping in a system, but amplify the existent damping of the system. In particular, it is revealed that damping by parametric excitation can be applied even to systems without self-excitation. Special attention is paid to symmetry properties of the parametric excitation matrices and their influence on the location of the parametric anti-resonance frequencies. The interaction between time-periodic stiffness and damping coefficients is investigated and the effect on the performance of vibration suppression is studied. It is revealed that a time-harmonic inertia variation is equivalent to a simultaneous stiffness and damping variation and that it is capable of being used as a tool for damping vibrations by parametric excitation as well.

By examining the energy flows, it is shown that by activating a parametric excitation energy is transferred into the system, which leads to a more effective dissipation of kinetic energy. The coupling of eigenvalues of the homogenous equation and the maximum possible damping achievable by introducing parametric excitation in a vibrating system are explored. An extension to systems with more degrees of freedom shows that under certain conditions a parametric excitation leads to a coupling of just two modes, while the remaining modes are not affected. The important consequence is that a parametric excitation with a single frequency can only stabilize one unstable mode.

An optimization is performed for a system where all stiffness coefficients are varied simultaneously, but with different amplitudes and phase angles. For a restricted overall parametric excitation, the optimum is found to be a configuration with just one maximum amplitude. The effect of additional constraints on the design of the amplitudes is investigated, too. Furthermore, for a system with a single time-periodic stiffness coefficient the optimum function shape in case of a restricted amplitude of the parametric excitation is found to be the simple rectangular shape, which corresponds to a bang-bang open-loop control. It is shown that only by adjusting the shape of the parametric excitation with fixed maximum amplitude the effective amplitude of the parametric excitation is enlarged significantly and leads to faster vibration suppression.

The results demonstrate that parametric excitation can be employed to extend significantly the area of stability in the parameter space of the system parameters. In comparison to a closed-loop control, the great advantage of the proposed open-loop control is that once a system identification is performed, no feedback from on-line measurements is necessary. The proposed method shows great potential in practical applications when a destabilization due to self-excitation occurs or when the damping of weakly damped systems shall be enhanced.

Kurzfassung

Das Ziel dieser Dissertation ist es, die Methode der Dämpfung mittels Parametererregung an mechanischen Systemen mit und ohne Selbsterregung zu untersuchen. Aus bisherigen Arbeiten ist bekannt, dass eine zeitlich periodische Steifigkeit eine zusätzliche Dämpfung in ein selbsterregtes System einbringen kann, wenn die Frequenz der Parametererregung in der Umgebung einer Parameterkombinationsfrequenz des ungedämpften Systems liegt – eine so genannte Parameterantiresonanzfrequenz. Diese Untersuchungen stellen den Ausgangspunkt der Arbeit dar. Die vorliegende Arbeit erweitert die vorgestellte Methode auf unterschiedliche Typen von periodischen Systemparametern und entwickelt einen tiefen Einblick in diese Methode der Schwingungsunterdrückung.

Diese Arbeit untersucht mechanische Systeme mit den allgemeinsten linearen Bewegungsgleichungen mit gleichzeitig periodischen Steifigkeiten, Dämpfungen und Trägheiten. Um Dämpfung mittels Parametererregung erreichen zu können, sind zumindest zwei Freiheitsgrade notwendig. Solche Systeme können Punktmassensysteme sein oder zwei Schwingungsmoden einer elastischen Struktur beschreiben. Ausgehend von einer Näherung erster Ordnung, wird eine sorgfältige Stabilitätsanalyse der vorgestellten Methode der Dämpfung mittels Parametererregung für generische Systeme mit zwei Freiheitsgraden durchgeführt. Die analytischen Formeln stellen ein zuverlässiges Werkzeug zur Beschreibung des durch Parametererregung zusätzlich erzeugten Stabilitätsgebietes in der Umgebung einer Antiresonanzfrequenz, als auch des Instabilitätsgebietes aufgrund einer Parametererregung in der Umgebung von Resonanzfrequenzen dar. Diese Arbeit kann als kompaktes Handbuch zur Auslegung einer Vorrichtung zur Schwingungsunterdrückung mittels Parametererregung verwendet werden.

Die Ergebnisse in zusammengefasster Form sind: Zeitlich periodische Steifigkeiten können keine zusätzliche Dämpfung im System erzeugen, sondern verstärken die bereits vorhandene Dämpfung. Es wird gezeigt, dass Dämpfung mittels Parametererregung auch auf Systeme ohne Selbsterregung angewendet werden kann. Besondere Aufmerksamkeit wird auf die Symmetrieeigenschaften der Parametererregungsmatrizen und deren Einfluss auf die Lage der Parameterantiresonanzfrequenz gelegt. Die Wechselwirkung zwischen periodischen Steifigkeiten und Dämpfungen wird analysiert und deren Auswirkung auf die Funktion der Schwingungsunterdrückung studiert. Es wird gezeigt, dass eine periodische Trägheitsänderung einer gleichzeitigen Steifigkeits- und Dämpfungsänderung entspricht und somit auch als Hilfsmittel zur Dämpfung von Schwingungen mittels Parametererregung eingesetzt werden können.

Die Analyse der Energieflüsse zeigt, dass durch Einsatz einer Parametererregung dem System Energie zugeführt wird, die zu einer effektiveren Dissipation der kinetischen Energie führt. Die Kopplung der Eigenwerte der homogenen Bewegungsgleichung und die maximal erreichbare Dämpfung mittels Parameterregung in einem schwingungsfähigen System werden untersucht. Eine Erweiterung auf Systeme mit mehreren Freiheitsgraden zeigt, dass eine Parametererregung unter bestimmten Bedingungen lediglich zwei Moden des Systems koppelt, während die verbleibenden Moden unbeeinflusst bleiben. Die wichtige Konsequenz daraus ist, dass eine einfrequente Parametererregung nur einen instabilen Mode stabilisieren kann.

Es wird eine Optimierung für ein System durchgeführt, in dem alle Steifigkeiten gleichzeitig aber mit unterschiedlichen Amplituden und Phasen veränderlich sind. Für eine begrenzte Gesamtenergie der Parametererregung ist eine Konfiguration mit nur einer maximalen Amplitude optimal. Der Einfluss von zusätzlichen Beschränkungen der Amplituden der Parametererregung wird ebenfalls untersucht. Weiters wird für ein System mit einer einzigen periodischen Steifigkeit als die optimale Funktionsform im Falle einer Parametererregung mit beschränkter Amplitude die simple Rechtecksfunktion bestimmt, die einer Zweipunktsteuerung entspricht. Es wird gezeigt, dass allein durch die Anpassung der Funktionsform der Paramatererregung die effektive Amplitude der Parametererregung erheblich vergrößert wird, das zu einer schnelleren Schwingungsunterdrückung führt.

Die Ergebnisse zeigen, dass der Einsatz einer Parametererregung das Stabilitätsgebiet im Parameterraum der Systemparameter wesentlich erweitert. Der große Vorteil der vorgestellten Steuerung im Vergleich zu einer Regelung ist, dass sobald eine Systemidentifikation durchgeführt wurde, keine Rückführung aus Messungen im Betrieb mehr notwendig ist. Die vorgestellte Methode zeigt ein großes Potential im praktischen Einsatz wo eine Destabilisierung aufgrund einer Selbsterregung auftritt oder um die Dämpfung von schwach gedämpften Systemen zu erhöhen.

Contents

Nomenclature ix

1. Introduction 1

2. Models 3
   2.1. Types of vibrations 3
   2.2. Equations of motion 4
   2.3. Transformation to normal form 5
      2.3.1. Transposed transformation matrix 6
      2.3.2. Inverted transformation matrix 7
   2.4. Equivalent minimum systems 10
      2.4.1. System 1 11
      2.4.2. System 2 13
      2.4.3. System 3 14
      2.4.4. System 4 15

I. Stability of normal forms – analytical approaches 17

3. Time-periodic stiffness 21
   3.1. Single harmonic stiffness variation 21
      3.1.1. Transformation and averaging 22
      3.1.2. Summary for small single harmonic stiffness variations 34
      3.1.3. Symmetry considerations 37
   3.2. General harmonic stiffness variation 39
      3.2.1. Transformation and averaging 39
      3.2.2. Summary for small general harmonic stiffness variations 46
      3.2.3. Symmetry considerations 53
      3.2.4. Literature review 55
   3.3. Periodic stiffness variation 57
      3.3.1. Transformation and averaging 58
      3.3.2. Summary for small general periodic stiffness variations 60

3.3.3. Symmetry considerations 61

4. Time-periodic damping   63
4.1. Transformation and averaging 64
4.2. Summary for small single harmonic damping variations 70
4.3. Symmetry considerations 73

5. Synchronous time-periodic damping and stiffness   75
5.1. Transformation and averaging 76
5.2. Summary for small variations 79
5.3. Symmetry considerations 81

6. Time-periodic inertia   85
6.1. Transformation and averaging 86
6.2. Summary for small single harmonic inertia variations 90
6.3. Symmetry considerations 91

7. Simultaneous time-periodic inertia, damping and stiffness   93
7.1. Transformation and averaging 94
7.2. Summary for small variations 98

II. Interpretations and optimization   101

8. Equivalent damping   103
8.1. Single harmonic stiffness variation 103
8.2. General harmonic stiffness variation 109
8.3. Adopting stability formulae 111
8.4. Systems with more degrees of freedom 112

9. Energy flow   115
9.1. Energy definition 116
9.2. Interaction of damping and parametric excitation 117
    9.2.1. Negative damping 118
    9.2.2. Positive damping 127

10. Optimization of periodic functions   131
10.1. Optimal multi-location harmonic stiffness variation 131
    10.1.1. Optimal phase angles 134
    10.1.2. Optimal amplification factors 147
    10.1.3. Summary for optimal phase and amplification factors 152

10.1.4. Generalization of results 158
 10.2. Optimal multi-location harmonic damping variation 161
 10.3. Optimal shape of functions for stiffness variation 162

III. Comparison with numerical results 169

11. Numerical methods of investigation 171
 11.1. Floquet theory 171
 11.2. Path following 173

12. Minimum systems 175
 12.1. Harmonic stiffness variation 175
  12.1.1. System 1 177
  12.1.2. System 2 191
  12.1.3. Coupled pendulum system 195
 12.2. Optimal shape of a stiffness variation 198
 12.3. Synchronous stiffness and damping variation 200
 12.4. Inertia variation 206

13. Conclusions and Outlook 209

Bibliography 212

Appendices 217

A. Routh-Hurwitz theorems 219

B. Trigonometric decomposition theorems 223

Nomenclature

General conventions

Variable	Example	Font
Scalar	a	italic letter (usually lowercase)
Vector	**a**	bold face lowercase letter
Matrix	**A**	bold face capital letter

Notation

Symbol	Description
$\mathbb{R}$	field of real numbers
$\mathbb{C}$	field of complex numbers
$\in$	belong to
$\approx$	approximately equal to
j	$\sqrt{-1}$, imaginary unit
$\mathbf{I}_k$	identity matrix of size $k \times k$
0	zero matrix or vector
$\mathbf{A}^T$	transpose of **A**
$\mathbf{A}^s$	symmetric part of **A**
$\mathbf{A}^a$	skew-symmetric part of **A**
$\mathbf{A}_0$	constant part of time-dependent **A**

(continued on next page)

Symbol	Description
$\mathbf{A}_c$	coefficient of cosine function
$\mathbf{A}_s$	coefficient of sine function
$\hat{\mathbf{a}}$	averaged $\mathbf{a}$

Symbols

Symbol	Description
m	inertia (mass) coefficient
c	damping coefficient
k	stiffness coefficient
t	time
τ	non-dimensional time
$\mathbf{x}$	position vector
$\mathbf{z}$	position vector of normal form
u_i, v_i	real-valued amplitudes
w_i	complex-valued amplitudes
$\hat{\mathbf{u}}$	averaged amplitude vector
$\mathbf{T}$	transformation matrix
$\mathbf{M}$	inertia matrix
$\mathbf{C}$	damping matrix
$\boldsymbol{\Theta}$	non-dimensional damping matrix of normal form
$\mathbf{K}$	stiffness matrix
$\boldsymbol{\Omega}^2$	diagonal stiffness matrix of normal form
Ω	natural frequency

(continued on next page)

Symbol	Description
ϖ	Ω/η, scaled natural frequency
λ	eigenvalue
ω	parametric excitation frequency
η	non-dimensional parametric excitation frequency
ε	small amplitude amplification factor
σ	detuning factor of η
α_{ij}	phase angle between different parametric excitations
s_i, c_i	abbreviation for $\sin(\Omega_i t)$, $\cos(\Omega_i t)$
δ_{ik}	Kronecker delta-function
Q	non-dimensional coefficient matrix of cosine part of stiffness matrix
P	non-dimensional coefficient matrix of sine part of stiffness matrix
R	non-dimensional coefficient matrix of cosine part of damping matrix
S	non-dimensional coefficient matrix of sine part damping matrix

1. Introduction

The destabilizing nature of parametric excitation in mechanical systems with two and more degrees of freedom has been investigated and well understood within the last fifty years. Some of the early main contributions to this field can be found in [6], [35], [49], [43] and [63] and the literature cited therein. The main focus of these works was to determine the intervals of the parametric excitation frequency, for which parametric excitation destabilizes the system and a parametric resonance occurs.

Opposite to these works A. Tondl showed in [53], [54], [55] and [56] that parametric excitation can be used to stabilize a dynamic system, a phenomenon that had been disregarded by other authors. He proved for the first time that systems with self-excitation can be stabilized by interaction with a parametric excitation due to a time-periodic stiffness variation. Vibrations can be suppressed successfully, if the frequency of the parametric excitation is tuned close to the parametric combination frequency of order n,

$$\frac{|\Omega_i - \Omega_k|}{n}, \quad i \neq k, \quad n \in \mathbb{N}, \tag{1.1}$$

where $\Omega_{i,k}$ are the natural frequencies of the linearized system without damping. This frequency is called parametric *anti*-resonance frequency. On the other hand, if the parametric excitation frequency is in the vicinity of

$$\frac{\Omega_i + \Omega_k}{n}, \quad n \in \mathbb{N}, \tag{1.2}$$

a classical parametric resonance occurs. For $j = k$ this is called parametric primary resonance and parametric combination resonance for $j \neq k$. Hence, tuning the frequency of the parametric excitation near to an anti-resonance frequency (1.1) enables the damping of mechanical vibrations by parametric excitation. Based on these findings, more specific results were derived in [58] and [27], which represent the starting point of the present work.

To achieve damping by parametric excitation in a mechanical system, at least two degrees of freedom are necessary. Such systems may be lumped mass systems or may represent two modes of a continuous structure. Systems with two degrees of freedom which are under the influence of parametric excitation due to time-periodic stiffness and/or damping and/or inertia coefficients are analyzed in this work. Although the original work [53] proposes vibration suppression by interaction of self-excitation and parametric excitation only, extended conditions for systems with and without self-excitation are developed.

The terms damping by parametric excitation and vibration suppression are used interchangeably in this work. The main questions posed and discussed are:

- Is it possible to suppress vibrations by time-periodic damping or inertia coefficients?
- How do different kinds of time-periodic coefficients influence each other?
- What is the optimum parametric excitation with respect to the location in the system and the shape of the time-periodic function?
- How is the energy flow in the system when vibration suppression occurs?

This book is structured as follows.

Starting with a short overview of different types of vibrations, the equations of motion of a generic mechanical system are presented and a transformation to its normal form is established. Four systems with two degrees of freedom, for which damping by parametric excitation can be achieved, are introduced.

The following content is subdivided in three main parts. In Part I, a rather general study is performed, that can be used as a comprehensive guide for designing a device for vibration suppression by parametric excitation, even in case when no self-excitation is present. Based on a perturbation technique of first order, an analytical study is applied to obtain explicit formulae for stability boundary curves. Revisiting systems with time-periodic stiffness coefficients by a cosine function as presented in [58] and [27], a general harmonic as well as a time-periodic stiffness variation are investigated. Additionally, systems with parametric excitation due to single harmonic damping coefficients are studied. Furthermore, the interaction between a parametric excitation due to a stiffness variation as well as a damping variation is analyzed. Finally, formulae of stability boundary curves are presented for a rather general parametric excitation based on simultaneous time-periodic stiffness, damping and inertia coefficient variation.

In Part II, the coupling of eigenvalues and the maximum possible damping achievable by introducing parametric excitation in a vibrating system are explored. A short outlook to systems with multi-degrees of freedom is given. Additionally, the occurring energy flows are examined in order to gain a deeper insight how a parametric excitation interacts with a system. For a system where all stiffness coefficients are varied simultaneously, an optimization of amplitudes and phase angles of these parametric excitations is performed. In case of a single time-harmonic stiffness coefficient an optimal function shape is derived.

In Part III, a numerical analysis of the proposed exemplary minimum systems is performed. These numerical results are compared with the analytical formulae derived in Part I. Finally, conclusions are drawn and future research is suggested.

2. Models

2.1. Types of vibrations

Vibrations that occur in a dynamical system can be classified with respect to their origin, e.g. see [34], [41] or [4]. Four different types are commonly distinguished:

1. free vibrations,

2. self-excited vibrations,

3. parametrically excited vibrations and

4. forced vibrations.

If a system performs *free vibrations* it starts from an initial state and continuous vibrating without external forces acting on the system.

Self-excited vibrations can occur if a system has access to an external reservoir of energy. Other than with forced vibrations where the frequency of the excitation is prescribed, here the system itself determines the frequency at which the energy is transferred into the system. The system absorbs energy at a rate as needed to maintain a certain level of vibration. A well known example of such a system is a pendulum clock, where the energy may be stored in the potential energy of a weight or a mechanical spring. If more energy per cycle flows into the system than energy is dissipated due to friction etc., unstable vibrations can occur, which lead to strongly increasing amplitudes as in a case of unstable feedback control loops.

Self-excited vibrations may also occur due to interaction between a structure and a fluid, as in the case of an airplane wing which tends to unstable vibrations when reaching a critical speed. Such vibrations are called flow-induced vibrations. The destruction of the Tacoma-Bridge in 1940 is an impressive example how dangerous self-excited vibrations by fluid-structure interaction can be. Other examples are unstable vibrations of turbo machinery coming from flow-induced vibrations, friction-induced vibrations in brakes (squeal noise) or stick-slip phenomena in tool machines leading to marks of the cutting tool at the surface of the workpiece. Also unstable bogie motion of rail vehicles at very high speeds belongs to the group of self-excited vibrations.

Models for a great class of self-excited vibrations are the Van der Pol- as well as the Rayleigh-oscillator, see [55]. They describe self-excitation using a non-linear force f_{se},

$$f_{se}^v = c\dot{x}\left(1 + \gamma x^2\right) \quad \text{and} \quad f_{se}^r = c\dot{x}\left(1 + \gamma \dot{x}^2\right), \tag{2.1}$$

respectively, where x represents a deflection and c is a negative damping coefficient. This work mainly analyzes the stable motion of a system, for which the equations of motion may be linearized around the equilibrium position $x, \dot{x} = 0$. The non-linear forces (2.1) yield $f_{se} = c\dot{x}$ as the linearized expression. Hence, this type self-excited vibrations can be described by a negative damping coefficient.

Parametrically excited vibrations occur if one or more coefficients of the differential equations are not constant but periodically time-varying. The frequency of the parameter change is prescribed explicitly as a function of time (independent on the motion of the system), e.g. by the rotational speed of a shaft. Examples are: pendulum with periodically varying length or periodically moving pivot point, rotating shaft with nonsymmetric cross-section, periodically varying stiffness of gear-wheels, etc.

Forced vibrations emerge from external forces and disturbances. The vibration frequency is determined by the frequency of those time-varying forces, which appear on the right hand side of the equations of motion.

2.2. Equations of motion

The most general equations of motion of a linear mechanical system with n degrees of freedom, which are under the influence of an external excitation $\mathbf{f}(t)$ are

$$\mathbf{M}(t)\ddot{\mathbf{x}}(t) + \varepsilon\mathbf{C}(t)\dot{\mathbf{x}}(t) + \mathbf{K}(t)\mathbf{x}(t) = \mathbf{f}(t), \tag{2.2}$$

with the inertia matrix $\mathbf{M}(t)$, the damping matrix $\mathbf{C}(t)$, the stiffness matrix $\mathbf{K}(t)$ and the position vector $\mathbf{x}(t) = (x_1(t), x_2(t))^T$, see [41] or [62]. These system matrices are time-dependent, but periodic matrices of the size $n \times n$. The equations (2.2) describe the dynamics of n modes and represent a system of n coupled linear differential equations with periodic coefficients. The main purpose of this work is to amplify the existent damping coefficients in a dynamical system. Therefore we investigate especially systems with small damping coefficients, which is expressed by the small factor $\varepsilon \ll 1$ introduced in (2.2).

The inertia matrix is assumed to be a symmetric matrix, while the damping and stiffness matrices can be of arbitrary shape. One should keep in mind, that only a closed system possesses symmetric system matrices, see [47] or [48]. Skew-symmetric system matrices occur in the presence of open systems. In this case gyroscopic forces may lead to a skew-symmetric part of the damping matrix, while non-conservative forces yield a skew-symmetric part in the stiffness matrix. Examples for such systems are any kind of rotors [54] or the cantilever with axial follower force [51], just to mention a few.

For the further analysis it is convenient to decompose harmonic matrices in the following way

$$A(t) = A_0 + \varepsilon A_s \sin(\omega t) + \varepsilon A_c \cos(\omega t) \quad \text{for} \quad A = M, C, K, \quad (2.3)$$

where the subscript 0 indicates the constant and time-independent part and the subscripts s or c the coefficient matrix of the sine or cosine function. The quadratic system matrices can be expressed as the sum of a symmetric part, with the superscript s, and a skew-symmetric (or anti-symmetric) part, with the superscript a,

$$A_i = \frac{1}{2}\left(A_i + A_i^T\right) + \frac{1}{2}\left(A_i - A_i^T\right) = A_i^s + A_i^a \quad \text{for} \quad i = 0, s, c. \quad (2.4)$$

Finally, the system matrices A can be decomposed to

$$A(t) = \{A_0^s + A_0^a\} + \varepsilon \{A_s^s + A_s^a\} \sin(\omega t) + \varepsilon \{A_c^s + A_c^a\} \cos(\omega t), \quad (2.5)$$

with the small amplification factor ε. Note that the matrices A_i allow the description of a time-periodic variation for more than a single system parameter. A system configuration with multiple time-periodic variation of one or more inertia, damping and/or stiffness coefficients is possible and will be investigated in the next chapters. The only restriction is, that these variations all occur with the same frequency ω. Phase shifts between variations of different parameters can be considered, too.

Replacing the coefficient matrices in (2.2) with the expressions from (2.5) yields for free vibrations, $f(t) = 0$,

$$M_0 \ddot{x}(t) + \varepsilon C_0 \dot{x}(t) + K_0 x(t) =$$
$$- \varepsilon M_s^s \ddot{x}(t) \sin(\omega t) - \varepsilon M_c^s \ddot{x}(t) \cos(\omega t)$$
$$- \varepsilon \{C_s^s + C_s^a\} \dot{x}(t) \sin(\omega t) - \varepsilon \{C_c^s + C_c^a\} \dot{x}(t) \cos(\omega t)$$
$$- \varepsilon \{K_s^s + K_s^a\} x(t) \sin(\omega t) - \varepsilon \{K_c^s + K_c^a\} x(t) \cos(\omega t). \quad (2.6)$$

On the left hand side the classical ordinary differential equations appear with constant coefficients, on the right hand side coupling terms of the differential equations through harmonic coefficients of order ε are arranged. Other than for stiffness and damping matrices, for physically meaningful system configurations the inertia matrix M_0 is always symmetric.

2.3. Transformation to normal form

This work deals with the mechanism of amplifying the existent damping in a system by introducing a single-frequency parametric excitation. To choose the right frequency of this parametric excitation is one of the most important tasks in order to achieve a damping effect at all. The sign of the product of two off-diagonal terms, the so-called parametric excitation term, as for example $Q_{ij}Q_{ji}$ in Tab. 3.2 on page 38, is crucial to decide whether this frequency is located at $|\Omega_i - \Omega_j|$ or at $\Omega_i + \Omega_j$. This will be more clear in the following chapters. The main question is, whether the sign changes by transforming the system into its normal form or not.

First, we review the case of a symmetric and constant stiffness matrix with the classical results for the case of using a transposed transformation matrix. We describe the conditions for symmetry conservation for the case of a general constant stiffness matrix using an inverted transformation matrix and then specialize on the case of a diagonal inertia matrix. A diagonal inertia matrix would arise for example from a model of coupled lumped masses. Finally, we give explicit expressions for some minimum systems that will be explored numerically in the last chapters.

2.3.1. Transposed transformation matrix

In physically meaningful systems the constant part $\mathbf{M}_0$ of the inertia matrix in (2.6),

$$\mathbf{M}_0\ddot{\mathbf{x}}(t) + \varepsilon\mathbf{C}_0\dot{\mathbf{x}}(t) + \mathbf{K}_0\mathbf{x}(t) = \varepsilon f\left(\mathbf{M}_i^s\ddot{\mathbf{x}}(t), (\mathbf{C}_i^s + \mathbf{C}_i^a)\dot{\mathbf{x}}(t), (\mathbf{K}_i^s + \mathbf{K}_i^a)\mathbf{x}(t), t\right), \quad (2.7)$$

is symmetric and positive definite. Considering the case where the constant part of the stiffness matrix is symmetric,

$$\mathbf{K}_0 = \mathbf{K}_0^s = \mathbf{K}_0^{s,T}, \qquad \mathbf{K}_0^a = \mathbf{0},$$

a transformation matrix $\mathbf{T}$ can be found that satisfies the following conditions

$$\begin{aligned}\tilde{\mathbf{M}}_0 &= \mathbf{T}^T\mathbf{M}_0\mathbf{T} = [m_i\delta_{ij}], \\ \tilde{\mathbf{K}}_0 &= \mathbf{T}^T\mathbf{K}_0\mathbf{T} = [m_i\Omega_i^2\delta_{ij}],\end{aligned} \quad (2.8)$$

where δ_{ij} is the discrete Kronecker delta function defined by

$$\delta_{ij} = \begin{cases} 0 & \text{for } i \neq j \\ 1 & \text{for } i = j. \end{cases}$$

Normalizing the eigenvectors to 1 the matrix becomes orthogonal,

$$\mathbf{T}\mathbf{T}^{-1} = \mathbf{T}\mathbf{T}^T = \mathbf{I}, \qquad \mathbf{T}^{-1} = \mathbf{T}^T.$$

The determinant is either +1 (rotation) or -1 (mirroring). By using a rescaled transformation matrix $\tilde{\mathbf{T}}$ (2.8) can be even simplified to

$$\begin{aligned}\tilde{\mathbf{M}}_0 &= \tilde{\mathbf{T}}^T\mathbf{M}_0\tilde{\mathbf{T}} = \mathbf{I} \\ \tilde{\mathbf{K}}_0 &= \tilde{\mathbf{T}}^T\mathbf{K}_0\tilde{\mathbf{T}} = [\Omega_i^2\delta_{ij}] = \mathbf{\Omega}^2.\end{aligned} \quad (2.9)$$

2 Transformation to normal form

The derivation of the special transformation matrix $\mathbf{T}$ in (2.8) or $\tilde{\mathbf{T}}$ in (2.9) is given for instance in [36].

The relations in (2.8) points out that using the transposed transformation matrix is sufficient to diagonalize the constant inertia matrix $\mathbf{M}_0$ and the constant part of the stiffness matrix $\mathbf{K}_0$ simultaneously. We obtain the following normal form of the equations of motion

$$\tilde{\mathbf{M}}_0 \ddot{\mathbf{z}}(t) + \varepsilon \tilde{\mathbf{C}}_0 \dot{\mathbf{z}}(t) + \tilde{\mathbf{K}}_0 \mathbf{z}(t) = \varepsilon f \left(\tilde{\mathbf{M}}_i^s \ddot{\mathbf{z}}(t), \left(\tilde{\mathbf{C}}_i^s + \tilde{\mathbf{C}}_i^a \right) \dot{\mathbf{z}}(t), \left(\tilde{\mathbf{K}}_i^s + \tilde{\mathbf{K}}_i^a \right) \mathbf{z}(t), t \right), \quad (2.10)$$

where $i = s, c$ and

$$\tilde{\mathbf{A}} = \mathbf{T}^T \mathbf{A} \mathbf{T} \quad \text{with} \quad \mathbf{A} = \mathbf{M}_j^s, \mathbf{C}_j^s, \mathbf{C}_j^a, \mathbf{K}_j^s, \mathbf{K}_j^a \quad \text{and} \quad j = 0, s, c. \quad (2.11)$$

Using the transposed transformation matrix, the symmetry property of the original matrix $\mathbf{A}$ is equivalent to the symmetry property of the transformed matrix $\tilde{\mathbf{A}}$, as can be seen from the following analysis. Decomposing $\mathbf{A}$ and $\tilde{\mathbf{A}}$ into a symmetric and an skew-symmetric part,

$$\mathbf{A} = \mathbf{A}^s + \mathbf{A}^a, \qquad \tilde{\mathbf{A}} = \tilde{\mathbf{A}}^s + \tilde{\mathbf{A}}^a,$$

applying the transformation (2.11)

$$\tilde{\mathbf{A}}^s + \tilde{\mathbf{A}}^a = \mathbf{T}^T \left(\mathbf{A}^s + \mathbf{A}^a \right) \mathbf{T} = \mathbf{T}^T \mathbf{A}^s \mathbf{T} + \mathbf{T}^T \mathbf{A}^a \mathbf{T},$$

yields

$$\tilde{\mathbf{A}}^s = \frac{1}{2} \left(\mathbf{T}^T \left(\mathbf{A}^s + \mathbf{A}^a \right) \mathbf{T} + \left(\mathbf{T}^T \left(\mathbf{A}^s + \mathbf{A}^a \right) \mathbf{T} \right)^T \right)$$

$$= \frac{1}{2} \left(\mathbf{T}^T \mathbf{A}^s \mathbf{T} + \mathbf{T}^T \mathbf{A}^a \mathbf{T} + \mathbf{T}^T \mathbf{A}^{s,T} \mathbf{T} + \mathbf{T}^T \mathbf{A}^{a,T} \mathbf{T} \right) = \mathbf{T}^T \mathbf{A}^s \mathbf{T},$$

$$\tilde{\mathbf{A}}^a = \frac{1}{2} \left(\mathbf{T}^T \mathbf{A}^s \mathbf{T} + \mathbf{T}^T \mathbf{A}^a \mathbf{T} - \left(\mathbf{T}^T \mathbf{A}^s \mathbf{T} + \mathbf{T}^T \mathbf{A}^a \mathbf{T} \right)^T \right)$$

$$= \frac{1}{2} \left(\mathbf{T}^T \mathbf{A}^s \mathbf{T} + \mathbf{T}^T \mathbf{A}^a \mathbf{T} - \mathbf{T}^T \mathbf{A}^{s,T} \mathbf{T} - \mathbf{T}^T \mathbf{A}^{a,T} \mathbf{T} \right) = \mathbf{T}^T \mathbf{A}^a \mathbf{T}. \quad (2.12)$$

As will be shown for instance in Tab. 3.2 on page 38 the sign of the product of two corresponding matrix entries, as for example the parametric excitation term $Q_{ij} Q_{ji}$, is decisive for the proper design of a parametric excitation. The sign of the parametric excitation term decides the location of the frequency of the parametric excitation that is capable of amplifying the existent system damping. The relations in (2.12) show that the symmetry, and therefore the sign of the matrix entries of $\tilde{\mathbf{M}}_i^s, \tilde{\mathbf{C}}_i^s, \tilde{\mathbf{C}}_i^a, \tilde{\mathbf{K}}_i^s, \tilde{\mathbf{K}}_i^a$, is conserved by performing the transformation in (2.11). Additionally, due to the conservation of symmetry, two corresponding off-diagonal expressions, as Q_{ij} and Q_{ji}, differ in their sign, but their absolute values are the same in case of $\mathbf{A}^a = \mathbf{0}$.

2.3.2. Inverted transformation matrix

In this section we investigate the influence of using an inverted transformation matrix on the symmetry property of the excitation matrices $\tilde{\mathbf{M}}_i^s, \tilde{\mathbf{C}}_i^s, \tilde{\mathbf{C}}_i^a, \tilde{\mathbf{K}}_i^s, \tilde{\mathbf{K}}_i^a$. In physically meaningful systems the constant part $\mathbf{M}_0$ of the inertia matrix in (2.6) is symmetric and positive definite. The determinant of a positive definite matrix is always positive, so a positive definite matrix is always nonsingular. Multiplying the original system equations (2.6) by the always existent inverted inertia matrix $\mathbf{M}_0^{-1}$ gives

$$\ddot{\mathbf{x}}(t) + \varepsilon \mathbf{M}_0^{-1}\mathbf{C}_0\dot{\mathbf{x}}(t) + \mathbf{M}_0^{-1}\mathbf{K}_0\mathbf{x}(t) = \varepsilon f\left(\ddot{\mathbf{x}}(t), \dot{\mathbf{x}}(t), \mathbf{x}(t), t\right). \tag{2.13}$$

If the matrix $\mathbf{K}_0$ is singular, then a similarity transformation that diagonalizes the original matrix cannot be provided. We assume that none of the eigenvalues of $\mathbf{M}_0^{-1}\mathbf{K}_0$ is zero. We introduce the constant and nonsingular linear transformation $\mathbf{x}(t) = \mathbf{T}\mathbf{z}(t)$, premultiply the entire equation by $\mathbf{T}^{-1}$ and obtain the following normal form

$$\ddot{\mathbf{z}}(t) + \varepsilon \tilde{\mathbf{C}}_0\dot{\mathbf{z}}(t) + \tilde{\mathbf{K}}_0\mathbf{z}(t) = \varepsilon f\left(\tilde{\mathbf{M}}_i^s\ddot{\mathbf{z}}(t), \tilde{\mathbf{C}}_i^s\dot{\mathbf{z}}(t), \tilde{\mathbf{C}}_i^a\dot{\mathbf{z}}(t), \tilde{\mathbf{K}}_i^s\mathbf{z}(t), \tilde{\mathbf{K}}_i^a\mathbf{z}(t), t\right), \tag{2.14}$$

where $i = s, c$ and

$$\tilde{\mathbf{A}} = \mathbf{T}^{-1}\mathbf{M}_0^{-1}\mathbf{A}\mathbf{T} \quad \text{with} \quad \mathbf{A} = \mathbf{M}_j^s, \mathbf{C}_j^s, \mathbf{C}_j^a, \mathbf{K}_j^s, \mathbf{K}_j^a \quad \text{and} \quad j = 0, s, c, \tag{2.15}$$

instead of (2.11). The inertia coefficient matrix remains a unity matrix because $\tilde{\mathbf{M}}_0 = \mathbf{T}^{-1}\mathbf{M}_0^{-1}\mathbf{M}_0\mathbf{T} = \mathbf{I}$. One can always choose $\mathbf{T}$ such that $\tilde{\mathbf{K}}_0 = \mathbf{T}^{-1}\mathbf{M}_0^{-1}\mathbf{K}_0\mathbf{T}$ is a Jordan canonical form. Assuming distinct and positive eigenvalues of $\tilde{\mathbf{K}}_0$

$$\Omega_1^2 < \Omega_2^2 < \cdots < \Omega_N^2,$$

a transformation matrix $\mathbf{T}$ can always be found such that the constant part of the stiffness coefficient will become diagonal

$$\tilde{\mathbf{K}}_0 = \mathbf{T}^{-1}\mathbf{M}_0^{-1}\mathbf{K}_0\mathbf{T} = \left[\Omega_i^2 \delta_{ij}\right] = \mathbf{\Omega}^2. \tag{2.16}$$

A similarity transformation does not affect the eigenvalues of $\mathbf{M}_0^{-1}\mathbf{K}_0$. The values of Ω_i^2 remain unchanged and are equivalent to the values in (2.9). Rewriting the equations of motion in expanded form yields

$$\ddot{\mathbf{z}}(t) + \varepsilon \mathbf{T}^{-1}\mathbf{M}_0^{-1}\mathbf{C}_0\mathbf{T}\dot{\mathbf{z}}(t) + \mathbf{\Omega}^2\mathbf{z}(t) =$$
$$-\varepsilon \mathbf{T}^{-1}\mathbf{M}_0^{-1}\mathbf{M}_s^s\mathbf{T}\ddot{\mathbf{z}}(t)\sin(\omega t) - \varepsilon \mathbf{T}^{-1}\mathbf{M}_0^{-1}\mathbf{M}_c^s\mathbf{T}\ddot{\mathbf{z}}(t)\cos(\omega t)$$
$$-\varepsilon \mathbf{T}^{-1}\mathbf{M}_0^{-1}\{\mathbf{C}_s^s + \mathbf{C}_s^a\}\mathbf{T}\dot{\mathbf{z}}(t)\sin(\omega t) - \varepsilon \mathbf{T}^{-1}\mathbf{M}_0^{-1}\{\mathbf{C}_c^s + \mathbf{C}_c^a\}\mathbf{T}\dot{\mathbf{z}}(t)\cos(\omega t)$$
$$-\varepsilon \mathbf{T}^{-1}\mathbf{M}_0^{-1}\{\mathbf{K}_s^s + \mathbf{K}_s^a\}\mathbf{T}\mathbf{z}(t)\sin(\omega t) - \varepsilon \mathbf{T}^{-1}\mathbf{M}_0^{-1}\{\mathbf{K}_c^s + \mathbf{K}_c^a\}\mathbf{T}\mathbf{z}(t)\cos(\omega t). \tag{2.17}$$

Applying a transposed transformation matrix on a system matrix $\mathbf{A}$ as in (2.8), the symmetry property is conserved, but performing the operations in (2.15) on a system matrix $\mathbf{A}$, in general, will destroy a previous symmetry. The symmetry property of $\mathbf{A}$ can be destroyed by two operations: first, by premultiplying by the inverted inertia matrix $\mathbf{M}_0^{-1}$ and second, by performing the similarity transformation using $\mathbf{T}$.

2 Transformation to normal form

Multiplication by inverted inertia matrix

Any real matrix $\mathbf{A}$ can be separated into a symmetric and an antisymmetric part. For the symmetric and constant inertia matrix

$$\left(\mathbf{M}_0^{-1}\right)^T = \left(\mathbf{M}_0^T\right)^{-1} = \mathbf{M}_0^{-1}$$

holds and multiplying by a parametric excitation matrix as in (2.17) yields the relations

$$\left(\tilde{\mathbf{A}}^s + \tilde{\mathbf{A}}^a\right) = \mathbf{M}_0^{-1}\left(\mathbf{A}^s + \mathbf{A}^a\right),$$
$$\tilde{\mathbf{A}}^s = \frac{1}{2}\left(\mathbf{M}_0^{-1}\left(\mathbf{A}^s + \mathbf{A}^a\right) + \left(\mathbf{A}^s - \mathbf{A}^a\right)\mathbf{M}_0^{-1}\right), \quad (2.18)$$
$$\tilde{\mathbf{A}}^a = \frac{1}{2}\left(\mathbf{M}_0^{-1}\left(\mathbf{A}^s + \mathbf{A}^a\right) - \left(\mathbf{A}^s - \mathbf{A}^a\right)\mathbf{M}_0^{-1}\right).$$

For a symmetric matrix $\mathbf{A} = \mathbf{A}^s$ the relations above simplify to

$$\tilde{\mathbf{A}}^s = \frac{1}{2}\left(\mathbf{M}_0^{-1}\mathbf{K}^s + \mathbf{K}^s\mathbf{M}_0^{-1}\right),$$
$$\tilde{\mathbf{A}}^a = \frac{1}{2}\left(\mathbf{M}_0^{-1}\mathbf{K}^s - \mathbf{K}^s\mathbf{M}_0^{-1}\right) \neq \mathbf{0}. \quad (2.19)$$

On one hand the eigenvalues of the original and the transformed matrix are the same, but on the other hand, an originally symmetric matrix, in general, is transformed into a matrix with a symmetric part $\tilde{\mathbf{A}}^s$ and, in addition, with a skew-symmetric part $\tilde{\mathbf{A}}^a$. In general, the existent (skew-)symmetry of any system matrix is broken and a multiplication of two (skew-)symmetric matrices does not imply the (skew-)symmetry of the resultant matrix, because the matrix multiplication $\mathbf{M}_0^{-1}\mathbf{K}^s$ is not commutative. Only for the special case of a diagonal inertia matrix the following holds

$$\mathbf{M}_0^{-1}\mathbf{A}^s = \mathbf{A}^s\mathbf{M}_0^{-1},$$
$$\tilde{\mathbf{A}}^s = \mathbf{M}_0^{-1}\mathbf{A}^s, \quad (2.20)$$
$$\tilde{\mathbf{A}}^a = \mathbf{0}.$$

Similarity transformation

The symmetry property of $\mathbf{A}$ may be destroyed by a similarity transformation as performed in (2.15), too. Examining a symmetric matrix $\mathbf{B}^s = \mathbf{M}_0^{-1}\mathbf{A}^s$ for a diagonal inertia matrix $\mathbf{M}_0$ yields

$$\mathbf{B}^s = \mathbf{B}^{s,T},$$
$$\tilde{\mathbf{B}} = \tilde{\mathbf{B}}^s + \tilde{\mathbf{B}}^a = \mathbf{T}^{-1}\mathbf{B}^s\mathbf{T},$$
$$\tilde{\mathbf{B}}^s = \frac{1}{2}\left(\mathbf{T}^{-1}\mathbf{B}^s\mathbf{T} + \left(\mathbf{T}^{-1}\mathbf{B}^s\mathbf{T}\right)^T\right)$$
$$= \frac{1}{2}\left(\mathbf{T}^{-1}\mathbf{B}^s\mathbf{T} + \mathbf{T}^T\mathbf{B}^s\mathbf{T}^{-T}\right),$$

$$\tilde{\mathbf{B}}^a = \frac{1}{2}\left(\mathbf{T}^{-1}\mathbf{B}^s\mathbf{T} - \left(\mathbf{T}^{-1}\mathbf{B}^s\mathbf{T}\right)^T\right)$$
$$= \frac{1}{2}\left(\mathbf{T}^{-1}\mathbf{B}^s\mathbf{T} - \mathbf{T}^T\mathbf{B}^s\mathbf{T}^{-T}\right) \neq \mathbf{0}.$$

The skew-symmetric part of this matrix does not vanish because, again, in general the matrix multiplication is not commutative

$$\left(\mathbf{TT}^T\right)^{-1}\mathbf{B}^s \neq \mathbf{B}^s\left(\mathbf{TT}^T\right)^{-1}.$$

This inequality does not hold for the case where the symmetric matrix $\mathbf{TT}^T$ is diagonal.

Recapitulating the case of the inverted transformation matrix, we can state that, in general, the symmetry property is lost by performing the transformations in (2.15). While in case of using a transposed transformation matrix as in (2.11) the absolute values of Q_{ij} and Q_{ji} are equal in case of $\mathbf{A}^a = \mathbf{0}$, now, these values differ in their size. Nevertheless, an originally positive (negative) definite matrix remains positive (negative) definite after transformation. Again, the sign of the parametric excitation term $Q_{ij}Q_{ji}$ is conserved.

2.4. Equivalent minimum systems

In general, a dynamical process can be modelled by an equivalent mechanical system. This section investigates the simplest possible systems for which damping by parametric excitation is achievable. These systems possess just two degrees of freedom and are called minimum systems.

The most general lumped mass system with two degrees of freedom and time-periodic coefficients is shown in Fig. 2.1. The corresponding equations of motion are given by (2.6). In the following paragraphs we give explicit expressions for the coefficient matrices of the normal form for four simple lumped mass systems, that possess only a single time-periodic stiffness and/or damping coefficient. Such systems have been investigated in [58], [27], [11] and [12].

As outlined in the preceding sections, both, applying the transformation in (2.11) or in (2.15), respectively, do not alter the sign of the determinant of the parametric excitation matrix, and consequently, the sign of the parametric excitation terms. Using an inverted transformation matrix as in (2.15), any symmetry is lost, but leads to simpler expressions than applying the transposed matrix as defined in (2.11). In order to obtain handsome expressions, we restrict our analysis to the transformation (2.15).

2 Equivalent minimum systems

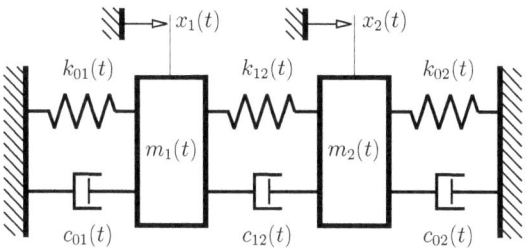

Figure 2.1.: Lumped mass system with two degrees of freedom.

2.4.1. System 1

The first system investigated is a simplified system for which only the stiffness k_{02} attached to the inertial reference frame is time-periodic and $k_{01}, c_{12} \equiv 0$, as presented in [58]. The corresponding non-vanishing system matrices of (2.6) are

$$\mathbf{M}_0 = \begin{bmatrix} m_1 & 0 \\ 0 & m_2 \end{bmatrix}, \quad \mathbf{K}_0^s = \begin{bmatrix} k_{12} & -k_{12} \\ -k_{12} & k_{12} + k_{02} \end{bmatrix},$$

$$\mathbf{C}_0^s = \begin{bmatrix} c_{01} & 0 \\ 0 & c_{02} \end{bmatrix}, \quad \mathbf{K}_c^s = \begin{bmatrix} 0 & 0 \\ 0 & k_{02} \end{bmatrix}. \quad (2.21)$$

Using the time transformation that is derived from the natural frequency of the subsystem m_1, k_{12},

$$\tau = \omega_1 t \quad \text{with} \quad \omega_1 = \sqrt{\frac{k_{12}}{m_1}}, \quad (2.22)$$

and defining the following characteristic parameters

$$\eta = \frac{\omega}{\omega_1}, \quad M = \frac{m_1}{m_2}, \quad \kappa_1 = \frac{c_{01}}{m_1 \omega_1}, \quad \kappa_2 = \frac{c_{02}}{m_2 \omega_1}, \quad Q^2 = \frac{k_{02}}{m_2 \omega_1^2} = \left(\frac{\omega_2}{\omega_1}\right)^2, \quad (2.23)$$

gives the non-dimensional system matrices

$$\bar{\mathbf{M}}_0 = \begin{bmatrix} 1 & 0 \\ 0 & 1 \end{bmatrix}, \quad \bar{\mathbf{K}}_0^s = \begin{bmatrix} 1 & -1 \\ -M & M + Q^2 \end{bmatrix},$$

$$\bar{\mathbf{C}}_0^s = \begin{bmatrix} \kappa_1 & 0 \\ 0 & \kappa_2 \end{bmatrix}, \quad \bar{\mathbf{K}}_c^s = \begin{bmatrix} 0 & 0 \\ 0 & Q^2 \end{bmatrix}. \quad (2.24)$$

The ratios defined in (2.23) represent relations between the dimensional physical system parameters. For a certain physical system specific values for some of the parameters have to be chosen additionally. The non-dimensional equations of motion are transformed into the normal form by applying the constant transformation matrix

$$\mathbf{x}(t) = \mathbf{T}\mathbf{z}(t) \quad \text{with} \quad \mathbf{T} = \begin{bmatrix} 1 & 1 \\ a_1 & a_2 \end{bmatrix}, \quad (2.25)$$

as in (2.14). The special form of the transformation matrix $\mathbf{T}$ is chosen in order to keep the following expressions simple. The coefficients of the transformation matrix $\mathbf{T}$ read for the system matrices (2.21)

$$a_1 = \frac{M}{Q^2 + M - \Omega_1^2}, \quad a_2 = \frac{M}{Q^2 + M - \Omega_2^2} \quad (2.26)$$

with the eigenvalues

$$\Omega_{1,2}^2 = \frac{1}{2}(1 + M + Q^2) \pm \sqrt{\frac{1}{4}(1 + M + Q^2)^2 - Q^2}. \quad (2.27)$$

Applying (2.25) on (2.21) the equations of motion are transformed into the normal form

$$\bar{\mathbf{K}}_0^s \mapsto \begin{bmatrix} \Omega_1^2 & 0 \\ 0 & \Omega_2^2 \end{bmatrix} = \mathbf{\Omega}^2, \quad \bar{\mathbf{C}}_0^s \mapsto [\Theta_{ij}], \quad \bar{\mathbf{K}}_c^s \mapsto [Q_{ij}]. \quad (2.28)$$

The modal damping coefficients yield

$$\Theta_{11} = \frac{-a_2\kappa_1 + a_1\kappa_2}{a_1 - a_2}, \quad \Theta_{12} = \frac{-a_2\kappa_1 + a_2\kappa_2}{a_1 - a_2},$$
$$\Theta_{21} = \frac{a_1\kappa_1 - a_1\kappa_2}{a_1 - a_2}, \quad \Theta_{22} = \frac{a_1\kappa_1 - a_2\kappa_2}{a_1 - a_2}, \quad (2.29)$$

and the coefficients of the parametric excitation result in

$$Q_{11} = Q^2 \frac{a_1}{a_1 - a_2}, \quad Q_{12} = Q^2 \frac{a_2}{a_1 - a_2},$$
$$Q_{21} = Q^2 \frac{-a_1}{a_1 - a_2}, \quad Q_{22} = Q^2 \frac{-a_2}{a_1 - a_2}. \quad (2.30)$$

Note that the determinant of the parametric excitation matrix $\mathbf{K}_c^s$ remains unchanged after applying the similarity transformation as well as introducing non-dimensional properties. The following relation is valid:

$$\det\{\mathbf{K}_c^s\} = 0 = \det\{[Q_{ij}]\} = Q_{11}Q_{22} - Q_{12}Q_{21}. \quad (2.31)$$

An approximation of the equations of motion up to the first order of ε is sufficient to describe the stability border in the entire parameter domain of the system parameters, as will be shown in the following part.

2.4.2. System 2

The second system of interest is a system where only the coupling stiffness k_{12} is time-periodic and $k_{01}, c_{01} \equiv 0$, as presented in [27]. The non-vanishing system matrices of (2.6) read

$$\mathbf{M}_0 = \begin{bmatrix} m_1 & 0 \\ 0 & m_2 \end{bmatrix}, \quad \mathbf{K}_0^s = \begin{bmatrix} k_{12} & -k_{12} \\ -k_{12} & k_{12} + k_{02} \end{bmatrix},$$

$$\mathbf{C}_0^s = \begin{bmatrix} c_{12} & -c_{12} \\ -c_{12} & c_{12} + c_{02} \end{bmatrix}, \quad \mathbf{K}_c^s = \begin{bmatrix} k_{12} & -k_{12} \\ -k_{12} & k_{12} \end{bmatrix}. \tag{2.32}$$

Applying the time transformation that is derived from the natural frequency of the subsystem m_2, k_{02},

$$\tau = \omega_2 t \quad \text{with} \quad \omega_2 = \sqrt{\frac{k_{02}}{m_2}} \quad \text{and} \quad \omega_1 = \sqrt{\frac{k_{12}}{m_1}} \tag{2.33}$$

and defining the following characteristic parameters

$$\eta = \frac{\omega}{\omega_2}, \quad M = \frac{m_1}{m_2}, \quad \kappa_1 = \frac{c_{12}}{m_1 \omega_2}, \quad \kappa_2 = \frac{c_{02}}{m_2 \omega_2}, \quad Q^2 = \frac{k_{12}}{m_1 \omega_2^2} = \left(\frac{\omega_1}{\omega_2}\right)^2, \tag{2.34}$$

the system matrices of the non-dimensional equations of motion yield

$$\bar{\mathbf{M}}_0 = \begin{bmatrix} 1 & 0 \\ 0 & 1 \end{bmatrix}, \quad \bar{\mathbf{K}}_0^s = \begin{bmatrix} Q^2 & -Q^2 \\ -MQ^2 & MQ^2 + 1 \end{bmatrix},$$

$$\bar{\mathbf{C}}_0^s = \begin{bmatrix} \kappa_1 & -\kappa_1 \\ -M\kappa_1 & M\kappa_1 + \kappa_2 \end{bmatrix}, \quad \bar{\mathbf{K}}_c^s = \begin{bmatrix} Q^2 & -Q^2 \\ -MQ^2 & MQ^2 \end{bmatrix}. \tag{2.35}$$

For the system in (2.32) the coefficients of the transformation matrix as defined in (2.25) become

$$a_1 = \frac{MQ^2}{1 + MQ^2 - \Omega_1^2}, \quad a_2 = \frac{MQ^2}{1 + MQ^2 - \Omega_2^2} \tag{2.36}$$

with the eigenvalues

$$\Omega_{1,2}^2 = \frac{1}{2}\left(1 + MQ^2 + Q^2\right) \pm \sqrt{\frac{1}{4}\left(1 + MQ^2 + Q^2\right)^2 - Q^2}. \tag{2.37}$$

The equations of motion are transformed into the normal form

$$\bar{\mathbf{K}}_0^s \mapsto \begin{bmatrix} \Omega_1^2 & 0 \\ 0 & \Omega_2^2 \end{bmatrix} = \Omega^2, \quad \bar{\mathbf{C}}_0^s \mapsto [\Theta_{ij}], \quad \bar{\mathbf{K}}_c^s \mapsto [Q_{ij}], \tag{2.38}$$

with the modal damping coefficients

$$\Theta_{11} = \frac{-(a_2 + M)(1 - a_1)\kappa_1 + a_1 \kappa_2}{a_1 - a_2}, \quad \Theta_{12} = \frac{-(a_2 + M)(1 - a_2)\kappa_1 + a_2 \kappa_2}{a_1 - a_2},$$

$$\Theta_{21} = \frac{(a_1 + M)(1 - a_1)\kappa_1 - a_2 \kappa_2}{a_1 - a_2}, \quad \Theta_{22} = \frac{(a_1 + M)(1 - a_2)\kappa_1 - a_2 \kappa_2}{a_1 - a_2}, \tag{2.39}$$

and the coefficients of the parametric excitation

$$Q_{11} = Q^2 \frac{-(a_2 + M)(1 - a_1)}{a_1 - a_2}, \quad Q_{12} = Q^2 \frac{-(a_2 + M)(1 - a_2)}{a_1 - a_2},$$
$$Q_{21} = Q^2 \frac{(a_1 + M)(1 - a_1)}{a_1 - a_2}, \quad Q_{22} = Q^2 \frac{(a_1 + M)(1 - a_2)}{a_1 - a_2}. \tag{2.40}$$

Using the relations $M = -a_1 a_2$ and $Q^2(1 - a_1)(1 - a_2) = 1$ simplifies the off-diagonal expressions in (2.40) to

$$Q_{21} = \frac{a_1}{a_1 - a_2}, \quad Q_{12} = \frac{-a_2}{a_1 - a_2}. \tag{2.41}$$

Note again that the determinant of the parametric excitation matrix $\mathbf{K}_c^s$ leads to the same relation as in (2.31).

2.4.3. System 3

The third system of interest is a modification of System 1 in (2.21), and has been discussed so far in [11]. The system possesses a time-periodic stiffness coefficient k_{02} and a time-periodic damping coefficient c_{02}. Both time-periodic coefficients are varied synchronously. The non-vanishing system matrices of (2.6) yield

$$\mathbf{M}_0 = \begin{bmatrix} m_1 & 0 \\ 0 & m_2 \end{bmatrix}, \quad \mathbf{K}_0^s = \begin{bmatrix} k_{12} & -k_{12} \\ -k_{12} & k_{12} + k_{02} \end{bmatrix},$$
$$\mathbf{C}_0^s = \begin{bmatrix} c_{01} & 0 \\ 0 & c_{02} \end{bmatrix}, \quad \mathbf{K}_c^s = \begin{bmatrix} 0 & 0 \\ 0 & k_{02} \end{bmatrix}, \quad \mathbf{C}_c^s = \begin{bmatrix} 0 & 0 \\ 0 & c_{02} \end{bmatrix}. \tag{2.42}$$

Using the characteristic parameters from (2.22) and (2.23) the following non-dimensional system matrices are obtained:

$$\bar{\mathbf{M}}_0 = \begin{bmatrix} 1 & 0 \\ 0 & 1 \end{bmatrix}, \quad \bar{\mathbf{K}}_0^s = \begin{bmatrix} 1 & -1 \\ -M & M + Q^2 \end{bmatrix},$$
$$\bar{\mathbf{C}}_0^s = \begin{bmatrix} \kappa_1 & 0 \\ 0 & \kappa_2 \end{bmatrix}, \quad \bar{\mathbf{K}}_c^s = \begin{bmatrix} 0 & 0 \\ 0 & Q^2 \end{bmatrix}, \quad \bar{\mathbf{C}}_c^s = \begin{bmatrix} 0 & 0 \\ 0 & \kappa_2 \end{bmatrix}. \tag{2.43}$$

The matrices indexed by 0 correspond to constant system properties and are equal to the matrices as defined in (2.24). The variables a_i of the transformation matrix in (2.25) and the eigenvalues Ω_i^2 are equal to the expressions presented in (2.26) and (2.27). The non-dimensional equations of motion are transformed into the normal form

$$\bar{\mathbf{K}}_0^s \mapsto \begin{bmatrix} \Omega_1^2 & 0 \\ 0 & \Omega_2^2 \end{bmatrix} = \mathbf{\Omega}^2, \quad \bar{\mathbf{C}}_0^s \mapsto [\Theta_{ij}], \quad \bar{\mathbf{K}}_c^s \mapsto [Q_{ij}], \quad \bar{\mathbf{C}}_c^s \mapsto [R_{ij}]. \tag{2.44}$$

2 Equivalent minimum systems   15

The modal damping coefficients Θ_{ij} and the coefficients Q_{ij}, corresponding to the time-periodic stiffness variation, are equal to the expressions in (2.29) and (2.30). The coefficients S_{ij} that corresponds to the time-periodic damping variation are

$$R_{11} = \kappa_2 \frac{a_1}{a_1 - a_2}, \quad R_{12} = \kappa_2 \frac{a_2}{a_1 - a_2},$$
$$R_{21} = \kappa_2 \frac{-a_1}{a_1 - a_2}, \quad R_{22} = \kappa_2 \frac{-a_2}{a_1 - a_2}. \tag{2.45}$$

As before, that the determinants of the parametric excitation matrices $\mathbf{K}_c^s$ and $\mathbf{C}_c^s$ lead to similar relations as in (2.31).

2.4.4. System 4

The fourth system discussed is a modification of System 1 in (2.21) and System 3 in (2.42), as investigated in [12]. It is a system with time-periodic inertia

$$m_1(t) = m_1(1 + \varepsilon \cos(\omega t)) \tag{2.46}$$

and $k_{01}, c_{12} \equiv 0$. Respecting Newton's law of motion the inertia force, the time derivative of the impulse, for a time dependent inertia yield

$$\frac{d}{dt}(m_1(t)\dot{y}_1(t)) = m_1(1 + \varepsilon \cos(\omega t))\ddot{y}_1(t) - \varepsilon m_1 \omega \sin(\omega t)\dot{y}_1(t). \tag{2.47}$$

Hence, a time-periodic inertia variation leads to an additional damping variation with a phase shift of $\pi/2$. With (2.47) the non-vanishing system matrices of (2.6) are

$$\mathbf{M}_0 = \begin{bmatrix} m_1 & 0 \\ 0 & m_2 \end{bmatrix}, \quad \mathbf{K}_0^s = \begin{bmatrix} k_{12} & -k_{12} \\ -k_{12} & k_{12} + k_{02} \end{bmatrix},$$
$$\mathbf{C}_0^s = \begin{bmatrix} c_{01} & 0 \\ 0 & c_{02} \end{bmatrix}, \quad \mathbf{M}_c^s = \begin{bmatrix} m_1 & 0 \\ 0 & 0 \end{bmatrix}, \quad \mathbf{C}_s^s = \begin{bmatrix} -m_1\omega & 0 \\ 0 & 0 \end{bmatrix}. \tag{2.48}$$

Using the characteristic parameters from (2.22) and (2.23) we obtain the following non-dimensional system matrices

$$\bar{\mathbf{M}}_0 = \begin{bmatrix} 1 & 0 \\ 0 & 1 \end{bmatrix}, \quad \bar{\mathbf{K}}_0^s = \begin{bmatrix} 1 & -1 \\ -M & M + Q^2 \end{bmatrix},$$
$$\bar{\mathbf{C}}_0^s = \begin{bmatrix} \kappa_1 & 0 \\ 0 & \kappa_2 \end{bmatrix}, \quad \bar{\mathbf{M}}_c^s = \begin{bmatrix} 1 & 0 \\ 0 & 0 \end{bmatrix}, \quad \bar{\mathbf{C}}_s^s = \eta \begin{bmatrix} -1 & 0 \\ 0 & 0 \end{bmatrix}. \tag{2.49}$$

The matrices with the subscript 0 correspond to constant system properties and are equal to the matrices defined in (2.24).

Dividing the first equation by the time periodic part $(1 + \varepsilon \cos \eta \tau)$ of the inertia coefficient in (2.46) we can expand the remaining coefficients into Taylor series

$$\frac{1}{1 + \varepsilon \cos \eta \tau} = 1 - \varepsilon \cos \eta \tau + \varepsilon^2 \cos^2 \eta \tau + \mathcal{O}\left(\varepsilon^3\right),$$

$$\eta \frac{\varepsilon \sin \eta \tau}{1 + \varepsilon \cos \eta \tau} = \eta \sin \eta \tau \left(\varepsilon - \varepsilon^2 \cos \eta \tau + \mathcal{O}\left(\varepsilon^3\right)\right), \qquad (2.50)$$

for small values of ε. Keeping only terms up to first order of ε results in

$$\bar{\mathbf{M}}_c^s \mapsto \bar{\mathbf{K}}_c^s = \begin{bmatrix} -1 & 1 \\ 0 & 0 \end{bmatrix} \quad \text{and} \quad \bar{\mathbf{C}}_s^s = \eta \begin{bmatrix} -1 & 0 \\ 0 & 0 \end{bmatrix}. \qquad (2.51)$$

The system matrix $\bar{\mathbf{M}}_c^s$ in (2.49), corresponding to a time-periodic inertia variation, leads to the system matrix $\bar{\mathbf{K}}_c^s$ that corresponds to a time-periodic stiffness variation and to the system matrix $\bar{\mathbf{C}}_s^s$ that corresponds to a time-periodic damping variation. Due to the occurrence of the two different subscripts c, s in $\bar{\mathbf{K}}_c^s$ and $\bar{\mathbf{C}}_s^s$, these variations are performed with the same frequency but with a phase shift of π. In this work we term such a variation asynchronous. Finally, in first order approximation, the non-dimensional equations of motion are transformed into the normal form

$$\bar{\mathbf{K}}_0^s \mapsto \begin{bmatrix} \Omega_1^2 & 0 \\ 0 & \Omega_2^2 \end{bmatrix} = \boldsymbol{\Omega}^2, \quad \bar{\mathbf{C}}_0^s \mapsto [\Theta_{ij}], \quad \bar{\mathbf{K}}_c^s \mapsto [Q_{ij}], \quad \bar{\mathbf{C}}_s^s \mapsto \eta [S_{ij}]. \qquad (2.52)$$

The modal damping coefficients Θ_{ij} are equal to the expressions presented in (2.29) and (2.30). The coefficients Q_{ij} that correspond to the time-periodic stiffness variation yield

$$Q_{11} = \frac{a_2 (1 - a_1)}{a_1 - a_2}, \quad Q_{12} = \frac{a_2 (1 - a_2)}{a_1 - a_2},$$
$$Q_{21} = -\frac{a_1 (1 - a_1)}{a_1 - a_2}, \quad Q_{22} = -\frac{a_1 (1 - a_2)}{a_1 - a_2}, \qquad (2.53)$$

and the coefficients R_{ij} that correspond to the time-periodic damping variation result in

$$S_{11} = \frac{a_2}{a_1 - a_2} = S_{12}, \quad S_{21} = -\frac{a_1}{a_1 - a_2} = S_{22}. \qquad (2.54)$$

The determinants of the parametric excitation matrices $\bar{\mathbf{K}}_c^s$ and $\bar{\mathbf{C}}_s^s$ remain unchanged. Consequently, similar relations as in (2.31) are valid for the coefficients in (2.53) and (2.54).

Part I.

Stability of normal forms – analytical approaches

2 Equivalent minimum systems

The two coupled linear differential equations similar to (2.6),

$$\mathbf{M}(t)\ddot{\mathbf{x}}(t) + \varepsilon \mathbf{C}(t)\dot{\mathbf{x}}(t) + \mathbf{K}(t)\mathbf{x}(t) = \mathbf{0}, \qquad (2.55)$$

with time-dependent, but periodic coefficients cannot be solved analytically exact. Sometimes an exact solution cannot be obtained for a differential equation and an approximate solution must be found. Other times, an approximate solution may convey more information than an exact solution. The exact solution is approximated by performing a perturbation technique as used in the literature in various fields of physics. Famous representatives of these techniques are: harmonic balance (or two-timing), Poincaré-Lindstedt method, averaging method, method of multiple scales or successive approximation to mention the most popular ones. A general survey of these methods can be found in [40] and [52]. The method of vibrational mechanics as proposed in [5] leads to demonstrative physical interpretations if the fast and the slow motions are coupled additively, but this method is not capable of provide a solution if the slow and the fast motion are coupled multiplicatively, as it is the case in (2.55). The method chosen here is the averaging method, which is a method with a strong mathematical background. The approximations to the solution will be derived for a first order perturbation and are only valid in a region and on the time scale $1/\varepsilon$. Only systems with distinct eigenvalues are considered. Performing the method of averaging leads to cumbersome expressions, therefore a software package for symbolic computations as MAPLE [46] is a valuable task.

The following studies are divided into five chapters. The first two chapters examine the stability and instability boundary curves of mechanical systems where only the variation of one type of physical property is examined, while the following chapters investigate boundary curves in case of a simultaneous variation of more than one physical property. The first chapter deals with the variation of stiffness coefficients with a cosine function, with a general harmonic variation and finally, with a general time-periodic variation that can be represented by Fourier series. In the second chapter, boundary curves in case of a harmonic damping variation are presented. The third chapter analyzes the case of a simultaneous stiffness and damping variation with the same phase, which corresponds to a synchronous variation, while the fourth chapter deals with an asynchronous variation. Finally, fifth chapter examines the boundary curves in case of a variation of all system parameters: a simultaneous variation with a single frequency of stiffness, damping and inertia coefficients. The analysis presented here represents a summary and extension of [8] that is based on [1] and [30].

There are several publications dealing with coupled equations having a time-periodic stiffness coefficient, a system of coupled and damped Mathieu equations. In respect to the method of damping by parametric excitation as proposed in this work, the main contributions can be found in [53], [55], [56], [58], [31], [1], [8] and [9]. These works mainly analyze stability boundary curves of systems with a negative damping coefficient. The main focus of these works is the stabilizing effect of a parametric excitation. Instability boundary curves of positively damped systems can be derived from these results, too, which lead to results as derived in [50], [28], [7] [37], [51]. The main focus of these works is the destabilizing effect of a parametric excitation.

The first works dealing with a time-periodic damping coefficient in combination with a negative damping coefficient can be found in [57] and [8]. Finally, stability analysis of systems where more than one type of system parameters are varied periodically in time cannot – to the author's knowledge – be found in the literature, see [11] and [12].

3. Time-periodic stiffness

3.1. Single harmonic stiffness variation

For the case of a single harmonic stiffness variation the mass and damping matrices are kept constant

$$\mathbf{M}(t) = \mathbf{M}_0, \quad \mathbf{C}(t) = \mathbf{C}_0,$$
$$\mathbf{K}(t) = \mathbf{K}_0 + \varepsilon \left\{ \mathbf{K}_c^s + \mathbf{K}_c^a \right\} \cos(\omega t).$$

In this case the general linear equations of motion (2.2) simplifies to

$$\ddot{\mathbf{z}}(t) + \varepsilon \tilde{\mathbf{C}}_0 \dot{\mathbf{z}}(t) + \tilde{\mathbf{K}}_0 \mathbf{z}(t) = -\varepsilon \left\{ \tilde{\mathbf{K}}_c^s + \tilde{\mathbf{K}}_c^a \right\} \cos(\omega t) \mathbf{z}(t).$$

Note that, due to the fact that $\tilde{\mathbf{K}}_c$ is a matrix, more than one stiffness parameter of the system can be varied. The only restrictions which apply are that these variations have the same frequency ω and the same phase 0.

The equations of motion in dimensional terms can be transformed ($\mathbf{z}(t) \longrightarrow \mathbf{z}(\tau)$) to a non-dimensional normal form

$$\mathbf{z}''(\tau) + \varepsilon \boldsymbol{\Theta} \mathbf{z}'(\tau) + \boldsymbol{\Omega}^2 \mathbf{z}(\tau) = -\varepsilon \left(\mathbf{Q}^s + \mathbf{Q}^a \right) \cos(\eta \tau) \mathbf{z}(\tau),$$
$$\text{where } \mathbf{Q} = \mathbf{Q}^s + \mathbf{Q}^a = \mathbf{T}^{-1} \mathbf{M}_0^{-1} \mathbf{K}_c \mathbf{T},$$

as shown in Section 2.3. For a mechanical system with two degrees of freedom this yields

$$\mathbf{z}''(\tau) + \boldsymbol{\Omega}^2 \mathbf{z}(\tau) = -\varepsilon \left\{ \boldsymbol{\Theta} \mathbf{z}'(\tau) + (\mathbf{Q}^s + \mathbf{Q}^a) \cos(\eta \tau) \mathbf{z}(\tau) \right\},$$

$$\mathbf{z}''(\tau) + \begin{bmatrix} \Omega_1^2 & 0 \\ 0 & \Omega_2^2 \end{bmatrix} \mathbf{z}(\tau) = -\varepsilon \left\{ \begin{bmatrix} \Theta_{11} & \Theta_{12} \\ \Theta_{21} & \Theta_{22} \end{bmatrix} \mathbf{z}'(\tau) + \begin{bmatrix} Q_{11} & Q_{12} \\ Q_{21} & Q_{22} \end{bmatrix} \cos(\eta \tau) \mathbf{z}(\tau) \right\},$$

and rewritten in the comprehensive index notation

$$z_i''(\tau) + \Omega_i^2 z_i(\tau) = -\varepsilon \left\{ \Theta_{ij} z_j'(\tau) + Q_{ij} z_j(\tau) \cos(\eta \tau) \right\}, \qquad (3.1)$$

with $i, j = 1, 2$, using Einstein summation, a convention that repeated indices are implicitly summed over. Herein i is the free index. The approach presented below follows the procedure in [1] and [30]. This procedure is included for the sake of completeness and clarity and will be extended to a detailed study in the frequency range η.

3.1.1. Transformation and averaging

Applying a time transformation to (3.1) in order to normalize the frequency η to become one on the chosen time gives

$$\eta \tau \to t, \quad ()' = \frac{d}{d\tau} = \eta \frac{d}{dt} = \eta\,()^{\cdot}, \quad \frac{d^2}{d\tau^2} = \eta^2 \frac{d^2}{dt^2}, \tag{3.2}$$

$$z_i(\tau) = z_i(t/\eta) = \bar{z}_i(t).$$

Substituting in (3.1), dividing by η^2 and omitting the bar yields

$$\ddot{z}_i + \frac{\Omega_i^2}{\eta^2} z_i = -\frac{\varepsilon}{\eta} \Theta_{ij} \dot{z}_j - \frac{\varepsilon}{\eta^2} Q_{ij} z_j \cos t, \quad \text{with } i,j = 1,2 \tag{3.3}$$

Allowing a small detuning of first order near η_0 of the form

$$\boxed{\eta = \eta_0 + \varepsilon\sigma + \mathcal{O}\left(\varepsilon^2\right)} \tag{3.4}$$

and expanding the coefficients to Taylor series for small values of parameter ε yields

$$\ddot{z}_i + \varpi_i^2 z_i = -\frac{\varepsilon}{\eta_0^2} \left\{ +\eta_0 \Theta_{ij} \dot{z}_j + Q_{ij} z_j \cos t - 2\Omega_i \varpi_i \sigma z_i \right\} + \mathcal{O}\left(\varepsilon^2\right), \tag{3.5}$$

with the abbreviations $\varpi_i = \Omega_i/\eta_0$. For a *first order approximation* all terms of higher order than ε are neglected. Similar to the classical method of estimating the particular solution from the homogenous solution by variation of parameters, we perform a coordinate transformation that fulfills the unperturbed equations, $\varepsilon = 0$,

$$z_i = u_i \cos \varpi_i t + v_i \sin \varpi_i t, \quad \dot{z}_i = -u_i \varpi_i \sin \varpi_i t + v_i \varpi_i \cos \varpi_i t, \tag{3.6}$$

and by using the abbreviations $s_i = \sin \varpi_i t$ and $c_i = \cos \varpi_i t$ we obtain:

1. Satisfying the transformation in (3.6) yields

$$\dot{z}_i: \quad \dot{u}_i c_i - u_i \varpi_i s_i + \dot{v}_i s_i + v_i \varpi_i c_i = -u_i \varpi_i s_i + v_i \varpi_i c_i,$$
$$\dot{u}_i c_i + \dot{v}_i s_i = 0. \tag{3.7}$$

2. By differentiating (3.6) follows the relation

$$\ddot{z}_i = -\dot{u}_i \varpi_i s_i - u_i \varpi_i^2 c_i + \dot{v}_i \varpi_i c_i - v_i \varpi_i^2 s_i,$$

and inserting into (3.5) cancels some terms on the left hand side and we obtain an equation which couples u_i, v_i on the right hand side

$$-\dot{u}_i \varpi_i s_i + \dot{v}_i \varpi_i c_i = \frac{\varepsilon}{\eta_0^2} F_i(u_1, v_1, u_2, v_2, t),$$

$$= -\frac{\varepsilon}{\eta_0^2} \left\{ \eta_0 \sum_j \Theta_{ij} \left(-u_j \varpi_j s_j + v_j \varpi_j c_j \right) + \right.$$

$$\left. + \sum_j Q_{ij} \left(u_j c_j + v_j s_j \right) \cos t - 2\Omega_i \varpi_i \sigma \left(u_i c_i + v_i s_i \right) \right\}. \tag{3.8}$$

3 Single harmonic stiffness variation

Inserting (3.7) in (3.8) and using the relation $s_i^2 + c_i^2 = 1$ yields

$$-\varpi_i s_i \left(-\frac{s_i}{c_i}\right) \dot{v}_i + \varpi_i c_i \dot{v}_i = \frac{\varpi_i \dot{v}_i}{c_i} = \frac{\varepsilon}{\eta_0^2} F_i(\mathbf{u},\, t),$$

with the state vector $\mathbf{u} = (u_1, v_1, u_2, v_2)^T$, and the equations finally become

$$\dot{u}_i = -\frac{\varepsilon}{\eta_0^2 \varpi_i} F_i(\mathbf{u},\, t) \quad s_i = \varepsilon \tilde{F}_i^s(\mathbf{u},\, t),$$
$$\dot{v}_i = \frac{\varepsilon}{\eta_0^2 \varpi_i} F_i(\mathbf{u},\, t) \quad c_i = \varepsilon \tilde{F}_i^c(\mathbf{u},\, t). \qquad (i=1,2) \qquad (3.9)$$

The function $\tilde{F}_i$ on the right hand side of this system of equations are quasi-periodic – they are not periodic but they can be split into a finite sum of different periodic terms of the following form

$$\tilde{F}_i^{s,c}(\mathbf{u},\, t) = \sum_{k=1}^{N} \tilde{F}_{i,k}^{s,c}(\mathbf{u},\, t),$$

with N fixed and $\tilde{F}_{i,k}(\mathbf{u},\, t)$ is T_k-periodic in t. For this case we can apply the *averaging method in the general case* from [60, p.145]

$$\dot{\hat{u}}_i = \varepsilon \sum_{k=1}^{N} \frac{1}{T_k} \int_0^{T_k} \tilde{F}_{i,k}^s(\hat{\mathbf{u}},\, t)\, dt,$$
$$\dot{\hat{v}}_i = \varepsilon \sum_{k=1}^{N} \frac{1}{T_k} \int_0^{T_k} \tilde{F}_{i,k}^c(\hat{\mathbf{u}},\, t)\, dt, \qquad (3.10)$$

where the difference between the solutions $\mathbf{u}$ of the original and $\hat{\mathbf{u}}$ the averaged system is of order ε, $\hat{u}_i(t) - u_i(t) = \mathcal{O}(\varepsilon)$ on the time scale $1/\varepsilon$. The integration over T_k is carried out for fixed average values $\hat{\mathbf{u}}$.

Hence, for averaging we first have to determine the periods of the right hand sides of (3.9). With the help of decomposition theorems, see Appendix B by substituting $n = 1$, the arising products of the trigonometric terms on the right hand sides of (3.8, 3.9) can be rearranged as a sum of basic trigonometric terms. For this simple system with two modes 12 different periods arise. Averaging over a basic trigonometric term yields always zero, except for the case where a term becomes resonant, i.e. the argument of a cosine function vanishes:

$$\Theta_{ii},\, \sigma: \quad \text{no resonant terms, constant factor } \tfrac{1}{2},$$
$$\Theta_{ij}: \quad \text{resonant terms for } \varpi_i \pm \varpi_j = 0,$$
$$Q_{ii}: \quad \text{resonant terms for } 2\varpi_i \pm 1 = 0,$$
$$Q_{ij}: \quad \text{resonant terms for } \varpi_i \pm \varpi_j \pm 1 = 0.$$

General case

If ϖ_1, ϖ_2 do not take on special values averaging over each simple trigonometric function gives zero and only the terms Θ_{ii}, σ remain. After averaging (3.9) we get

$$\dot{\hat{u}}_i = \frac{\varepsilon}{\eta_0^2 \varpi_i} \left\{ -\frac{\eta_0}{2} \Theta_{ii} \varpi_i \hat{u}_i - \Omega_i \varpi_i \sigma \hat{v}_i \right\},$$

$$\dot{\hat{v}}_i = \frac{\varepsilon}{\eta_0^2 \varpi_i} \left\{ -\frac{\eta_0}{2} \Theta_{ii} \varpi_i \hat{v}_i + \Omega_i \varpi_i \sigma \hat{u}_i \right\}.$$

(3.11)

Using the abbreviations

$$\hat{w}_i = \hat{u}_i + j\hat{v}_i,$$

where $j = \sqrt{-1}$ is the complex unit, we obtain from (3.11)

$$\dot{\hat{w}}_i = \frac{\varepsilon}{\eta_0^2} \left\{ -\frac{\eta_0}{2} \Theta_{ii} + j\Omega_i \sigma \right\} \hat{w}_i.$$

Rewritten in matrix notation, keeping in mind that i is the free index, we obtain equations of the form

$$\dot{\hat{\mathbf{w}}}(t) = \frac{\varepsilon}{\eta_0^2} \begin{bmatrix} -\frac{\eta_0}{2}\Theta_{11} + j\Omega_1\sigma & 0 \\ 0 & -\frac{\eta_0}{2}\Theta_{22} + j\Omega_2\sigma \end{bmatrix} \hat{\mathbf{w}}(t) = \frac{\varepsilon}{\eta_0^2} \mathbf{A} \hat{\mathbf{w}}(t),$$

with the complex state vector $\hat{\mathbf{w}} = (\hat{w}_1, \hat{w}_2)^T$. After rescaling time by ε/η_0^2 we obtain the characteristic equation of the coefficient matrix

$$\det(\lambda \mathbf{I}_2 - \mathbf{A}) = 0,$$

(3.12)

that determines the stability of the trivial solution, where $\mathbf{I}_k$ is the unity matrix of dimension k. Of course we could calculate the stability border from two simple complex polynomials of order one. Nevertheless, for a comparison and deeper understanding of the relations between this general case and the following special cases the stability is calculated from the more complicated complex polynomial of order two.

First we analyze the case of $\sigma = 0$, checking whether the system is stable at the frequency η_0 or not according to (3.4). Looking for a *stable point* yields

$$\lambda^2 - \text{trace}(\mathbf{A})\lambda + \det(\mathbf{A}) = 0,$$

$$\lambda^2 + \left(\tfrac{1}{2}\eta_0 (\Theta_{11} + \Theta_{22})\right)\lambda + \tfrac{1}{4}\eta_0^2 \Theta_{11}\Theta_{22} = 0.$$

(3.13)

Applying the Routh-Hurwitz criterion for real valued polynomials according to (A.1) this polynomial is stable and all eigenvalues have only negative real parts, if the corresponding determinants satisfy the following necessary and sufficient conditions

$$\Delta_1 : \tfrac{1}{2}\eta_0 (\Theta_{11} + \Theta_{22}) > 0,$$

$$\Delta_2 : \tfrac{1}{4}\eta_0^2 (\Theta_{11} + \Theta_{22})^2 \Theta_{11}\Theta_{22} > 0.$$

3 Single harmonic stiffness variation

For physically meaningful systems η_0 is positive, and the conditions simplify to

$$\Theta_{11} + \Theta_{22} > 0,$$

$$\Theta_{11}\Theta_{22} > 0.$$

The second inequality means that Θ_{11}, Θ_{22} need to have the same sign. Respecting the first inequality demands that both coefficients have to be positive

$$\boxed{\Theta_{11} > 0 \quad \text{and} \quad \Theta_{22} > 0.} \qquad (3.14)$$

These are the general conditions for stability, where the system motion is damped until it comes to rest. This result is the classical result that can be found in every common textbook on dynamical systems: A non-conservative, non-gyroscopic system is stable if both damping coefficients are positive, and therefore kinetic energy is dissipated. This result is already clear from (3.11) for $\sigma = 0$, where the equations are decoupled. Note that positive damping parameters for the non-dimensional modal-transformed system do *not* imply that all damping coefficients of the physical system are positive, too. There are parameter sets where, although some the physical damping parameters are negative, the modal damping parameters Θ_{ii} are all positive, which represents a so-called *stability reserve* of the physical system. This will become more clear in the sections dealing with mechanical examples and their numerical analysis.

Analyzing (3.12) for $\sigma \neq 0$ we check whether a stability border exists. Looking for a *stability interval* σ according to (3.4) yields

$$\lambda^2 + \left(\tfrac{1}{2}\eta_0 \left(\Theta_{11} + \Theta_{22}\right) - j\left(\Omega_1 + \Omega_2\right)\sigma\right)\lambda$$
$$+ \tfrac{1}{4}\eta_0^2 \Theta_{11}\Theta_{22} - \Omega_1\Omega_2\sigma^2 + j\tfrac{1}{2}\eta_0 \left(\Theta_{11}\Omega_2 + \Theta_{22}\Omega_1\right)\sigma = 0.$$

Applying the extended Routh-Hurwitz criterion for polynomials with complex coefficients from (A.2) this complex polynomial is stable iff

$$\Delta_1: \quad \tfrac{1}{2}\eta_0 \left(\Theta_{11} + \Theta_{22}\right) > 0,$$

$$\Delta_2: \begin{cases} \tfrac{1}{4}\eta_0^2 \left(\tfrac{1}{4}\eta_0^2 \left(\Theta_{11} + \Theta_{22}\right)^2 + \left(\Omega_1 - \Omega_2\right)^2 \sigma^2\right)\Theta_{11}\Theta_{22} > 0, \\ \tfrac{1}{4}\eta_0^2 \left(\sigma - \sigma_1\right)\left(\sigma - \sigma_2\right)\Theta_{11}\Theta_{22} > 0, \\ a_0 \sigma^2 + a_2 > 0, \\ \text{critical values:} \\ \sigma_{1,2} = \pm \dfrac{\eta_0 \left(\Theta_{11} + \Theta_{22}\right)}{2\left(\Omega_1 - \Omega_2\right)}\sqrt{-1}. \end{cases}$$

The critical values $\sigma_{1,2}$ border a frequency interval where the inequality Δ_2 is either satisfied or violated. Looking at the different cases for the incomplete set of possible parameter combinations

$$\{a_i\} = \left\{ \begin{Bmatrix} \Theta_{11} > 0 \\ \Theta_{22} > 0 \end{Bmatrix}, \begin{Bmatrix} \Theta_{11} > 0 \\ \Theta_{22} < 0 \end{Bmatrix}, \begin{Bmatrix} \Theta_{11} < 0 \\ \Theta_{22} > 0 \end{Bmatrix}, \begin{Bmatrix} \Theta_{11} < 0 \\ \Theta_{22} < 0 \end{Bmatrix} \right\},$$

a_1 is always satisfied; a_2, a_3 do not satisfy the condition Δ_2, and a_4 does not satisfy both conditions in Δ_1 and Δ_2. In all other cases, where one of the damping terms vanishes, the system is on the stability limit, the vibrations are neither damped nor excited.

The critical values $\sigma_{1,2}$ determine the boundaries between stable and unstable frequency intervals. For purely complex σ-values a stability boundary does not exist. Globally the system is either always stable or always unstable depending on conditions in (3.14). The term "always" means for any chosen frequency η_0! However, there can exist *local* stability around certain frequency values η_0 where (3.11) is not longer valid and the strict stability conditions in (3.14) are relaxed. This will be shown in the next paragraphs.

Some resonant cases

If ϖ_1, ϖ_2 take on special values, some terms become resonant and influence the previous solution in (3.11), specifically the averaging process. In physically meaningful systems the two natural frequencies Ω_1, Ω_2 are *positive*. The case, where one frequency is equal to zero, which is the case for systems that are not connected to the inertia reference frame, is not relevant for the study of systems with two degrees of freedom.

1. Simple parametric resonance frequencies of first kind

By setting

$$\boxed{\eta_0 = 2\Omega_k \quad \text{or} \quad 2\varpi_k = 1, \quad \text{with } k = 1 \text{ or } 2}$$

some terms become resonant and produce additional terms after averaging. Instead of (3.11) we obtain from (3.9)

$$\dot{\hat{u}}_i = \frac{\varepsilon}{\eta_0^2} \left\{ -\frac{\eta_0}{2}\Theta_{ii}\hat{u}_i - \left(\frac{Q_{ii}}{4\varpi_i}\delta_{ik} + \Omega_i\sigma\right)\hat{v}_i \right\},$$

$$\dot{\hat{v}}_i = \frac{\varepsilon}{\eta_0^2} \left\{ -\frac{\eta_0}{2}\Theta_{ii}\hat{v}_i - \left(\frac{Q_{ii}}{4\varpi_i}\delta_{ik} - \Omega_i\sigma\right)\hat{u}_i \right\},$$

(3.15)

where δ_{ik} is the Kronecker delta-function. This case cannot be rewritten in complex notation like in the previous general case. By rearranging the equations and representing them in matrix notation one gets equations of the form

$$\dot{\hat{\mathbf{u}}}(t) = \frac{\varepsilon}{\eta_0^2} \begin{bmatrix} \mathbf{L}_1 & \mathbf{0} \\ \mathbf{0} & \mathbf{L}_2 \end{bmatrix} \hat{\mathbf{u}}(t) = \frac{\varepsilon}{\eta_0^2}\mathbf{A}_4\hat{\mathbf{u}}(t),$$

with $\hat{\mathbf{u}} = (\hat{u}_1, \hat{v}_1, \hat{u}_2, \hat{v}_2)^T$ and

$$\mathbf{L}_i = \begin{bmatrix} -\dfrac{\eta_0}{2}\Theta_{ii} & -\Omega_i\sigma - \dfrac{Q_{ii}}{4\varpi_i}\delta_{ik} \\ \Omega_i\sigma - \dfrac{Q_{ii}}{4\varpi_i}\delta_{ik} & -\dfrac{\eta_0}{2}\Theta_{ii} \end{bmatrix}.$$

3 Single harmonic stiffness variation

Thus the states $(\hat{u}_1, \hat{v}_1)$ and $(\hat{u}_2, \hat{v}_2)$ are decoupled. After rescaling time by ε/η_0^2 we obtain the characteristic equation

$$\det(\lambda \mathbf{I}_4 - \mathbf{A}_4) = \det(\lambda \mathbf{I}_2 - \mathbf{L}_1)\det(\lambda \mathbf{I}_2 - \mathbf{L}_2) = 0.$$

At least one of the determinants has to be zero. Without restriction of generality we choose the first determinant to be zero, this means we are looking just for stable behavior of the states $(\hat{u}_1, \hat{v}_1)$

$$\det(\lambda \mathbf{I}_2 - \mathbf{L}_1) = 0 = \lambda^2 + b_1\lambda + b_0.$$

Looking for a *stable point*, means analyzing (3.15) for $\sigma = 0$, and yields

$$\lambda^2 + \eta_0 \Theta_{11}\lambda + \frac{1}{4}\eta_0^2\Theta_{11}^2 - \frac{1}{4}\eta_0^2\frac{Q_{11}^2}{4\Omega_1^2}\delta_{1k} = 0. \tag{3.16}$$

Applying the Routh-Hurwitz criterion according to (A.1) this polynomial is stable iff

$$\Delta_1 : \eta_0\Theta_{11} > 0,$$

$$\Delta_2 : \frac{1}{16\Omega_1^2}\eta_0^3\Theta_{11}(2\Omega_1\Theta_{11} - Q_{11}\delta_{1k})(2\Omega_1\Theta_{11} + Q_{11}\delta_{1k}) > 0,$$

which simplifies for $\eta_0, \Omega_1 > 0$ (meaningful system)

$$\Theta_{11} > 0,$$

$$2\Omega_1\Theta_{11} > |Q_{11}|\delta_{1k}.$$

For the second determinant, looking just for stable behavior of $(\hat{u}_2, \hat{v}_2)$, we get exactly the same result by changing the indices from $1 \longmapsto 2$. Thus, for stability at $\eta_0 = 2\Omega_k$ the conditions

$$\eta_0 = 2\Omega_1: \quad \Theta_{11} > \frac{|Q_{11}|}{2\Omega_1} \quad \text{and} \quad \Theta_{22} > 0,$$

$$\eta_0 = 2\Omega_2: \quad \Theta_{11} > 0 \quad \text{and} \quad \Theta_{22} > \frac{|Q_{22}|}{2\Omega_2},$$

have to be satisfied, which can be written in short notation as

$$\boxed{\Theta_{ii} > \frac{|Q_{ii}|}{2\Omega_i}\delta_{ik} \geq 0.} \tag{3.17}$$

Contrary to (3.14) this result demands a more restrictive condition for Θ_{kk} in the case of a parametric resonance frequency of first kind at $2\Omega_k$.

These results hold only for $\sigma = 0$. For the case $\sigma \neq 0$, instead of (3.16), we obtain

$$\lambda^2 + \eta_0\Theta_{11}\lambda + \frac{1}{4}\eta_0^2\Theta_{11}^2 + \Omega_1^2\sigma^2 - \frac{1}{16\Omega_1^2}\eta_0^2 Q_{11}^2\delta_{1k} = 0. \tag{3.18}$$

Applying the Routh-Hurwitz criterion according to (A.1) this polynomial is stable if and only if

$$\Delta_1: \quad \eta_0 \Theta_{11} > 0,$$

$$\Delta_2: \quad \frac{1}{16\Omega_1^2} \eta_0^3 \Theta_{11} \left(\frac{16\Omega_1^2}{\eta_0^2} \sigma^2 + 4\Omega_1^2 \Theta_{11}^2 - Q_{11}^2 \delta_{1k} \right) > 0.$$

This yields for $\eta_0, \Omega_1 > 0$ the critical values, the boundaries between stable and unstable frequency intervals

$$\Theta_{11} > 0, \quad \sigma_{1,2} = \pm \frac{\eta_0}{2\Omega_1} \sqrt{-\Theta_{11}^2 + \frac{Q_{11}^2}{4\Omega_1^2} \delta_{1k}}.$$

For the second determinant, looking just for stable behavior of the states $(\hat{u}_2, \hat{v}_2)$, we obtain the same result by changing the indices from $1 \longmapsto 2$. Finally, for stability at $\eta_0 = 2\Omega_k$ the conditions

$$\boxed{\Theta_{ii} > 0, \quad \sigma_{1,2} = \pm \frac{\eta_0}{2\Omega_i} \sqrt{-\Theta_{ii}^2 + \frac{Q_{ii}^2}{4\Omega_i^2} \delta_{ik}}} \quad (3.19)$$

have to be satisfied. We see from (3.19) that, if the damping coefficients have different signs, there is no parameter combination possible for which we can stabilize the system. Or in other words: introducing one negative modal damping Θ_{ii} *always destabilizes* the system at the frequencies $\eta_0 = 2\Omega_k$. The same conclusion can be made if both modal damping coefficients are negative.

If we ask for a stable states $(\hat{u}_i, \hat{v}_i)$, then (3.17) has to hold and the stability borders in (3.19) become purely imaginary. Thus, a stability boundary for a stable behavior of $(\hat{u}_i, \hat{v}_i)$ does not exist. In this case the system is always stable in the vicinity of the frequencies $\eta_0 = 2\Omega_k$ as well as in the whole frequency range of η_0 according to (3.14).

If we choose the coefficients such that the conditions (3.14) are satisfied, but conditions (3.17) are not, then

$$\frac{|Q_{ii}|}{2\Omega_i} \delta_{ik} > \Theta_{ii} > 0$$

holds. Now the critical value σ_1 for the stability boundary in (3.19) becomes real valued. Although both damping coefficients are positive, the additional restriction in (3.17) is violated and the system is unstable at the frequency $\eta = \eta_0 = 2\Omega_k$ as well as within a distance of $\pm \varepsilon \sigma_1$ from η_0. Outside this stability area the system is stable due to the conditions in (3.14). These results are sketched in Fig. 3.1.

The figure shows the stability boundary as a function of an arbitrary system parameter p for the case when the conditions in (3.17) are not satisfied. The modal coefficients are more or less complicated functions of this system parameter, i.e the undamped natural frequency $\Omega_i = \Omega_i(p)$. Therefore, the frequency line $\eta = \eta_0 = 2\Omega_k$ is drawn as a curved line in this diagram. If the conditions in (3.17) are not satisfied, then the critical values σ calculated from (3.19) are real valued. There exist stability boundaries at $\eta = \eta_0 + \varepsilon \sigma_1$ and $\eta = \eta_0 + \varepsilon \sigma_2$ that

3 Single harmonic stiffness variation

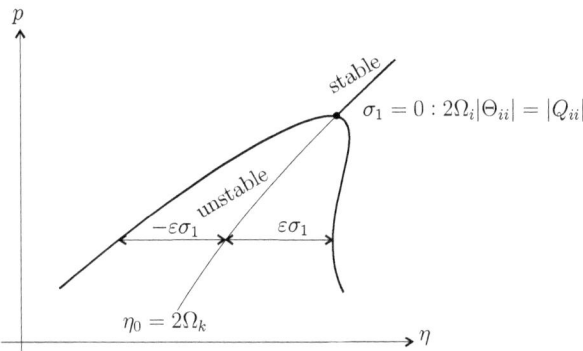

Figure 3.1.: Stability boundary from (3.19) in dependency of conditions in (3.17).

separate the parameter sets for unstable and stable system behavior. Since the conditions in (3.17) are violated, the system is *unstable inside* the stability boundary and according to the conditions in (3.14) the system is stable outside. On the other hand, if the conditions in (3.17) are satisfied then the critical value σ_1 for the stability boundary is purely imaginary. In this case the frequency line $\eta = \eta_0$ itself is stable as well as the whole frequency range according to (3.14). For the case where the value σ_1 vanishes, the relation

$$\Theta_{kk} = \frac{|Q_{kk}|}{2\Omega_k}$$

holds, which is the critical relation for (3.17).

For the case where one modal damping parameter is negative the system is always unstable due to the general case conditions in (3.14) and unstable due to the special case conditions in (3.17), although (3.19) gives a real valued σ_1.

Recapitulating this special case it is not possible to stabilize the system for a negative modal damping as well as for a moderate positive modal damping, which does not satisfy the conditions in (3.17).

2. Combination parametric resonance of difference type and first kind

By setting

$$\boxed{\eta_0 = |\Omega_1 - \Omega_2| > 0 \quad \text{or} \quad \varpi_1 - \varpi_2 = \mp 1}$$

some terms become resonant and produce additional terms after averaging and we get from (3.9)

$$\dot{\hat{u}}_i = \frac{\varepsilon}{\eta_0^2 \varpi_i} \left\{ -\frac{\eta_0}{2} \Theta_{ii} \varpi_i \hat{u}_i + \frac{Q_{ij}}{4} \hat{v}_j - \Omega_i \varpi_i \sigma \hat{v}_i \right\},$$

$$\dot{\hat{v}}_i = \frac{\varepsilon}{\eta_0^2 \varpi_i} \left\{ -\frac{\eta_0}{2} \Theta_{ii} \varpi_i \hat{v}_i - \frac{Q_{ij}}{4} \hat{u}_j + \Omega_i \varpi_i \sigma \hat{u}_i \right\},$$

(3.20)

instead of (3.11). Rewritten in matrix notation yields

$$
\begin{pmatrix} \dot{\hat{u}}_1 \\ \dot{\hat{v}}_1 \\ \dot{\hat{u}}_2 \\ \dot{\hat{v}}_2 \end{pmatrix} = \frac{\varepsilon}{\eta_0^2} \begin{bmatrix} -\frac{\eta_0}{2}\Theta_{11} & -\Omega_1\sigma & 0 & \frac{\eta_0}{4\Omega_1}Q_{12} \\ \Omega_1\sigma & -\frac{\eta_0}{2}\Theta_{11} & -\frac{\eta_0}{4\Omega_1}Q_{12} & 0 \\ 0 & \frac{\eta_0}{4\Omega_2}Q_{21} & -\frac{\eta_0}{2}\Theta_{22} & -\Omega_2\sigma \\ -\frac{\eta_0}{4\Omega_2}Q_{21} & 0 & \Omega_2\sigma & -\frac{\eta_0}{2}\Theta_{22} \end{bmatrix} \begin{pmatrix} \hat{u}_1 \\ \hat{v}_1 \\ \hat{u}_2 \\ \hat{v}_2 \end{pmatrix} = \frac{\varepsilon}{\eta_0^2} \mathbf{A}_4 \hat{\mathbf{u}},
$$

the same equation as in [1, p.69]. These equations are fully coupled and cannot be split as in the previous section. After rescaling time by ε/η_0^2 we obtain the characteristic equation

$$\det(\lambda \mathbf{I}_4 - \mathbf{A}_4) = 0 = \lambda^4 + b_1\lambda^3 + b_2\lambda^2 + b_3\lambda + b_4,$$

a polynomial of order four. Using the abbreviations

$$\hat{w}_i = \hat{u}_i + j\hat{v}_i,$$

where $j = \sqrt{-1}$ is the complex unit, (3.20) is equivalent to

$$\dot{\hat{w}}_i = \frac{\varepsilon}{\eta_0^2} \left\{ \left(-\frac{\eta_0}{2}\Theta_{ii} + j\Omega_i\sigma \right) \hat{w}_i - j\frac{\eta_0}{4\Omega_i} Q_{ik} \hat{w}_k \right\}, \quad k \neq i.$$

Rewritten in matrix notation we obtain equations of the form

$$\dot{\hat{\mathbf{w}}}(t) = \frac{\varepsilon}{\eta_0^2} \begin{bmatrix} -\frac{\eta_0}{2}\Theta_{11} + j\Omega_1\sigma & -j\frac{\eta_0}{4\Omega_1}Q_{12} \\ -j\frac{\eta_0}{4\Omega_2}Q_{21} & -\frac{\eta_0}{2}\Theta_{22} + j\Omega_2\sigma \end{bmatrix} \hat{\mathbf{w}}(t) = \frac{\varepsilon}{\eta_0^2} \mathbf{A}_2 \hat{\mathbf{w}}(t), \quad (3.21)$$

with the complex state vector $\hat{\mathbf{w}} = (\hat{w}_1, \hat{w}_2)^T$. After rescaling time by ε/η_0^2 the characteristic equation of the coefficient matrix is reduced from a real polynomial of order four to a complex polynomial of order two

$$\det(\lambda \mathbf{I}_2 - \mathbf{A}_2) = 0,$$

$$\lambda^2 + \left(\tfrac{1}{2}\eta_0(\Theta_{11} + \Theta_{22}) - j(\Omega_1 + \Omega_2)\sigma \right) \lambda \\ + \left(\tfrac{1}{2}\eta_0\Theta_{11} - j\Omega_1\sigma \right)\left(\tfrac{1}{2}\eta_0\Theta_{22} - j\Omega_2\sigma \right) + \frac{\eta_0^2}{16\Omega_1\Omega_2}Q_{12}Q_{21} = 0. \quad (3.22)$$

Comparing with (3.13) reveals that the determinant of $\mathbf{A}$ is extended by the parametric excitation term $Q_{12}Q_{21}$ while the trace of $\mathbf{A}$ remains unchanged, see (3.13).

First analyzing the case for $\sigma = 0$, looking for a stable point, by applying the extended Routh-Hurwitz criterion according to (A.2) this polynomial is stable iff the conditions

$$\Delta_1 : \tfrac{1}{2}\eta_0(\Theta_{11} + \Theta_{22}) > 0,$$
$$\Delta_2 : \tfrac{1}{16}\eta_0^4 (\Theta_{11} + \Theta_{22})^2 \left(\Theta_{11}\Theta_{22} + \frac{1}{4\Omega_1\Omega_2}Q_{12}Q_{21} \right) > 0,$$

3 Single harmonic stiffness variation

are fulfilled. These conditions can be simplified for $\eta_0 > 0$ and read

$$\Theta_{11} + \Theta_{22} > 0,$$
$$\Theta_{11}\Theta_{22} + \frac{1}{4\Omega_1\Omega_2}Q_{12}Q_{21} > 0. \tag{3.23}$$

For the case $\sigma \neq 0$, looking for a stability interval gives after some lengthy calculations the following stability conditions

$$\Delta_1 : \tfrac{1}{2}\eta_0\left(\Theta_{11} + \Theta_{22}\right) > 0,$$

$$\Delta_2 : \begin{cases} \tfrac{1}{16}\eta_0^4\left(\Theta_{11} + \Theta_{22}\right)^2\left(\Theta_{11}\Theta_{22} + \dfrac{1}{4\Omega_1\Omega_2}Q_{12}Q_{21} - \dfrac{4}{\eta_0^2}\Omega_1\Omega_2\sigma^2\right) + \\ +\tfrac{1}{4}\eta_0^2\left(\Theta_{11}\Omega_2 + \Theta_{22}\Omega_1\right)\left(\Theta_{11}\Omega_1 + \Theta_{22}\Omega_2\right)\sigma^2 > 0, \\ a_2\sigma^2 + a_0 > 0, \\ \text{critical values:} \\ \sigma_{1,2} = \pm\dfrac{\eta_0\left(\Theta_{11} + \Theta_{22}\right)}{2\left(\Omega_1 - \Omega_2\right)}\sqrt{-\dfrac{\Theta_{11}\Theta_{22} + \dfrac{1}{4\Omega_1\Omega_2}Q_{12}Q_{21}}{\Theta_{11}\Theta_{22}}}. \end{cases} \tag{3.24}$$

Note that the critical case $\sigma_1 = 0 = \sigma_2$ in (3.24) is an equivalent formulation for the critical case of the second condition in (3.23).

In combination with (3.23) a classification of the critical values $\sigma_{1,2}$, which are the boundaries between the stable and unstable parameter domains, for distinct modal damping parameters can be made:

If both modal damping coefficients are positive, then $\Theta_{11}\Theta_{22} > 0$ holds. Hence, the general stability conditions in (3.14) are fulfilled. Although the system is generally stable, if the conditions in (3.23) are not satisfied, the system becomes unstable in the vicinity of the parametric combination frequency $\eta_0 = |\Omega_1 - \Omega_2|$. The first condition in (3.23) is trivially fulfilled for positive damping parameters, but the second condition demands the further restriction

$$\Theta_{11}\Theta_{22} > \frac{-1}{4\Omega_1\Omega_2}Q_{12}Q_{21}$$

if the term $Q_{12}Q_{21}$ is *negative*. Resulting from (3.24) the calculated critical values $\sigma_{1,2}$ are real-valued. Hence, the stable parameter space is overlapped by an instability domain near η_0 with a width of $\sigma_{1,2}$ according to (3.4). In this case the frequency η_0 is called a parametric resonance frequency. On the other hand, if the conditions in (3.23) are fulfilled, then the calculated critical values $\sigma_{1,2}$ are purely imaginary and the system is stable at η_0, too.

In general a system is unstable once one modal damping parameter becomes negative. Hence, the inequality $\Theta_{11}\Theta_{22} < 0$ is valid and the general stability conditions in (3.14) cannot be satisfied. In the case where the frequency of the single harmonic parametric excitation is near to the parametric combination resonance frequency $\eta_0 = |\Omega_1 - \Omega_2|$ the system remains unstable as long as the inequalities in (3.23) are not fulfilled. The stability width $\sigma_{1,2}$ in (3.24), the critical values for stability, are purely imaginary. Thus no stability change occurs, the system remains unstable near η_0. On the other hand, if the conditions from (3.23)

$$\Theta_{11} + \Theta_{22} > 0,$$

$$0 > \Theta_{11}\Theta_{22} > \frac{-1}{4\Omega_1\Omega_2} Q_{12} Q_{21},$$

are satisfied, then the system becomes stable, although a negative modal damping is present. In this case the frequency η_0 is called a parametric *anti*-resonance frequency. From the above conditions it can be easily concluded that a parametric anti-resonance at $\eta_0 = |\Omega_1 - \Omega_2|$ can only occur if the term $Q_{12}Q_{21}$ is *positive*. This classification is summarized in Tab. 3.1 on page 36. A more detailed study on the symmetry property of the parametric excitation term $Q_{12}Q_{21}$ is investigated in Section 3.1.3.

3. Combination parametric resonance of summation type and first kind
If we assume that

$$\boxed{\eta_0 = \Omega_i + \Omega_j \quad \text{or} \quad \varpi_i + \varpi_j = 1,}$$

some terms become resonant and produce additional terms after averaging. Instead of (3.11) we obtain from (3.9)

$$\dot{\hat{u}}_i = \frac{\varepsilon}{\eta_0^2 \varpi_i} \left\{ -\frac{\eta_0}{2}\Theta_{ii}\varpi_i \hat{u}_i - \frac{Q_{ij}}{4}\hat{v}_j - \Omega_i \varpi_i \sigma \hat{v}_i \right\},$$

$$\dot{\hat{v}}_i = \frac{\varepsilon}{\eta_0^2 \varpi_i} \left\{ -\frac{\eta_0}{2}\Theta_{ii}\varpi_i \hat{v}_i - \frac{Q_{ij}}{4}\hat{u}_j + \Omega_i \varpi_i \sigma \hat{u}_i \right\}.$$

(3.25)

Comparing with (3.20) reveals that only the sign of the parametric excitation terms Q_{ij} in the first set of equations changes. Using now the abbreviations

$$\hat{w}_i = \hat{u}_i + j(-1)^{i-1}\hat{v}_i,$$

$$\hat{w}_i = \hat{u}_i + j\hat{v}_i, \quad \text{for } i = \text{even} : \hat{w}_i \to \overline{\hat{w}_i} \text{ (c.c.)}$$

where $j = \sqrt{-1}$ is the complex unit and i is the free index, (3.25) is equivalent to

$$\dot{\hat{w}}_i = \frac{\varepsilon}{\eta_0^2} \left\{ \left(-\frac{\eta_0}{2}\Theta_{ii} + j(-1)^{i-1}\Omega_i \sigma \right) \hat{w}_i - j(-1)^{i-1}\frac{\eta_0}{4\Omega_i} Q_{ik} \hat{w}_k \right\} \quad k \neq i.$$

Rewritten in matrix notation we get equations of the form

$$\dot{\hat{\mathbf{w}}}(t) = \frac{\varepsilon}{\eta_0^2} \begin{bmatrix} -\frac{\eta_0}{2}\Theta_{11} + j\Omega_1 \sigma & -j\frac{\eta_0}{4\Omega_1} Q_{12} \\ j\frac{\eta_0}{4\Omega_2} Q_{21} & -\frac{\eta_0}{2}\Theta_{22} - j\Omega_2 \sigma \end{bmatrix} \hat{\mathbf{w}}(t) = \frac{\varepsilon}{\eta_0^2} \mathbf{A}_2 \hat{\mathbf{w}}(t), \quad (3.26)$$

3 Single harmonic stiffness variation

with the complex state vector $\hat{\mathbf{w}} = (\hat{w}_1, \hat{w}_2)^T$. After rescaling time by ε/η_0^2 the characteristic equation of the coefficient matrix is a polynomial of order two

$$\det(\lambda \mathbf{I}_2 - \mathbf{A}_2) = 0,$$

$$\lambda^2 + \left(\tfrac{1}{2}\eta_0 \left(\Theta_{11} + \Theta_{22}\right) - j\left(\Omega_1 - \Omega_2\right)\sigma\right)\lambda$$
$$+ \left(\tfrac{1}{2}\eta_0\Theta_{11} - j\Omega_1\sigma\right)\left(\tfrac{1}{2}\eta_0\Theta_{22} + j\Omega_2\sigma\right) - \frac{\eta_0^2}{16\Omega_1\Omega_2}Q_{12}Q_{21} = 0. \quad (3.27)$$

Comparing with results of the averaging process for the general case (3.11) the determinant of the system matrix is extended by the expression $Q_{12}Q_{21}$ while the trace remains unchanged, according to (3.13). Note the different sign of terms containing Ω_2 compared with the previous section where $\eta_0 = |\Omega_1 - \Omega_2|$. Hence we can conclude the stability conditions for $\eta_0 = \Omega_1 + \Omega_2$ by performing the substitution $\Omega_2 \mapsto -\Omega_2$ in the resultant stability conditions of $\eta_0 = |\Omega_1 - \Omega_2|$.

First analyzing the case for $\sigma = 0$, checking whether the system is stable at the frequency $\eta_0 = \Omega_1 + \Omega_2$ or not, the necessary and sufficient conditions are

$$\Theta_{11} + \Theta_{22} > 0,$$
$$\Theta_{11}\Theta_{22} - \frac{1}{4\Omega_1\Omega_2}Q_{12}Q_{21} > 0, \quad (3.28)$$

according to (3.23). Note the different sign of the $Q_{12}Q_{21}$ term compared with the previous section where $\eta_0 = |\Omega_1 - \Omega_2|$. For the case $\sigma \neq 0$, looking for a *stability interval* σ in the vicinity of the frequency η_0 according to (3.4), the following stability conditions

$$\Delta_1: \tfrac{1}{2}\eta_0\left(\Theta_{11} + \Theta_{22}\right) > 0,$$

$$\Delta_2: \begin{cases} \tfrac{1}{16}\eta_0^4\left(\Theta_{11} + \Theta_{22}\right)^2 \left(\Theta_{11}\Theta_{22} - \dfrac{1}{4\Omega_1\Omega_2}Q_{12}Q_{21} + \dfrac{4}{\eta_0^2}\Omega_1\Omega_2\sigma^2\right) - \\ \quad - \tfrac{1}{4}\eta_0^2\left(\Theta_{11}\Omega_2 - \Theta_{22}\Omega_1\right)\left(\Theta_{11}\Omega_1 - \Theta_{22}\Omega_2\right)\sigma^2 > 0, \\ a_2\sigma^2 + a_0 > 0, \\ \text{critical values:} \\ \sigma_{1,2} = \pm \dfrac{\eta_0\left(\Theta_{11} + \Theta_{22}\right)}{2\left(\Omega_1 + \Omega_2\right)}\sqrt{-\dfrac{\Theta_{11}\Theta_{22} - \dfrac{1}{4\Omega_1\Omega_2}Q_{12}Q_{21}}{\Theta_{11}\Theta_{22}}}, \end{cases} \quad (3.29)$$

are obtained from (3.24). Note the change in the sign not only for the parametric excitation term but also in the denominator of the last equation. Note further that again the critical case $\sigma_1 = 0$ in (3.29) is an equivalent formulation for the critical case of the second condition in (3.28).

In combination with (3.28) a classification can be made for distinct modal damping parameters, equal to the previous section, by respecting the different sign of the parametric excitation term $Q_{12}Q_{21}$:

If both modal damping coefficients are positive, and hence $\Theta_{11}\Theta_{22} > 0$ is valid, the second condition demands the further restriction

$$\Theta_{11}\Theta_{22} > \frac{1}{4\Omega_1\Omega_2}Q_{12}Q_{21}$$

if the term $Q_{12}Q_{21}$ is *positive*.
If one modal damping parameter becomes negative and the conditions from (3.28)

$$\Theta_{11} + \Theta_{22} > 0,$$

$$0 > \Theta_{11}\Theta_{22} > \frac{1}{4\Omega_1\Omega_2}Q_{12}Q_{21},$$

are satisfied, then the system becomes stable. From this conditions it can be easily concluded that a parametric anti-resonance at $\eta_0 = \Omega_1 + \Omega_2$ can only occur if the term $Q_{12}Q_{21}$ is *negative*. This result is summarized in Tab. 3.1 on page 36. A more detailed study on the symmetry property of the parametric excitation term $Q_{12}Q_{21}$ is investigated in Section 3.1.3.

3.1.2. Summary for small single harmonic stiffness variations

Whether a system is stable or not depends mostly on the frequency η of its stiffness variation. In general the system is stable for arbitrary frequency values η iff

$$\boxed{\Theta_{11} > 0 \quad \text{and} \quad \Theta_{22} > 0}$$

according to (3.14). If the frequency η is close to a certain combination frequency the system might be locally stable:

Case 1. For $\eta \approx 2\Omega_k$ if the conditions in (3.17) are satisfied

$$\boxed{\Theta_{ii} > \frac{|Q_{ii}|}{2\Omega_i}\delta_{ik}}$$

In this case the critical values $\sigma_{1,2}$ in (3.19) are purely imaginary, no stability boundary occurs.

Cases 2. and 3. For $\eta \approx |\Omega_1 \mp \Omega_2| = \eta_0$ if the following conditions are satisfied:

$$\begin{array}{|ll|}
\hline
(3.23, 3.28),\ \Delta_1 > 0: & \Theta_{11} + \Theta_{22} > 0 \\
(3.23, 3.28),\ \text{stable } \eta_0: & \Theta_{11}\Theta_{22} \pm \dfrac{1}{4\Omega_1\Omega_2}Q_{12}Q_{21} > 0 \\
(3.24, 3.29),\ \Delta_2 = 0: & \sigma_1 = \dfrac{\Theta_{11} + \Theta_{22}}{2}\sqrt{-\dfrac{\Theta_{11}\Theta_{22} \pm \dfrac{1}{4\Omega_1\Omega_2}Q_{12}Q_{21}}{\Theta_{11}\Theta_{22}}} \\
(3.24),\ \Delta_2 > 0: & \Theta_{11}\Theta_{22} < 0 \qquad\qquad \Theta_{11}\Theta_{22} > 0 \\
& \sigma_1 \text{ is real-valued} \qquad \sigma_1 \text{ is purely imaginary} \\
& -|\sigma_1| < \sigma < |\sigma_1| \qquad \text{no boundary} \\
\hline
\end{array}$$

3 Single harmonic stiffness variation

Note the different signs of the $Q_{12}Q_{21}$ terms in the case of $\eta_0 = |\Omega_1 - \Omega_2|$ and $\eta_0 = \Omega_1 + \Omega_2$, respectively. If we consider the possibility of stability at $\eta_0 = |\Omega_1 - \Omega_2|$ and $\eta_0 = \Omega_1 + \Omega_2$ simultaneously, we obtain from above, (3.24) and (3.29), the following conditions for stability

$$\left. \begin{array}{l} \eta_0 = |\Omega_1 - \Omega_2| : \quad \Theta_{11}\Theta_{22} + \dfrac{1}{4\Omega_1\Omega_2}Q_{12}Q_{21} > 0 \\[2mm] \eta_0 = \Omega_1 + \Omega_2 : \quad \Theta_{11}\Theta_{22} - \dfrac{1}{4\Omega_1\Omega_2}Q_{12}Q_{21} > 0 \end{array} \right\} \Rightarrow 2\Theta_{11}\Theta_{22} > 0. \qquad (3.30)$$

This means that in the case $\Theta_{11}\Theta_{22} < 0$ it is *not* possible that the system is stable at the parametric excitation frequency of the difference type $\eta_0 = |\Omega_1 - \Omega_2|$ and the summation type $\eta_0 = \Omega_1 + \Omega_2$ simultaneously!

The critical values $\sigma_{1,2}$ in (3.24) or (3.29) represent the stability width of an interval according to (3.4). Plotting these critical values as a function of a system parameter p, the stability width changes. The possible values for $\sigma_{1,2}$ are either real-valued or purely imaginary because the variables in these formulaes appear always purely real-valued. Hence, the condition $\sigma_{1,2} = 0$, that is equivalent to (3.23) or (3.28), represents a switch between these two completely different qualities.

Table 3.1 shows the development of the stability border in dependency of an arbitrary system parameter p. Read this table by either using always the upper sign or always the lower sign. Mixing up cases is not allowed. Plotting the frequency $\eta_0 = |\Omega_1 \mp \Omega_2|$ as function of a system parameter p results in a frequency line. If the conditions in (3.23) or (3.28) are violated then the system becomes unstable for parameter values p lying on the frequency line η_0. This is shown in the left column in Tab. 3.1. If, additionally, the stability conditions in the general case from (3.14) are violated, then the system is unstable on the frequency line and in the remaining parameter domain too, see left bottom cell. The critical values $\sigma_{1,2}$ are purely imaginary, the system is unstable in the whole parameter domain. On the other hand, if for an unstable frequency line η_0 the conditions in (3.14) are satisfied, the system is unstable in the vicinity of the frequency line η_0 and stable in the remaining parameter domain, see left top cell. In this case we call the frequency η_0 a *parametric resonance frequency*, because it disturbs an otherwise stable system.

For a stable frequency line η_0 the conditions in (3.23) or (3.28) are satisfied. This is shown in the right column in Tab. 3.1. Now, if the general stability conditions in (3.14) are satisfied, then the system is stable in the whole parameter domain, see right top cell. On the other hand, if the general stability conditions are violated, the system remains stable in the vicinity of the frequency line but becomes unstable in the remaining parameter domain, see right bottom cell. In this case we call the frequency η_0 a *parametric anti-resonance frequency*, because it stabilizes an otherwise unstable system.

If a parameter combination p fulfills the switching condition $\sigma_{1,2} = 0$ then the stability on the frequency line η_0 switches from stable to unstable or vice versa from unstable to stable.

Table 3.1.: Summary for single harmonic stiffness variation at parametric combination resonance frequencies.

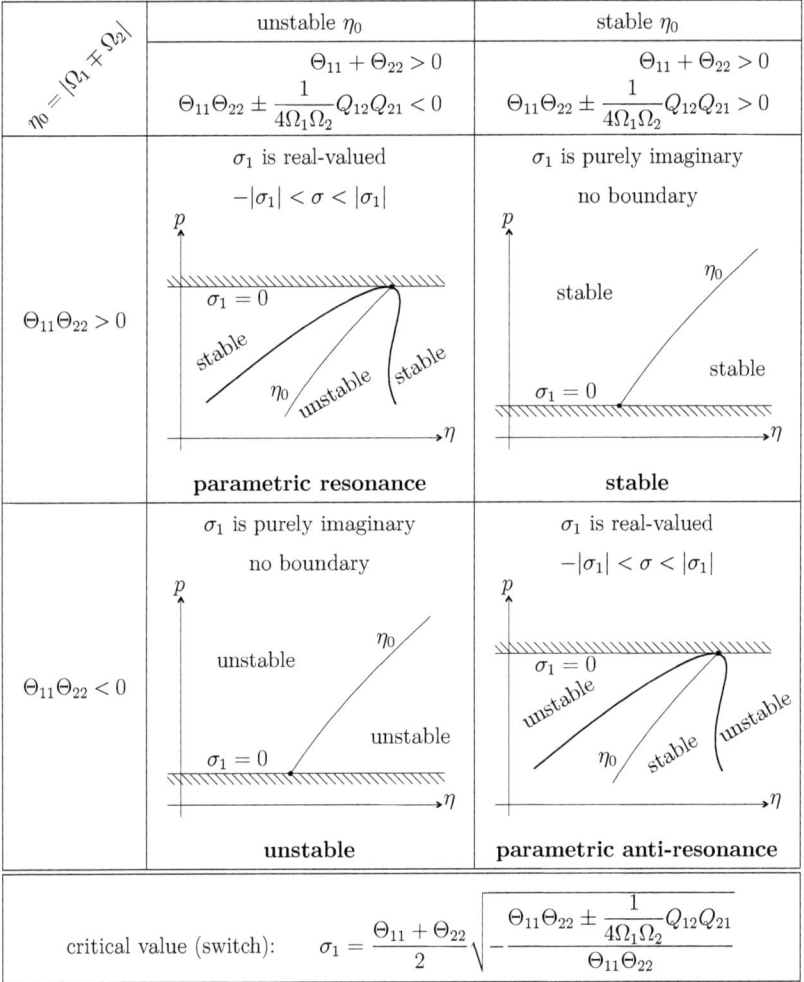

This is indicated by a hatched border. Reaching this border for such a critical parameter value we have to switch the corresponding column. These results are valid on the time scale $1/\varepsilon$.

The case where $\Theta_{11} + \Theta_{22} < 0$ is valid leads always to an unstable system. In this case the stability map for a certain parameter p looks like in the sketch in the left bottom corner in Tab. 3.1. A detailed discussion is therefore omitted.

3 Single harmonic stiffness variation

3.1.3. Symmetry considerations

The sign of the parametric excitation term $Q_{12}Q_{21}$ is crucial for the position of the parametric anti-resonance frequency, whether the anti-resonance is located at $\eta_0 = |\Omega_1 - \Omega_2|$ or at $\eta_0 = \Omega_1 + \Omega_2$.

As it was shown in the previous section the conditions for a system with partially negative modal damping ($0 > \Theta_{11}\Theta_{22}$) to become stable at $\eta_0 = |\Omega_1 - \Omega_2|$ are

$$\Theta_{11} + \Theta_{22} > 0,$$

$$0 > \Theta_{11}\Theta_{22} > \frac{-1}{4\Omega_1\Omega_2}Q_{12}Q_{21}.$$

If the parametric excitation matrix

$$Q = \begin{bmatrix} Q_{11} & Q_{12} \\ Q_{21} & Q_{22} \end{bmatrix}$$

is a symmetric matrix then the product $Q_{12}Q_{21}$ is positive and the conditions above are satisfiable. On the other hand, if the parametric excitation matrix is skew-symmetric, the product $Q_{12}Q_{21}$ is negative. For $Q_{12}Q_{21} < 0$ the second inequality results in

$$0 > \Theta_{11}\Theta_{22} > 0.$$

In this case it is impossible to fulfill the second inequality for non-vanishing modal damping coefficients.

The same results are obtained for a desired stability at the frequency $\eta_0 = \Omega_1 + \Omega_2$ where the conditions

$$\Theta_{11} + \Theta_{22} > 0,$$

$$0 > \Theta_{11}\Theta_{22} > \frac{1}{4\Omega_1\Omega_2}Q_{12}Q_{21},$$

have to be fulfilled. Now if the parametric excitation matrix is symmetric, $Q_{12}Q_{21} > 0$ results again in

$$0 > \Theta_{11}\Theta_{22} > 0.$$

Hence, the second inequality cannot be satisfied. These results are summarized in Tab. 3.2.

Table 3.2.: Symmetry considerations of parametric excitation matrix at frequencies $\eta_0 = |\Omega_1 \mp \Omega_2|$ for single harmonic stiffness variation.

$\eta_0 = |\Omega_1 - \Omega_2|$	(3.14) is fulfilled: $\Theta_{11} + \Theta_{22} > 0$ $\Theta_{11}\Theta_{22} > 0$	(3.14) is violated: $\Theta_{11} + \Theta_{22} > 0$ $\Theta_{11}\Theta_{22} < 0$
$Q_{12}Q_{21} < 0$ (skew-sym.)	additional restriction: **parametric resonance** for $0 < \Theta_{11}\Theta_{22} < -\dfrac{1}{4\Omega_1\Omega_2}Q_{12}Q_{21}$, $\sigma_{1,2}$ is real-valued: $-|\sigma_{1,2}| < \sigma < |\sigma_{1,2}|$	(3.23) is violated: **unstable** $\sigma_{1,2}$ is purely imaginary: no boundary
$Q_{12}Q_{21} > 0$ (sym.)	(3.23) is fulfilled: **stable** $\sigma_{1,2}$ is purely imaginary: no boundary	additional restriction: **parametric anti-resonance** for $0 < \Theta_{11}\Theta_{22} < -\dfrac{1}{4\Omega_1\Omega_2}Q_{12}Q_{21}$, $\sigma_{1,2}$ is real-valued: $-|\sigma_{1,2}| < \sigma < |\sigma_{1,2}|$
$\eta_0 = |\Omega_1 + \Omega_2|$	(3.14) is fulfilled: $\Theta_{11} + \Theta_{22} > 0$ $\Theta_{11}\Theta_{22} > 0$	(3.14) is violated: $\Theta_{11} + \Theta_{22} > 0$ $\Theta_{11}\Theta_{22} < 0$
$Q_{12}Q_{21} < 0$ (skew-sym.)	(3.28) is fulfilled: **stable** $\sigma_{1,2}$ is purely imaginary: no boundary	additional restriction: **parametric anti-resonance** for $0 > \Theta_{11}\Theta_{22} > -\dfrac{1}{4\Omega_1\Omega_2}Q_{12}Q_{21}$, $\sigma_{1,2}$ is real-valued: $-|\sigma_{1,2}| < \sigma < |\sigma_{1,2}|$
$Q_{12}Q_{21} > 0$ (sym.)	additional restriction: **parametric resonance** for $0 < \Theta_{11}\Theta_{22} < \dfrac{1}{4\Omega_1\Omega_2}Q_{12}Q_{21}$, $\sigma_{1,2}$ is real-valued: $-|\sigma_{1,2}| < \sigma < |\sigma_{1,2}|$	(3.28) is violated: **unstable** $\sigma_{1,2}$ is purely imaginary: no boundary

3.2. General harmonic stiffness variation

For the case of a general harmonic stiffness variation, the mass and damping matrices are kept constant

$$\mathbf{M}(t) = \mathbf{M}_0, \quad \mathbf{C}(t) = \mathbf{C}_0,$$
$$\mathbf{K}(t) = \mathbf{K}_0 + \varepsilon \left\{ \mathbf{K}_c^s + \mathbf{K}_c^a \right\} \cos(\omega t) + \varepsilon \left\{ \mathbf{K}_s^s + \mathbf{K}_s^a \right\} \sin(\omega t).$$

In this case, the general linear equations of motion (2.2) simplify to

$$\ddot{\mathbf{z}}(t) + \varepsilon \tilde{\mathbf{C}}_0 \dot{\mathbf{z}}(t) + \tilde{\mathbf{K}}_0 \mathbf{z}(t) = -\varepsilon \left\{ \left\{ \tilde{\mathbf{K}}_c^s + \tilde{\mathbf{K}}_c^a \right\} \cos(\omega t) + \left\{ \mathbf{K}_s^s + \mathbf{K}_s^a \right\} \sin(\omega t) \right\} \mathbf{z}(t).$$

We mentioned already that, due to the fact that $\tilde{\mathbf{K}}_{c/s}$ are matrices, more than one stiffness parameter in the system can be varied. In Section 3.1 we restricted our study to variations with the same frequency ω and phase relation of zero. In this section we skip the restrictions on the phase relations.

The equations of motion can be transformed ($\mathbf{z}(t) \longrightarrow \mathbf{z}(\tau)$) to a non-dimensional normal form

$$\mathbf{z}''(\tau) + \varepsilon \boldsymbol{\Theta} \mathbf{z}'(\tau) + \boldsymbol{\Omega}^2 \mathbf{z}(\tau) = -\varepsilon \left\{ (\mathbf{Q}^s + \mathbf{Q}^a) \cos(\eta \tau) + (\mathbf{P}^s + \mathbf{P}^a) \sin(\eta \tau) \right\} \mathbf{z}(\tau),$$
$$\text{where } \mathbf{Q} = \mathbf{Q}^s + \mathbf{Q}^a = \mathbf{T}^{-1} \mathbf{M}_0^{-1} \mathbf{K}_c \mathbf{T},$$
$$\text{and } \mathbf{P} = \mathbf{P}^s + \mathbf{P}^a = \mathbf{T}^{-1} \mathbf{M}_0^{-1} \mathbf{K}_s \mathbf{T},$$

as it is shown in Section 2.3, or rewritten for a 2dof-system in the comprehensive index notation

$$z_i''(\tau) + \Omega_i^2 z_i(\tau) = -\varepsilon \left\{ \Theta_{ij} z_j'(\tau) + Q_{ij} z_j(\tau) \cos(\eta \tau) + P_{ij} z_j(\tau) \sin(\eta \tau) \right\}, \quad (3.31)$$

with $i, j = 1, 2$. The approach presented below follows the analysis in Section 3.1.

3.2.1. Transformation and averaging

Applying the procedure for a *first order approximation* according to Section 3.1 extends the right hand sides of (3.8) by the general harmonic parametric excitation term $P_{ij} \sin t$ to

$$F_i(\mathbf{u}, t) = -\eta_0 \sum_j \Theta_{ij} \left(-u_j \varpi_j s_j + v_j \varpi_j c_j \right) +$$
$$- \sum_j (Q_{ij} \cos t + P_{ij} \sin t)(u_j c_j + v_j s_j) + 2\Omega_i \varpi_i \sigma (u_i c_i + v_i s_i), \quad (3.32)$$

with the abbreviations $s_i = \sin \varpi_i t$ and $c_i = \cos \varpi_i t$ and the state vector $\mathbf{u} = (u_1, v_1, u_2, v_2)^T$. Comparing (3.32) with (3.8) points out the difference between a single harmonic stiffness variation (Q_{ij}) and a general harmonic stiffness variation (Q_{ij}, P_{ij}).

The final system of equations for the slow amplitudes $\mathbf{u}$ are equal to (3.9), with quasi-periodic right hand sides. For this simple two mode system we obtain again 12 different periods, according to Appendix B, by substituting $n = 1$. Averaging over these different periodic terms yields always zero, except for the case where a term becomes resonant:

$\Theta_{ii}, \sigma :$ no resonant terms, constant factor $\frac{1}{2}$,

$\Theta_{ij} :$ resonant terms for $\varpi_i \pm \varpi_j = 0$,

$Q_{ii}, P_{ii} :$ resonant terms for $2\varpi_i \pm 1 = 0$,

$Q_{ij}, P_{ij} :$ resonant terms for $\varpi_i \pm \varpi_j \pm 1 = 0$.

General Case

If ϖ_1, ϖ_2 do not take on special values only the terms Θ_{ii}, σ remain after averaging and we obtain

$$\dot{\hat{u}}_i = \frac{\varepsilon}{\eta_0^2 \varpi_i} \left\{ -\frac{\eta_0}{2} \Theta_{ii} \varpi_i \hat{u}_i - \Omega_i \varpi_i \sigma \hat{v}_i \right\},$$

$$\dot{\hat{v}}_i = \frac{\varepsilon}{\eta_0^2 \varpi_i} \left\{ -\frac{\eta_0}{2} \Theta_{ii} \varpi_i \hat{v}_i + \Omega_i \varpi_i \sigma \hat{u}_i \right\},$$
(3.33)

which is identical to the system of equations in (3.11). The general conditions for vibration suppression for $\eta_0 > 0$ results from (3.14) to

$$\boxed{\Theta_{11} > 0 \quad \text{and} \quad \Theta_{22} > 0,}$$
(3.34)

and from the previous discussion of this result in Section 3.1 we know that σ_1 always is purely imaginary.

Some resonant cases

1. Combination parametric resonance of difference type and first kind

By setting

$$\boxed{\eta_0 = \Omega_2 - \Omega_1 > 0 \quad \text{or} \quad \varpi_1 - \varpi_2 = -1}$$

some terms become resonant and produce additional terms after averaging at η_0

$$\langle s_1 c_2 \sin t \rangle|_{\eta_0} = -\frac{1}{4}, \quad \langle s_2 c_1 \sin t \rangle|_{\eta_0} = \frac{1}{4}.$$

We obtain from (3.32)

$$\dot{\hat{u}}_i = \frac{\varepsilon}{\eta_0^2 \varpi_i} \left\{ -\frac{\eta_0}{2} \Theta_{ii} \varpi_i \hat{u}_i + \frac{Q_{ij}}{4} \hat{v}_j + (-1)^i \frac{P_{ij}}{4} \hat{u}_j - \Omega_i \varpi_i \sigma \hat{v}_i \right\},$$

$$\dot{\hat{v}}_i = \frac{\varepsilon}{\eta_0^2 \varpi_i} \left\{ -\frac{\eta_0}{2} \Theta_{ii} \varpi_i \hat{v}_i - \frac{Q_{ij}}{4} \hat{u}_j + (-1)^i \frac{P_{ij}}{4} \hat{v}_j + \Omega_i \varpi_i \sigma \hat{u}_i \right\},$$
(3.35)

3 General harmonic stiffness variation

instead of (3.33) or (3.20). Rewritten in matrix notation yields

$$\begin{pmatrix} \dot{\hat{u}}_1 \\ \dot{\hat{v}}_1 \\ \dot{\hat{u}}_2 \\ \dot{\hat{v}}_2 \end{pmatrix} = \frac{\varepsilon}{\eta_0^2} \begin{bmatrix} -\frac{\eta_0}{2}\Theta_{11} & -\Omega_1\sigma & -\frac{\eta_0}{4\Omega_1}P_{12} & \frac{\eta_0}{4\Omega_1}Q_{12} \\ \Omega_1\sigma & -\frac{\eta_0}{2}\Theta_{11} & -\frac{\eta_0}{4\Omega_1}Q_{12} & -\frac{\eta_0}{4\Omega_1}P_{12} \\ \frac{\eta_0}{4\Omega_2}P_{21} & \frac{\eta_0}{4\Omega_2}Q_{21} & -\frac{\eta_0}{2}\Theta_{22} & -\Omega_2\sigma \\ -\frac{\eta_0}{4\Omega_2}Q_{21} & \frac{\eta_0}{4\Omega_2}P_{21} & \Omega_2\sigma & -\frac{\eta_0}{2}\Theta_{22} \end{bmatrix} \begin{pmatrix} \hat{u}_1 \\ \hat{v}_1 \\ \hat{u}_2 \\ \hat{v}_2 \end{pmatrix} = \frac{\varepsilon}{\eta_0^2} \mathbf{A}_4 \hat{\mathbf{u}}.$$

In contrary to the coefficient matrix resulting from (3.20) the existing coefficient matrix is now fully occupied, which is a direct generalization of [1, p.69]. After rescaling time by ε/η_0^2 we obtain the characteristic equation

$$\det(\lambda \mathbf{I}_4 - \mathbf{A}_4) = 0 = \lambda^4 + b_1\lambda^3 + b_2\lambda^2 + b_3\lambda + b_4,$$

a polynomial of order four. Using the abbreviations

$$\hat{w}_i = \hat{u}_i + j\hat{v}_i,$$

where $j = \sqrt{-1}$ is the complex unit, (3.35) is equivalent to

$$\dot{\hat{w}}_i = \frac{\varepsilon}{\eta_0^2} \left\{ \left(-\frac{\eta_0}{2}\Theta_{ii} + j\Omega_i\sigma\right)\hat{w}_i - \frac{\eta_0}{4\Omega_i}\left((-1)^{i+1}P_{ik} + jQ_{ik}\right)\hat{w}_k \right\} \qquad k \neq i.$$

Rewritten in matrix notation we get equations of the form

$$\dot{\hat{\mathbf{w}}} = \frac{\varepsilon}{\eta_0^2} \begin{bmatrix} -\frac{\eta_0}{2}\Theta_{11} + j\Omega_1\sigma & -\frac{\eta_0}{4\Omega_1}(P_{12} + jQ_{12}) \\ -\frac{\eta_0}{4\Omega_2}(-P_{21} + jQ_{21}) & -\frac{\eta_0}{2}\Theta_{22} + j\Omega_2\sigma \end{bmatrix} \hat{\mathbf{w}} = \frac{\varepsilon}{\eta_0^2} \mathbf{A}_2 \hat{\mathbf{w}}, \qquad (3.36)$$

with the complex state vector $\hat{\mathbf{w}} = (\hat{w}_1, \hat{w}_2)^T$. Compared to a single harmonic stiffness variation the off-diagonal terms are extended by real-valued terms P_{ik}. After rescaling time by ε/η_0^2 the characteristic equation of the coefficient matrix is reduced from a real polynomial of order four to a complex polynomial of order two

$$\det(\lambda \mathbf{I}_2 - \mathbf{A}_2) = 0,$$

$$\lambda^2 + \left(\tfrac{1}{2}\eta_0(\Theta_{11} + \Theta_{22}) - j(\Omega_1 + \Omega_2)\sigma\right)\lambda \\ + \left(\tfrac{1}{2}\eta_0\Theta_{11} - j\Omega_1\sigma\right)\left(\tfrac{1}{2}\eta_0\Theta_{22} - j\Omega_2\sigma\right) - \frac{\eta_0^2}{16\Omega_1\Omega_2}(P_{12} + jQ_{12})(-P_{21} + jQ_{21}) = 0. \qquad (3.37)$$

Because the coefficients P_{ij}, Q_{ij} are real-valued, the parametric excitation term in this polynomial is a product of two general complex values and therefore consists of both real- and imaginary valued terms. Applying the extended Routh-Hurwitz criterion from (A.2) demands a separation of real and imaginary parts. But the parametric excitation coefficients contribute to both parts, therefore a simple mapping

$$Q_{12} \mapsto (P_{12} + jQ_{12}), \qquad Q_{21} \mapsto (-P_{21} + jQ_{21}),$$

of (3.23) similar to (4.11) in a following chapter about single harmonic damping variation is *not* allowed and we have to perform a formal analysis.

First analyzing the case for $\sigma = 0$ this polynomial is stable according to (A.2) iff

$$\Delta_1 : \tfrac{1}{2}\eta_0 \left(\Theta_{11} + \Theta_{22}\right) > 0,$$

$$\Delta_2 : \tfrac{1}{16}\eta_0^4 \left(\left(\Theta_{11} + \Theta_{22}\right)^2 \left(\Theta_{11}\Theta_{22} + \frac{Q_{12}Q_{21}}{4\Omega_1\Omega_2} + \frac{P_{12}P_{21}}{4\Omega_1\Omega_2}\right) - \left(\frac{P_{12}Q_{21}}{4\Omega_1\Omega_2} - \frac{Q_{12}P_{21}}{4\Omega_1\Omega_2}\right)^2\right) > 0,$$

are fulfilled. These conditions can be rewritten for $\eta_0 > 0$ as

$$\Theta_{11} + \Theta_{22} > 0,$$

$$\Theta_{11}\Theta_{22} + \frac{Q_{12}Q_{21} + P_{12}P_{21}}{4\Omega_1\Omega_2} > \left(\frac{Q_{12}P_{21} - P_{12}Q_{21}}{4\Omega_1\Omega_2 \left(\Theta_{11} + \Theta_{22}\right)}\right)^2 \geq 0. \qquad (3.38)$$

Note the additional restriction compared to the stability conditions in (3.23). Now the left hand terms have not only to be positive but have to exceed a certain value. Terms resulting from a product of coefficients that correspond to different kinds of harmonic stiffness excitation $(Q_{ij}P_{ji})$ are named *interaction terms*, because they arise only if both a cosine and a sine excitation operate simultaneously. Hence, the occurrence of phase shifted parametric excitations leads to a more restricted stability conditions at the frequency $\eta = \eta_0$.

Note also that only in the case where the coefficients fulfill the condition (see Section 2.3)

$$Q_{ij}\sin\varphi = -P_{ij}\cos\varphi, \qquad (3.39)$$

where φ is a constant phase for all ij, the stability conditions in (3.38) are identical with the simpler conditions for a single harmonic stiffness variation as in (3.23). By using the abbreviations

$$Q_{ij} = \hat{Q}_{ij}\cos\varphi, \qquad P_{ij} = -\hat{Q}_{ij}\sin\varphi$$

the conditions (3.39) above are satisfied and the two parametric excitation matrices in (3.32) can be combined to

$$Q_{ij}\cos\omega t + P_{ij}\sin\omega t = \hat{Q}_{ij}\cos\varphi\cos\omega t - \hat{Q}_{ij}\sin\varphi\sin\omega t = \hat{Q}_{ij}\cos\left(\omega t + \varphi\right).$$

Furthermore, we can rewrite the parametric excitation term as

$$Q_{12}Q_{21} + P_{12}P_{21} = \left(\hat{Q}_{12}\cos\varphi\right)\left(\hat{Q}_{21}\cos\varphi\right) + \left(-\hat{Q}_{12}\sin\varphi\right)\left(-\hat{Q}_{21}\sin\varphi\right) = \hat{Q}_{12}\hat{Q}_{21}$$

and the interaction term as

$$Q_{12}P_{21} - P_{12}Q_{21} = \left(\hat{Q}_{12}\cos\varphi\right)\left(-\hat{Q}_{21}\sin\varphi\right) - \left(-\hat{Q}_{12}\sin\varphi\right)\left(\hat{Q}_{21}\cos\varphi\right) = 0.$$

Inserting into (3.38) gives finally the stability conditions

$$\Theta_{11} + \Theta_{22} > 0,$$

$$\Theta_{11}\Theta_{22} + \frac{1}{4\Omega_1\Omega_2}\hat{Q}_{12}\hat{Q}_{21} > 0,$$

3 General harmonic stiffness variation

in analogy to (3.23).

For the case $\sigma \neq 0$, looking for a *stability interval* σ in the vicinity of the frequency η_0 according to (3.4), the following stability conditions are obtained:

$\Delta_1 : \quad \frac{1}{2}\eta_0 \left(\Theta_{11} + \Theta_{22}\right) > 0,$

$$\Delta_2 : \begin{cases} \frac{1}{16}\eta_0^4 \left(\Theta_{11} + \Theta_{22}\right)^2 \left(\Theta_{11}\Theta_{22} + \dfrac{Q_{12}Q_{21} + P_{12}P_{21}}{4\Omega_1\Omega_2} - \dfrac{4}{\eta_0^2}\Omega_1\Omega_2\sigma^2\right) + \\ + \frac{1}{4}\eta_0^2 \left(\left(\Theta_{11}\Omega_2 + \Theta_{22}\Omega_1\right)\sigma - \dfrac{\eta_0\left(Q_{12}P_{21} - P_{12}Q_{21}\right)}{8\Omega_1\Omega_2}\right) \\ \left(\left(\Theta_{11}\Omega_1 + \Theta_{22}\Omega_2\right)\sigma + \dfrac{\eta_0\left(Q_{12}P_{21} - P_{12}Q_{21}\right)}{8\Omega_1\Omega_2}\right) > 0, \end{cases} \quad (3.40)$$

$$\Delta_2 : \begin{cases} a_0\sigma^2 + a_1\left(P_{ij}\right)\sigma + a_2\left(P_{ij}\right) > 0, \\ \text{critical values:} \\ \sigma_{1,2} = \sigma_s \pm \tilde{\sigma}_{1,2}, \\ \sigma_s = \dfrac{\eta_0\left(\Theta_{11} - \Theta_{22}\right)}{2\Theta_{11}\Theta_{22}\left(\Omega_1 - \Omega_2\right)} \left(\dfrac{Q_{12}P_{21} - P_{12}Q_{21}}{8\Omega_1\Omega_2}\right), \\ \tilde{\sigma}_{1,2} = \dfrac{\eta_0\left(\Theta_{11} + \Theta_{22}\right)}{2\Theta_{11}\Theta_{22}\left(\Omega_1 - \Omega_2\right)}\sqrt{d}, \\ \text{with} \quad d = -\Theta_{11}\Theta_{22}\left(\Theta_{11}\Theta_{22} + \dfrac{Q_{12}Q_{21} + P_{12}P_{21}}{4\Omega_1\Omega_2}\right) + \left(\dfrac{Q_{12}P_{21} - P_{12}Q_{21}}{8\Omega_1\Omega_2}\right)^2, \\ \text{or} \quad d = -\left(\Theta_{11}\Theta_{22} + \dfrac{Q_{12}Q_{21} + P_{12}P_{21}}{8\Omega_1\Omega_2}\right)^2 + \dfrac{\left(Q_{12}^2 + P_{12}^2\right)\left(Q_{21}^2 + P_{21}^2\right)}{64\Omega_1^2\Omega_2^2}. \end{cases}$$

If the correlation in (3.39) is fulfilled the stability conditions (3.40) simplifies to the stability conditions presented in (3.24). Note that, contrary to the case of a single harmonic stiffness variation in the previous chapter, the condition $\sigma_1 = 0$ in (3.40) is *not* an equivalent formulation for the critical case of the second condition in (3.38). This will become more clear in Section 3.2.2.

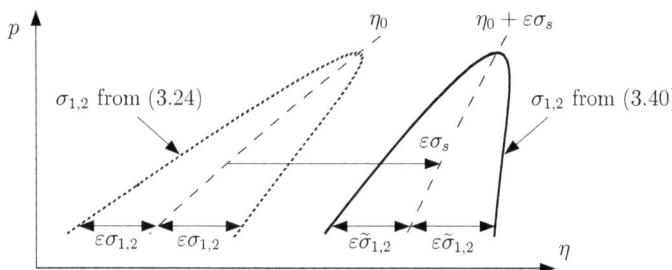

Figure 3.2.: Shift of stability boundary due to the interaction term in (3.40) compared to the stability boundary (3.24).

Note the occurrence of an additional frequency shift σ_s. Inserting (3.40) into the original frequency approximation in (3.4) the first order approximation of the stability boundary is determined by

$$\eta_0 + \varepsilon\sigma_s - \varepsilon\tilde{\sigma}_{1,2} \leq \eta \leq \eta_0 + \varepsilon\sigma_s + \varepsilon\tilde{\sigma}_{1,2}. \tag{3.41}$$

Hence, the skeleton line η_0 of the stability boundary curve in (3.40) is shifted by $+\varepsilon\sigma_s$ compared to the skeleton line from (3.24). A graphical interpretation is shown in Fig. 3.2. The diagram shows stability boundary curves and their corresponding skeleton line depending on a specific system parameter p. The dotted line indicates the stability boundary curve without an additional shift from the combination resonance frequency $\eta_0 = \Omega_2 - \Omega_1$ that would occur from the stability conditions in (3.24). The solid line indicates the final stability boundary curve respecting the additional shift $+\varepsilon\sigma_s$ according to (3.40, 3.41). The skeleton line η_0 of a single harmonic stiffness variation is moved to the skeleton line $\eta_0 + \varepsilon\sigma_s$ due to the interaction term in the case of a general harmonic stiffness variation.

Changing signs

In a physically meaningful system the frequency η_0 would be positive. The analysis shown above is carried out for $\eta_0 = \Omega_2 - \Omega_1$ which assumes implicitly that $\Omega_2 > \Omega_1$. If $\Omega_2 < \Omega_1$ holds then determining the corresponding stability conditions by simple mapping $\eta_0 \mapsto -\eta_0$ in (3.38) is *not* allowed and would lead to the wrong stability condition

$$\Theta_{11} + \Theta_{22} < 0.$$

We need to perform a formal analysis by setting

$$\boxed{\eta_0 = \Omega_1 - \Omega_2 > 0 \quad \text{or} \quad \varpi_1 - \varpi_2 = 1}$$

Now the non-vanishing terms after averaging are

$$\langle s_1 c_2 \sin t \rangle_{\eta_0} = \frac{1}{4}, \quad \langle s_2 c_1 \sin t \rangle_{\eta_0} = -\frac{1}{4},$$

and we obtain from (3.32)

$$\dot{\hat{u}}_i = \frac{\varepsilon}{\eta_0^2 \varpi_i} \left\{ -\frac{\eta_0}{2} \Theta_{ii} \varpi_i \hat{u}_i + \frac{Q_{ij}}{4} \hat{v}_j + (-1)^{i+1} \frac{P_{ij}}{4} \hat{u}_j - \Omega_i \varpi_i \sigma \hat{v}_i \right\},$$

$$\dot{\hat{v}}_i = \frac{\varepsilon}{\eta_0^2 \varpi_i} \left\{ -\frac{\eta_0}{2} \Theta_{ii} \varpi_i \hat{v}_i - \frac{Q_{ij}}{4} \hat{u}_j + (-1)^{i+1} \frac{P_{ij}}{4} \hat{v}_j + \Omega_i \varpi_i \sigma \hat{u}_i \right\}. \tag{3.42}$$

Comparing these equations after averaging for $\eta_0 = \Omega_1 - \Omega_2$ with the equations averaged for $\eta_0 = \Omega_2 - \Omega_1$ as obtained in (3.35) gives the following relations

$$P_{ij} \mapsto -P_{ij}.$$

Performing this transformation the stability conditions in (3.38, 3.40) remain unchanged for the case of $\Omega_2 < \Omega_1$.

3 General harmonic stiffness variation

2. Combination parametric resonance of summation type and first kind

By defining
$$\boxed{\eta_0 = \Omega_1 + \Omega_2 \quad \text{or} \quad \varpi_1 + \varpi_2 = 1}$$

the resonant terms are

$$\langle s_1 c_2 \sin t \rangle_{\eta_0} = -\frac{1}{4}, \qquad \langle s_2 c_1 \sin t \rangle_{\eta_0} = \frac{1}{4},$$

and we obtain from (3.32)

$$\dot{\hat{u}}_i = \frac{\varepsilon}{\eta_0^2 \varpi_i} \left\{ -\frac{\eta_0}{2} \Theta_{ii} \varpi_i \hat{u}_i - \frac{Q_{ij}}{4} \hat{v}_j + \frac{P_{ij}}{4} \hat{u}_j - \Omega_i \varpi_i \sigma \hat{v}_i \right\},$$
$$\dot{\hat{v}}_i = \frac{\varepsilon}{\eta_0^2 \varpi_i} \left\{ -\frac{\eta_0}{2} \Theta_{ii} \varpi_i \hat{v}_i - \frac{Q_{ij}}{4} \hat{u}_j - \frac{P_{ij}}{4} \hat{v}_j + \Omega_i \varpi_i \sigma \hat{u}_i \right\},$$
(3.43)

instead of (3.25). Using the abbreviations

$$\hat{w}_i = \hat{u}_i + j(-1)^{i-1} \hat{v}_i,$$

where $j = \sqrt{-1}$ is the complex unit, (3.43) is equivalent to

$$\dot{\hat{w}}_i = \frac{\varepsilon}{\eta_0^2} \left\{ \left(-\frac{\eta_0}{2} \Theta_{ii} + j\Omega_i \sigma \right) \hat{w}_i - \frac{\eta_0}{4\Omega_i} \left((-1)^{i+1} P_{ik} + jQ_{ik} \right) \hat{w}_k \right\} \qquad k \neq i.$$

Rewritten in matrix notation we get equations of the form

$$\dot{\hat{\mathbf{w}}} = \frac{\varepsilon}{\eta_0^2} \begin{bmatrix} -\frac{\eta_0}{2} \Theta_{11} + j\Omega_1 \sigma & \frac{\eta_0}{4\Omega_1} (P_{12} - jQ_{12}) \\ \frac{\eta_0}{4\Omega_2} (P_{21} + jQ_{21}) & -\frac{\eta_0}{2} \Theta_{22} - j\Omega_2 \sigma \end{bmatrix} \hat{\mathbf{w}} = \frac{\varepsilon}{\eta_0^2} \mathbf{A}_2 \hat{\mathbf{w}},$$
(3.44)

with the complex state vector $\hat{\mathbf{w}} = (\hat{w}_1, \hat{w}_2)^T$. Again, compared to a single harmonic stiffness variation the off-diagonal terms are extended by real-valued terms P_{ik}. Comparing with the averaging process at the frequency $\eta_0 = |\Omega_1 - \Omega_2|$ the following correlations can be identified

$$\Omega_2 \mapsto -\Omega_2, \qquad P_{ij} \mapsto -P_{ij}.$$

First analyzing the case of $\sigma = 0$ the necessary and sufficient conditions for stability are

$$\Theta_{11} + \Theta_{22} > 0,$$
$$\Theta_{11}\Theta_{22} - \frac{Q_{12}Q_{21} + P_{12}P_{21}}{4\Omega_1\Omega_2} > \left(\frac{Q_{12}P_{21} - P_{12}Q_{21}}{4\Omega_1\Omega_2 (\Theta_{11} + \Theta_{22})} \right)^2 \geq 0,$$
(3.45)

according to (3.38). Note the different sign of the parametric excitation term $(Q_{12}Q_{21} + P_{12}P_{21})$ compared to the previous section where $\eta_0 = |\Omega_1 - \Omega_2|$. For the case $\sigma \neq 0$, looking for a stability interval σ in the vicinity of the frequency η_0 according to (3.4), the following stability conditions

$\Delta_1: \frac{1}{2}\eta_0 (\Theta_{11} + \Theta_{22}) > 0,$

$\Delta_2: \begin{cases} a_0 \sigma^2 + \tilde{a}_1 (P_{ij}) \sigma + \tilde{a}_2 (P_{ij}) > 0, \\ \text{critical values:} \\ \sigma_{1,2} = \sigma_s \pm \tilde{\sigma}_{1,2}, \\ \sigma_s = \dfrac{\eta_0 (\Theta_{11} - \Theta_{22})}{2\Theta_{11}\Theta_{22}(\Omega_1 + \Omega_2)} \left(\dfrac{Q_{12}P_{21} - P_{12}Q_{21}}{8\Omega_1\Omega_2} \right), \\ \tilde{\sigma}_{1,2} = \dfrac{\eta_0 (\Theta_{11} + \Theta_{22})}{2\Theta_{11}\Theta_{22}(\Omega_1 + \Omega_2)} \sqrt{d}, \\ \text{with} \quad d = -\Theta_{11}\Theta_{22}\left(\Theta_{11}\Theta_{22} - \dfrac{Q_{12}Q_{21} + P_{12}P_{21}}{4\Omega_1\Omega_2}\right) + \left(\dfrac{Q_{12}P_{21} - P_{12}Q_{21}}{8\Omega_1\Omega_2}\right)^2, \\ \text{or} \quad d = -\left(\Theta_{11}\Theta_{22} - \dfrac{Q_{12}Q_{21} + P_{12}P_{21}}{8\Omega_1\Omega_2}\right)^2 + \dfrac{(Q_{12}^2 + P_{12}^2)(Q_{21}^2 + P_{21}^2)}{64\Omega_1^2\Omega_2^2}, \end{cases}$

(3.46)

are obtained from (3.40). Note the change in the sign not only for the parametric excitation term but also in the denominator. Furthermore, note that, in contrary to the case of a single harmonic stiffness variation in the previous chapter, the condition $\sigma_1 = 0$ in (3.46) is again *not* an equivalent formulation for the critical case of the second condition in (3.45). This will become more clear in the following section.

3.2.2. Summary for small general harmonic stiffness variations

Whether a system is stable or not depends mainly on the frequency η of its stiffness variation. Globally the system is stable for arbitrary frequency values η iff

$\boxed{\Theta_{11} > 0 \quad \text{and} \quad \Theta_{22} > 0,}$

according to (3.34). If the frequency η is close to a certain combination frequency the system might be locally stable:

Cases 1. and 2. For $\eta \approx |\Omega_1 \mp \Omega_2| = \eta_0$ if the following conditions are satisfied:

$\boxed{\begin{array}{l} (3.38, 3.45), \Delta_1 > 0: \quad \Theta_{11} + \Theta_{22} > 0 \\ (3.38, 3.45), \text{stable } \eta_0: \quad \Theta_{11}\Theta_{22} \pm \dfrac{Q_{12}Q_{21} + P_{12}P_{21}}{4\Omega_1\Omega_2} > \dfrac{\beta^2}{(\Theta_{11} + \Theta_{22})^2} \geq 0 \\ (3.40, 3.46), \Delta_2 = 0: \quad \sigma_1 = \sigma_s - \tilde{\sigma}_1, \quad \sigma_2 = \sigma_s + \tilde{\sigma}_1 \end{array}}$

3 General harmonic stiffness variation

with the abbreviations

$$\beta = \left(\frac{Q_{12}P_{21} - P_{12}Q_{21}}{4\Omega_1\Omega_2}\right),$$

$$\sigma_s = \frac{(\Theta_{11} - \Theta_{22})}{2\Theta_{11}\Theta_{22}}\frac{\beta}{2}, \quad \tilde{\sigma}_1 = \frac{(\Theta_{11} + \Theta_{22})}{2\Theta_{11}\Theta_{22}}\sqrt{d},$$

$$\text{with} \quad d = -\Theta_{11}\Theta_{22}\left(\Theta_{11}\Theta_{22} \pm \frac{Q_{12}Q_{21} + P_{12}P_{21}}{4\Omega_1\Omega_2}\right) + \left(\frac{\beta}{2}\right)^2,$$

$$\text{or} \quad d = -\left(\Theta_{11}\Theta_{22} \pm \frac{Q_{12}Q_{21} + P_{12}P_{21}}{8\Omega_1\Omega_2}\right)^2 + \frac{(Q_{12}^2 + P_{12}^2)(Q_{21}^2 + P_{21}^2)}{64\Omega_1^2\Omega_2^2}.$$

Note that the frequency shift σ_s is independent whether the frequency of the stiffness variation is near to a parametric combination frequency of the difference type $\eta_0 = |\Omega_1 - \Omega_2|$, or near to the summation type $\eta_0 = \Omega_1 + \Omega_2$. Further note the different signs of the parametric excitation term $(Q_{12}Q_{21} + P_{12}P_{21})$ in the case of $\eta_0 = |\Omega_1 - \Omega_2|$ and $\eta_0 = \Omega_1 + \Omega_2$, respectively. If we consider the case of stability at $\eta_0 = |\Omega_1 - \Omega_2|$ and at $\eta_0 = \Omega_1 + \Omega_2$ *simultaneously*, we obtain from above, (3.40) and (3.46), the following conditions for stability

$$\left.\begin{array}{l} \eta_0 = |\Omega_1 - \Omega_2| : \quad \Theta_{11}\Theta_{22} + \dfrac{Q_{12}Q_{21} + P_{12}P_{21}}{4\Omega_1\Omega_2} > \dfrac{\beta^2}{(\Theta_{11} + \Theta_{22})^2} \geq 0 \\[2ex] \eta_0 = \Omega_1 + \Omega_2 : \quad \Theta_{11}\Theta_{22} - \dfrac{Q_{12}Q_{21} + P_{12}P_{21}}{4\Omega_1\Omega_2} > \dfrac{\beta^2}{(\Theta_{11} + \Theta_{22})^2} \geq 0 \end{array}\right\} \Rightarrow 2\Theta_{11}\Theta_{22} > 0.$$

(3.47)

This means that in the case $\Theta_{11}\Theta_{22} < 0$ it is *not* possible that the system is stable at the parametric excitation frequency of the difference type $\eta_0 = |\Omega_1 - \Omega_2|$ and the summation type $\eta_0 = \Omega_1 + \Omega_2$ simultaneously. This is the same result as in the case of a single harmonic stiffness variation!

The additional frequency shift σ_s changes its sign at least if a modal damping term changes its sign, too. The critical case where one modal damping parameter is equal to zero results in an infinite frequency shift σ_s. Hence, a change in the sign of σ_s never occurs by crossing the frequency line η_0. This circumstance is pointed out in Tab. 3.3 for the case of $\Theta_{11} > \Theta_{22}$.

The critical values $\sigma_{1,2}$ in (3.40) or (3.46) represent the width of the stability interval according to (3.4). Plotting these critical values as a function of a system parameter p, this stability width changes. The possible values for $\sigma_{1,2}$ are either real-valued or purely imaginary because the variables in these formulaes appear always real-valued. For the case of a single harmonic stiffness variation the stability conditions in (3.23, 3.28) guarantee that the radical in the second stability condition in (3.24, 3.29) is positive and hence the stability width $\sigma_{1,2}$ is real-valued, as long as system is stable at the frequency η_0 is stable. In the case of a general harmonic stiffness variation the qualitative picture changes due to the occurrence of the interaction term $(Q_{12}P_{21} - P_{12}Q_{21})$. To make similar statements from the stability conditions (3.38, 3.45) and (3.40, 3.46) the additional frequency shift σ_s from the frequency η_0 has to be considered.

48   3 General harmonic stiffness variation

Table 3.3.: Summary for frequency shift σ_s for general harmonic stiffness variation at parametric combination resonance frequencies.

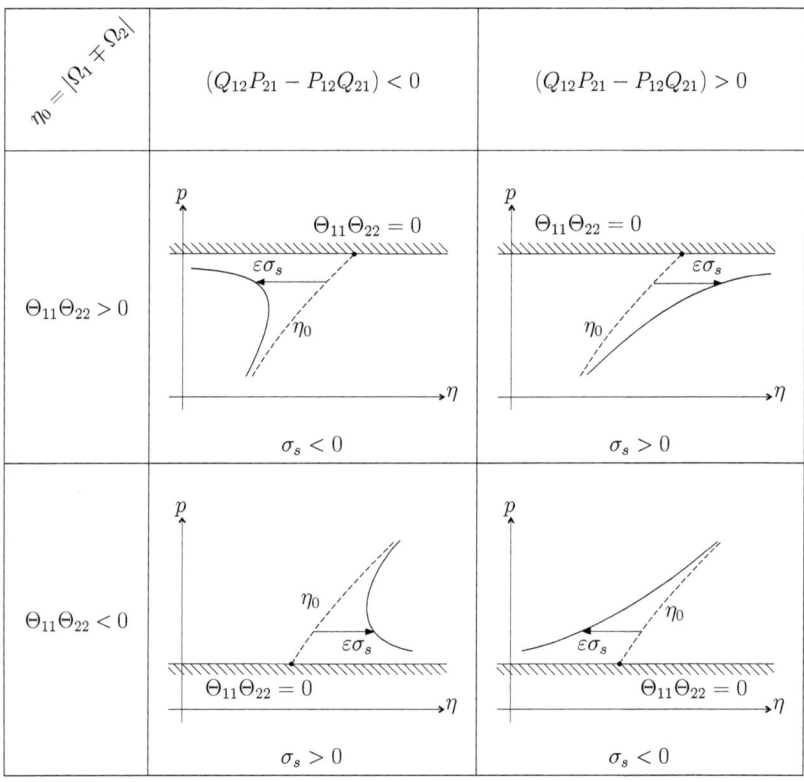

The sign of the radical d in (3.40, 3.46) determines whether the stability width $\sigma_{1,2}$ is real-valued or purely imaginary. If the special condition

$$Q_{12}P_{21} - P_{12}Q_{21} = 0$$

is satisfied, then the qualitative behavior of the system is equivalent to the single harmonic stiffness variation in Tab. 3.1. On the other hand, for a general harmonic stiffness variation where $Q_{12}P_{21} - P_{12}Q_{21} \neq 0$ the sign of the radical d is non-trivial. The deduction is shown in the following:

1. For
$$\Theta_{11} + \Theta_{22} > 0,$$
$$\Theta_{11}\Theta_{22} \pm \frac{Q_{12}Q_{21} + P_{12}P_{21}}{4\Omega_1\Omega_2} < 0 \leq \left(\frac{Q_{12}P_{21} - P_{12}Q_{21}}{4\Omega_1\Omega_2(\Theta_{11} + \Theta_{22})}\right)^2,$$

3 General harmonic stiffness variation

the system is unstable at frequency η_0. A positive modal damping, $\Theta_{11}\Theta_{22} > 0$, trivially implies that the radical d is positive, $d > 0$. If one of the modal damping terms is negative, $\Theta_{11}\Theta_{22} < 0$, then the sign of the radical can be determined by using the second inequality:

$$d \geq 0: \quad \underbrace{\Theta_{11}\Theta_{22}}_{<0}\underbrace{\left(\Theta_{11}\Theta_{22} \pm \frac{Q_{12}Q_{21} + P_{12}P_{21}}{4\Omega_1\Omega_2}\right)}_{<0} \leq \left(\frac{Q_{12}P_{21} - P_{12}Q_{21}}{8\Omega_1\Omega_2}\right)^2,$$

$$\frac{-4\Theta_{11}\Theta_{22}}{(\Theta_{11} + \Theta_{22})^2}\left(\frac{Q_{12}P_{21} - P_{12}Q_{21}}{8\Omega_1\Omega_2}\right)^2 \leq \left(\frac{Q_{12}P_{21} - P_{12}Q_{21}}{8\Omega_1\Omega_2}\right)^2,$$

$$-4\Theta_{11}\Theta_{22} < 0 \leq (\Theta_{11} + \Theta_{22})^2,$$

$$\Theta_{11} \leq \left(-3 - 2\sqrt{2}\right)\Theta_{22} \quad \text{or} \quad \Theta_{11} \geq \left(-3 + 2\sqrt{2}\right)\Theta_{22} \tag{3.48}$$

If (3.48) is satisfied then d is positive. Otherwise a general statement about the sign of d is not possible.

2. For

$$\Theta_{11} + \Theta_{22} > 0,$$

$$0 < \Theta_{11}\Theta_{22} \pm \frac{Q_{12}Q_{21} + P_{12}P_{21}}{4\Omega_1\Omega_2} < \left(\frac{Q_{12}P_{21} - P_{12}Q_{21}}{4\Omega_1\Omega_2(\Theta_{11} + \Theta_{22})}\right)^2 \geq 0,$$

the system is unstable at the frequency η_0. For positive modal damping $\Theta_{11}\Theta_{22} > 0$ the sign of the radical can be determined by the same procedure as for the previous case:

$$d \geq 0: \quad \underbrace{\Theta_{11}\Theta_{22}}_{>0}\underbrace{\left(\Theta_{11}\Theta_{22} \pm \frac{Q_{12}Q_{21} + P_{12}P_{21}}{4\Omega_1\Omega_2}\right)}_{>0} \leq \left(\frac{Q_{12}P_{21} - P_{12}Q_{21}}{8\Omega_1\Omega_2}\right)^2,$$

$$\frac{4\Theta_{11}\Theta_{22}}{(\Theta_{11} + \Theta_{22})^2}\left(\frac{Q_{12}P_{21} - P_{12}Q_{21}}{8\Omega_1\Omega_2}\right)^2 \leq \left(\frac{Q_{12}P_{21} - P_{12}Q_{21}}{8\Omega_1\Omega_2}\right)^2,$$

$$0 \leq (\Theta_{11} - \Theta_{22})^2.$$

This implies that the radical d is always positive, $d > 0$. If one of the modal damping terms is negative, $\Theta_{11}\Theta_{22} < 0$, then it is trivial that the radical d is positive, $d > 0$, too.

3. For

$$\Theta_{11} + \Theta_{22} > 0,$$

$$\Theta_{11}\Theta_{22} \pm \frac{Q_{12}Q_{21} + P_{12}P_{21}}{4\Omega_1\Omega_2} > \left(\frac{Q_{12}P_{21} - P_{12}Q_{21}}{4\Omega_1\Omega_2(\Theta_{11} + \Theta_{22})}\right)^2 \geq 0,$$

the system is stable at frequency η_0 and a general statement for the sign of d is not possible.

A positive radical d yields a real-valued stability width $\tilde{\sigma}_{1,2}$, and a negative radical a purely imaginary stability width, respectively. For real-valued critical values $\tilde{\sigma}_{1,2}$ the stability width is according to (3.41)

$$\sigma_s - |\tilde{\sigma}_1| < \sigma < \sigma_s + |\tilde{\sigma}_1|.$$

3 General harmonic stiffness variation

For a single harmonic stiffness variation as in Chapter 3.1, it is possible to characterize the different system qualities by two crucial inequalities, as in Tab. 3.1. These inequalities arise quite naturally from the stability consideration at the frequency line $\eta_0 = |\Omega_1 \mp \Omega_2|$ in (3.23, 3.28), because exactly the same expression occurs in the radical of the stability width $\sigma_{1,2}$ in (3.24, 3.29). In case of a general harmonic stiffness variation, this advantageous correlation is destroyed by the interaction term $(Q_{12}P_{21} - P_{12}Q_{21})$. As a consequence, the skeleton line of the determined stability interval $\tilde{\sigma}_{1,2}$ in (3.40, 3.46) is shifted by σ_s. This shift leads to an intersection between the stability boundary curve and the frequency line η_0, as illustrated in Fig. 3.3 for real-valued $\tilde{\sigma}_{1,2}$. Which part of this frequency line lies in the stable and which in the unstable parameter domain is determined by (3.38, 3.45).

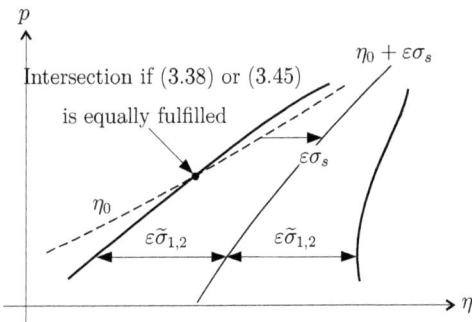

Figure 3.3.: Intersection of frequency line η_0 with shifted stability boundary curve.

Therefore the conditions in (3.38, 3.45) are helpful to determine the stability on the frequency line η_0, but in general, contrary to the results for a single harmonic stiffness variation, they are irrelevant for classification of the stability domain in the parameter space according to (3.40, 3.46), as the following short derivation shows:
The critical stability condition for η_0 is

$$\eta_0: \quad \Theta_{11}\Theta_{22} \pm \frac{Q_{12}Q_{21} + P_{12}P_{21}}{4\Omega_1\Omega_2} - \left(\frac{Q_{12}P_{21} - P_{12}Q_{21}}{4\Omega_1\Omega_2(\Theta_{11} + \Theta_{22})}\right)^2 = 0,$$

according to (3.38, 3.45). The critical condition where $\tilde{\sigma}_{1,2}$ switches from real-valued to purely imaginary is, $\tilde{\sigma}_{1,2} = 0$,

$$\tilde{\sigma}_{1,2}: \quad d = \Theta_{11}\Theta_{22}\left(\Theta_{11}\Theta_{22} \pm \frac{Q_{12}Q_{21} + P_{12}P_{21}}{4\Omega_1\Omega_2}\right) - \left(\frac{P_{12}Q_{21} - Q_{12}P_{21}}{8\Omega_1\Omega_2}\right)^2 = 0,$$

according to (3.40, 3.46). Respecting both equation gives

$$\frac{4\Theta_{11}\Theta_{22}}{(\Theta_{11} + \Theta_{22})^2}\left(\frac{Q_{12}P_{21} - P_{12}Q_{21}}{8\Omega_1\Omega_2}\right)^2 = \left(\frac{P_{12}Q_{21} - Q_{12}P_{21}}{8\Omega_1\Omega_2}\right)^2.$$

3 General harmonic stiffness variation

For a non-vanishing interaction term, $(P_{12}Q_{21} - Q_{12}P_{21}) \neq 0$, we obtain finally

$$(\Theta_{11} - \Theta_{22})^2 = 0,$$
$$\Theta_{11} = \Theta_{22}.$$

This means, that for a non-vanishing interaction term the stability conditions (3.38, 3.45) are capable of classifying the conditions (3.40, 3.46) only in the case when both modal damping terms are equal. Otherwise these conditions are inapplicable for a classification. On the other hand, for vanishing interaction term, $(P_{12}Q_{21} - Q_{12}P_{21}) = 0$, the relations simplify to equivalent conditions in (3.23, 3.28). In this special case the stability conditions (3.38, 3.45) are the adequate critical conditions for (3.40, 3.46), which enables the result in Tab. 3.1.

The adequate critical condition for non-vanishing interaction term follows from a stability analysis of the original characteristic polynomial in (3.37) by setting $\sigma = \sigma_s$. The shifted frequency line $\eta_0 + \varepsilon \sigma_s$ is stable iff

$$\Delta_1 : \tfrac{1}{2}\eta_0 (\Theta_{11} + \Theta_{22}) > 0,$$
$$\Delta_2 : \tfrac{1}{16}\eta_0^4 (\Theta_{11} + \Theta_{22})^2 \left(\Theta_{11}\Theta_{22} \pm \frac{Q_{12}Q_{21} + P_{12}P_{21}}{4\Omega_1\Omega_2} - \frac{1}{\Theta_{11}\Theta_{22}} \left(\frac{P_{12}Q_{21} - Q_{12}P_{21}}{8\Omega_1\Omega_2} \right)^2 \right) > 0,$$

are fulfilled, according to (A.2). These conditions can be rewritten for $\eta_0 > 0$ as

$$\Theta_{11} + \Theta_{22} > 0,$$
$$-\frac{d}{\Theta_{11}\Theta_{22}} > 0. \tag{3.49}$$

Using the inequalities in (3.49) enables a classification of the conditions (3.40, 3.46) for a general harmonic stiffness variation similar to Tab. 3.1, even if the additional frequency shift σ_s of the skeleton line η_0 occurs. This result is summarized in Tab. 3.4. Additionally, if we ask for simultaneous stability at the frequency $\eta_0 = |\Omega_1 - \Omega_2| + \varepsilon \sigma_s$ and $\eta_0 = \Omega_1 + \Omega_2 + \varepsilon \sigma_s$, respectively, we obtain the same result as in (3.47).

Table 3.4 shows the development of the stability border in dependency of an arbitrary system parameter p for non-vanishing interaction term $(Q_{12}P_{21} - P_{12}Q_{21})$. Read this table by either using always the upper sign or always the lower sign in analogy to Tab. 3.1. Plotting the frequency $\eta_0 = |\Omega_1 \mp \Omega_2|$ as function of a system parameter p results in a frequency line. The frequency shift σ_s is drawn for positive interaction term and $\Theta_{11} > \Theta_{22}$. If the condition in (3.49) is violated then the system becomes unstable for parameter values p lying on the shifted frequency line $\eta_0 + \varepsilon \sigma_s$. This is shown in the left column in Tab. 3.4. If additionally the stability conditions in the general case from (3.34) are violated, then the system is unstable on the shifted frequency line as well as in the remaining parameter domain and the stability width $\tilde{\sigma}_{1,2}$ is purely imaginary, see left bottom cell. On the other hand, if the conditions in (3.34) are satisfied, the system is unstable in the vicinity of the shifted frequency line $\eta_0 + \sigma_s$ and stable

Table 3.4.: Summary for general harmonic stiffness variation at parametric combination resonance frequencies for $Q_{12}P_{21} - P_{12}Q_{21} \neq 0$.

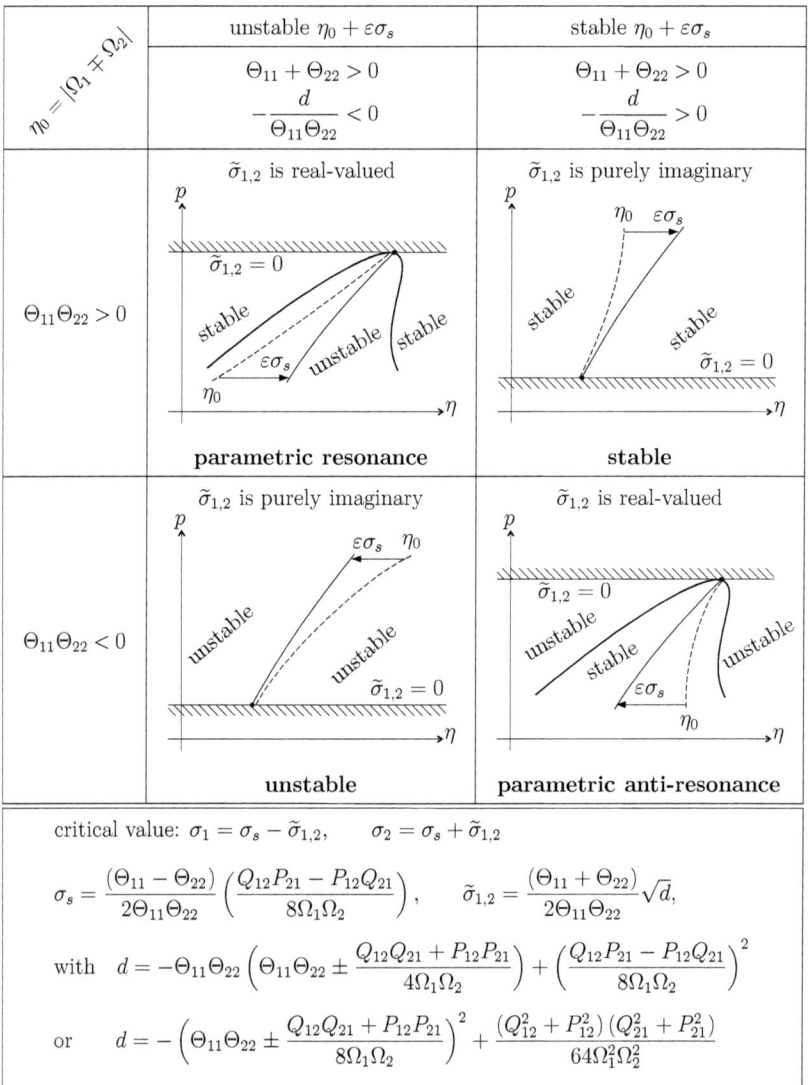

in the remaining parameter domain, see left top cell. In this case we call the frequency η_0 a parametric resonance frequency, because it disturbs an otherwise stable system.

3 General harmonic stiffness variation

For a stable frequency line $\eta_0 + \varepsilon\sigma_s$ the conditions in (3.38) or (3.45) are satisfied. This is shown in the right column in Tab. 3.4. Now if the general stability conditions in (3.34) are satisfied then the system is stable in the whole parameter domain, see right top cell. On the other hand, if the general stability conditions are violated, the system remains stable in the vicinity of the frequency line but becomes unstable in the remaining parameter domain, see right bottom cell. In this case we call the frequency η_0 a parametric *anti*-resonance frequency, because it stabilizes an otherwise unstable system.

Comparing the two left cells reveals, that passing the critical condition $\Theta_{11}\Theta_{22} = 0$ from $\Theta_{11}\Theta_{22} > 0$ to $\Theta_{11}\Theta_{22} < 0$ by varying a system parameter p, leads to a change of the sign of the frequency shift σ_s. Additionally, a positive radical d, corresponding to a real-valued stability width $\tilde{\sigma}_{1,2}$, switches to a negative radical, corresponding to a purely imaginary stability width. Generally, if a parameter combination p fulfills this switching condition $\Theta_{11}\Theta_{22} = 0$, then the stability on the shifted frequency line $\eta_0 + \varepsilon\sigma_s$ switches from stable to unstable or vice versa from unstable to stable. This is indicated by a hatched border. Reaching this border for such a critical parameter value, we have to switch the corresponding column.

Table 3.4 shows the qualitative stability map for a positive frequency shift $\sigma_s > 0$. If for specific parameter values p the frequency shift changes its sign the frequency shift σ_s has to be modified according to Tab. 3.3. An intersection of the frequency line η_0 with the stability border as in Fig. 3.3 is not considered.

The case where $\Theta_{11} + \Theta_{22} < 0$ is valid leads always to an unstable system. In this case the stability map for a certain parameter looks like in the sketch in the left bottom corner in Tab. 3.1. A detailed listing is therefore omitted.

The results in Tab. 3.4 hold as long as the interaction term does not vanish, $(Q_{12}P_{21} - P_{12}Q_{21}) \neq 0$. If the interaction term disappears, then the qualitative behavior is equivalent to that in Tab. 3.1. In this case the additional frequency shift σ_s vanishes and the general stability width $\tilde{\sigma}_{1,2}$ simplifies to $\sigma_{1,2}$ in Tab. 3.1.

3.2.3. Symmetry considerations

The sign of the parametric excitation term $(Q_{12}Q_{21} + P_{12}P_{21})$ is crucial for the position of the parametric anti-resonance frequency, whether the anti-resonance occurs at $\eta_0 = |\Omega_1 - \Omega_2|$ or at $\eta_0 = \Omega_1 + \Omega_2$.

As shown in the previous section, the conditions for stability of a system with negative modal damping $(0 > \Theta_{11}\Theta_{22})$ at the frequency $\eta_0 + \varepsilon\sigma_s$ for $\eta_0 = |\Omega_1 - \Omega_2|$ are

$$\Theta_{11} + \Theta_{22} > 0,$$
$$0 > \alpha > -\frac{Q_{12}Q_{21} + P_{12}P_{21}}{4\Omega_1\Omega_2},$$

Table 3.5.: Symmetry considerations of parametric excitation matrix at frequencies $\eta_0 = |\Omega_1 \mp \Omega_2|$ for general harmonic stiffness variation.

$\eta_0 = \|\Omega_1 - \Omega_2\|$	(3.34) is fulfilled: $\Theta_{11} + \Theta_{22} > 0$ $\alpha > 0$	(3.34) is violated: $\Theta_{11} + \Theta_{22} > 0$ $\alpha < 0$
$Q_{12}Q_{21}+$ $+P_{12}P_{21} < 0$	additional restriction: **parametric resonance** for $0 < \alpha < -\dfrac{Q_{12}Q_{21} + P_{12}P_{21}}{4\Omega_1\Omega_2}$, $\tilde{\sigma}_{1,2}$ is real-valued: $\sigma_s - |\tilde{\sigma}_{1,2}| < \sigma < \sigma_s + |\tilde{\sigma}_{1,2}|$	(3.38) is violated: **unstable** no boundary
$Q_{12}Q_{21}+$ $+P_{12}P_{21} > 0$	(3.38) is fulfilled: **stable** no boundary	additional restriction: **parametric anti-resonance** for $0 > \alpha > -\dfrac{Q_{12}Q_{21} + P_{12}P_{21}}{4\Omega_1\Omega_2}$, $\tilde{\sigma}_{1,2}$ is real-valued: $\sigma_s - |\tilde{\sigma}_{1,2}| < \sigma < \sigma_s + |\tilde{\sigma}_{1,2}|$
$\eta_0 = \Omega_1 + \Omega_2$	(3.34) is fulfilled: $\Theta_{11} + \Theta_{22} > 0$ $\alpha > 0$	(3.34) is violated: $\Theta_{11} + \Theta_{22} > 0$ $\alpha < 0$
$Q_{12}Q_{21}+$ $+P_{12}P_{21} < 0$	(3.45) is fulfilled: **stable** no boundary	additional restriction: **parametric anti-resonance** for $0 > \alpha > \dfrac{Q_{12}Q_{21} + P_{12}P_{21}}{4\Omega_1\Omega_2}$, $\tilde{\sigma}_{1,2}$ is real-valued: $\sigma_s - |\tilde{\sigma}_{1,2}| < \sigma < \sigma_s + |\tilde{\sigma}_{1,2}|$
$Q_{12}Q_{21}+$ $+P_{12}P_{21} > 0$	additional restriction: **parametric resonance** for $0 < \alpha < \dfrac{Q_{12}Q_{21} + P_{12}P_{21}}{4\Omega_1\Omega_2}$, $\tilde{\sigma}_{1,2}$ is real-valued: $\sigma_s - |\tilde{\sigma}_{1,2}| < \sigma < \sigma_s + |\tilde{\sigma}_{1,2}|$	(3.45) is violated: **unstable** no boundary

using the abbreviation

$$\alpha = \Theta_{11}\Theta_{22} - \frac{\beta^2}{(\Theta_{11} + \Theta_{22})^2}.$$

3 General harmonic stiffness variation

If the parametric excitation matrices

$$\mathbf{Q} = \begin{bmatrix} Q_{11} & Q_{12} \\ Q_{21} & Q_{22} \end{bmatrix}, \quad \mathbf{P} = \begin{bmatrix} P_{11} & P_{12} \\ P_{21} & P_{22} \end{bmatrix}$$

are symmetric matrices, then the product $(Q_{12}Q_{21} + P_{12}P_{21})$ is positive and the aforementioned inequalities can be satisfied. On the other hand, if the parametric excitation matrices are skew-symmetric, the product $(Q_{12}Q_{21} + P_{12}P_{21})$ is negative. For $(Q_{12}Q_{21} + P_{12}P_{21}) < 0$ the second inequality results in

$$0 > \alpha > 0.$$

In this case it is impossible to fulfill the second inequality for non-vanishing modal damping coefficients.

The same results are obtained for a desired stability at the frequency $\eta_0 + \varepsilon\sigma_s$ for $\eta_0 = \Omega_1 + \Omega_2$ where the conditions

$$\Theta_{11} + \Theta_{22} > 0,$$

$$0 > \alpha > \frac{Q_{12}Q_{21} + P_{12}P_{21}}{4\Omega_1\Omega_2},$$

have to be fulfilled. Now, if the parametric excitation matrices are symmetric, $(Q_{12}Q_{21} + P_{12}P_{21}) > 0$ results again in

$$0 > \alpha > 0.$$

Hence, it is not possible to satisfy the second inequality. These results are summarized in Tab. 3.5.

3.2.4. Literature review

Only very few research has been published dealing with analytical approximations of stability domains in case of a general harmonic stiffness variation as described. Valuable works to mention are [28] and [50]. Eicher [28] used the method of successive approximation in combination with an integral version of Fredholm's alternative. This approach restricts the natural frequencies Ω_1, Ω_2 to positive integer numbers n_1, n_2. Expressed in the notation as used here, the stability formulae derived in [28, p.363, (12.44)] yield

$$\eta = m + \frac{\Theta_{11} - \Theta_{22}}{4\Theta_{11}\Theta_{22}n_1n_2}\left(\frac{Q_{12}P_{21} - P_{12}Q_{21}}{4}\right) \pm \frac{1}{2}\sqrt{b}, \qquad (3.50)$$

where $m = n_1 - n_2$ and

$$b = +\left(\frac{Q_{12}P_{21} - P_{12}Q_{21}}{4}\right)^2\left(\frac{\Theta_{11} + \Theta_{22}}{2\Theta_{11}\Theta_{22}n_1n_2}\right)^2$$
$$-\left(\frac{Q_{12}P_{21} + P_{12}Q_{21}}{4}\right)\frac{(\Theta_{11} + \Theta_{22})^2}{\Theta_{11}\Theta_{22}n_1n_2} - (\Theta_{11} + \Theta_{22})^2.$$

Using the relationship

$$\left(Q_{12}^2 + P_{12}^2\right)\left(Q_{21}^2 + P_{21}^2\right) = \left(Q_{12}Q_{21} + P_{12}P_{21}\right)^2 + \left(Q_{12}P_{21} - Q_{12}P_{21}\right)^2$$

gives the more handsome expressions

$$b = \left(\frac{\Theta_{11} + \Theta_{22}}{2\Theta_{11}\Theta_{22}}\right)^2 \left(-\Theta_{11}\Theta_{22}\left(\Theta_{11}\Theta_{22} + \frac{Q_{12}P_{21} + P_{12}Q_{21}}{4n_1 n_2}\right) + \left(\frac{Q_{12}P_{21} - P_{12}Q_{21}}{8n_1 n_2}\right)^2\right).$$

Now it is easy to compare (3.50) with the condition derived in (3.40). Both results are the same. The only difference is the valid frequency scale. While the formulae derived by Eicher are only valid for natural-valued frequencies n_i, the conditions presented in this work are valid for real-valued frequencies Ω_i, which suggests the mapping relation

$$n_1,\ n_2,\ m = n_1 - n_2 \ \in \mathbb{N} \quad \Longleftrightarrow \quad \Omega_1,\ \Omega_2,\ \eta_0 = \Omega_1 - \Omega_2 \ \in \mathbb{R}.$$

Hence, the method of successive approximation and the averaging method produce the same results on their *own* scale, but the averaging method is valid on a denser scale.

3.3. Periodic stiffness variation

In the previous Sections 3.1 and 3.2 we analyzed the stability boundary curves of a mechanical system with a single harmonic variation of its stiffness parameters. In some cases the variation of a stiffness parameter is not a harmonic process, or even may be a non-smooth switching operation. In this section we relax the restriction of a single harmonic parametric excitation. We demand, that the parametric excitation is periodic and therefore can be expanded into a Fourier series, a sum of single harmonic functions. For instance, employing a symmetric rectangular variation in time, a special form of a so-called bang-bang control of a stiffness parameter,

$$\text{rect}(t) = \begin{cases} +1 & 0 < t \leq \frac{T}{2} \\ -1 & \frac{T}{2} < t \leq T \end{cases}, \quad T = \frac{2\pi}{\eta},$$

we obtain, instead of a single harmonic variation, a sum of single harmonic variations

$$k(t) = k(1 + \varepsilon \, \text{rect}(\eta\tau)) \approx k\left(1 + \varepsilon \frac{4}{\pi} \sum_{n=1}^{\infty} \frac{(-1)^{n+1}}{2n-1} \cos((2n-1)\eta\tau)\right). \tag{3.51}$$

For the case of a general periodic stiffness variation the mass and damping matrices are kept constant

$$\mathbf{M}(t) = \mathbf{M}_0, \quad \mathbf{C}(t) = \mathbf{C}_0,$$

$$\mathbf{K}(t) = \mathbf{K}_0 + \varepsilon \sum_{n=1}^{N} \left\{\mathbf{K}_{c,n}^s + \mathbf{K}_{c,n}^a\right\} \cos(n\omega t) + \varepsilon \left\{\mathbf{K}_{s,n}^s + \mathbf{K}_{s,n}^a\right\} \sin(n\omega t),$$

where N indicates the highest order term of the Fourier series considered for the analysis. In this case the general linear equations of motion (2.2) simplifies to

$$\ddot{\mathbf{z}}(t) + \varepsilon \tilde{\mathbf{C}}_0 \dot{\mathbf{z}}(t) + \tilde{\mathbf{K}}_0 \mathbf{z}(t) = -\varepsilon \sum_{n=1}^{N} \left\{\left\{\tilde{\mathbf{K}}_{c,n}^s + \tilde{\mathbf{K}}_{c,n}^a\right\} \cos(n\omega t) + \left\{\tilde{\mathbf{K}}_{s,n}^s + \tilde{\mathbf{K}}_{s,n}^a\right\} \sin(n\omega t)\right\} \mathbf{z}(t). \tag{3.52}$$

Note again, that due to the fact that $\tilde{\mathbf{K}}_{c/s,n}$ are matrices, more than one stiffness in the system can be varied. In Section 3.2 we restricted our study to variations with the same frequency ω and fixed phase relations. In this section we allow variation variations with the frequency ω and its multiples ω_n, a multi-frequency variation with integer-valued frequency ratios. The phase relations between these harmonic variations are arbitrary, but fixed. A general multi-frequency variation with arbitrary frequency ratios is *not* considered.

The equations of motion can be transformed ($\mathbf{z}(t) \longrightarrow \mathbf{z}(\tau)$) to a non-dimensional normal form

$$\mathbf{z}''(\tau) + \varepsilon \Theta \mathbf{z}'(\tau) + \Omega^2 \mathbf{z}(\tau) = -\varepsilon \sum_{n=1}^{N} \left\{(\mathbf{Q}_n^s + \mathbf{Q}_n^a) \cos(n\eta\tau) + (\mathbf{P}_n^s + \mathbf{P}_n^a) \sin(n\eta\tau)\right\} \mathbf{z}(\tau),$$

$$\text{where } \mathbf{Q}_n = \mathbf{Q}_n^s + \mathbf{Q}_n^a = \mathbf{T}^{-1} \mathbf{M}_0^{-1} \mathbf{K}_{c,n} \mathbf{T},$$

$$\text{and } \mathbf{P}_n = \mathbf{P}_n^s + \mathbf{P}_n^a = \mathbf{T}^{-1} \mathbf{M}_0^{-1} \mathbf{K}_{s,n} \mathbf{T},$$

as already shown in Section 2.3, or rewritten for a 2dof-system in the comprehensive index notation

$$z_i''(\tau) + \Omega_i^2 z_i(\tau) = -\varepsilon \left\{ \Theta_{ij} z_j'(\tau) + z_j(\tau) \sum_{n=1}^{N} Q_{ijn} \cos(n\eta\tau) + P_{ijn} \sin(n\eta\tau) \right\}, \quad (3.53)$$

with $i, j = 1, 2$. The approach presented below follows the procedure in Section 3.1 and is a direct generalization of Section 3.2.

3.3.1. Transformation and averaging

Applying the procedure for a *first order approximation* in analogy to Sections 3.1 and 3.2 extends the right hand sides of (3.8) by the sine parametric excitation term P_{ij}

$$F_i(\mathbf{u}, t) = -\eta_0 \sum_j \Theta_{ij} \left(-u_j \varpi_j s_j + v_j \varpi_j c_j \right) +$$

$$- \sum_j \left(\sum_{n=1}^{N} Q_{ijn} \cos nt + P_{ijn} \sin nt \right) (u_j c_j + v_j s_j) + 2\Omega_i \varpi_i \sigma (u_i c_i + v_i s_i), \quad (3.54)$$

with the abbreviations $s_i = \sin \varpi_i t$ and $c_i = \cos \varpi_i t$ and the state vector $\mathbf{u} = (u_1, v_1, u_2, v_2)^T$. Comparing (3.54) with (3.32) points out the difference between a harmonic stiffness variation (Q_{ij}, P_{ij}) and a periodic stiffness variation (Q_{ijn}, P_{ijn}).

The final equations for the slow amplitudes $\mathbf{u}$ are equal to (3.9), with quasi-periodic right hand sides. For averaging we first have to determine the periods of the right hand sides of (3.9). With the help of decomposition theorems in Appendix B the arising products of the trigonometric terms on the right hand sides of (3.54) can be rearranged as a sum of basic trigonometric terms. For this system with two modes we obtain $(3 + 9 \times N)$ different periods. Averaging over a basic trigonometric term yields always zero, except for the case where a term becomes resonant, i.e. the argument of a cosine function vanishes:

Θ_{ii}, σ : no resonant terms, constant factor $\frac{1}{2}$,

Θ_{ij} : resonant terms for $\varpi_i \pm \varpi_j = 0$,

Q_{iin}, P_{ijn} : resonant terms for $2\varpi_i \pm 1 = 0$,

Q_{ijn}, P_{ijn} : resonant terms for $\varpi_i \pm \varpi_j \pm 1 = 0$.

General case

If ϖ_1, ϖ_2 are arbitrary, averaging over each trigonometric function gives zero, only the terms Θ_{ii}, σ remain and we obtain the same stability conditions as in (3.14):

$$\boxed{\Theta_{11} > 0 \quad \text{and} \quad \Theta_{22} > 0.} \quad (3.55)$$

3 Periodic stiffness variation 59

The system is stable if both damping coefficient has the same sign. The boundaries between stable and unstable frequency intervals are always purely imaginary.

Some resonant cases

If ϖ_1, ϖ_2 take on special values, some terms become resonant and influence the previous solution.

1. Combination parametric resonance of difference type and first kind and order n
If we assume that
$$\boxed{\eta_{0n} = \frac{|\Omega_2 - \Omega_1|}{n} > 0 \quad \text{or} \quad \varpi_j - \varpi_i = \mp n}$$
with
$$n = 1, 2, 3, \ldots \quad \text{and} \quad \eta_0 = \eta_{01} = |\Omega_2 - \Omega_1|$$
some terms become resonant and produce additional terms after averaging. We get from (3.54)
$$\begin{aligned}\dot{\hat{u}}_i &= \frac{\varepsilon}{\eta_{0n}^2 \varpi_i}\left\{-\frac{\eta_0}{2}\Theta_{ii}\varpi_i\hat{u}_i - \Omega_i\varpi_i\sigma_n\hat{v}_i + (-1)^i\frac{P_{ijn}}{4}\hat{u}_j + \frac{Q_{ijn}}{4}\hat{v}_j\right\}, \\ \dot{\hat{v}}_i &= \frac{\varepsilon}{\eta_{0n}^2 \varpi_i}\left\{-\frac{\eta_0}{2}\Theta_{ii}\varpi_i\hat{v}_i + \Omega_i\varpi_i\sigma_n\hat{u}_i + (-1)^i\frac{P_{ijn}}{4}\hat{v}_j - \frac{Q_{ijn}}{4}\hat{u}_j\right\}.\end{aligned} \quad (3.56)$$
For $n = 1$ this system of equations is equivalent to a single harmonic stiffness variation as in (3.35). The stability conditions are obtained by performing the substitutions
$$Q_{ij} \mapsto Q_{ijn}, \quad P_{ij} \mapsto P_{ijn}, \quad \eta_0 \mapsto \eta_{0,n},$$
for $\sigma_n = 0$ in (3.38) and for $\sigma_n \neq 0$ in (3.40).

2. Combination parametric resonance of summation type and first kind and order n
By setting
$$\boxed{\eta_{0n} = \frac{\Omega_1 + \Omega_2}{n} > 0 \quad \text{or} \quad \varpi_i + \varpi_j = n}$$
with
$$n = 1, 2, 3, \ldots \quad \text{and} \quad \eta_0 = \eta_{01} = \Omega_1 + \Omega_2$$
some terms become resonant and produce additional terms after averaging. We get from (3.54)
$$\begin{aligned}\dot{\hat{u}}_i &= \frac{\varepsilon}{\eta_{0n}^2 \varpi_i}\left\{-\frac{\eta_0}{2}\Theta_{ii}\varpi_i\hat{u}_i - \Omega_i\varpi_i\sigma_n\hat{v}_i + \frac{P_{ijn}}{4}\hat{u}_j - \frac{Q_{ijn}}{4}\hat{v}_j\right\}, \\ \dot{\hat{v}}_i &= \frac{\varepsilon}{\eta_{0n}^2 \varpi_i}\left\{-\frac{\eta_0}{2}\Theta_{ii}\varpi_i\hat{v}_i + \Omega_i\varpi_i\sigma_n\hat{u}_i - \frac{P_{ijn}}{4}\hat{v}_j - \frac{Q_{ijn}}{4}\hat{u}_j\right\}.\end{aligned} \quad (3.57)$$
As in the previous paragraph, for $n = 1$ this system of equations is equivalent to a single harmonic stiffness variation as in (3.43). The stability conditions for $\sigma_n = 0$ are obtained from (3.45) and for $\sigma_n \neq 0$ from (3.46) by performing the substitutions
$$Q_{ij} \mapsto Q_{ijn}, \quad P_{ij} \mapsto P_{ijn}, \quad \eta_0 \mapsto \eta_{0,n}.$$

3.3.2. Summary for small general periodic stiffness variations

Whether a considered system is stable or not, depends mostly on the frequency η of the stiffness variation. In general the system is stable for arbitrary values of η iff

$$\boxed{\Theta_{11} > 0 \quad \text{and} \quad \Theta_{22} > 0,}$$

are satisfied. On the other hand, if η is close to a certain combination frequencies, the system might be locally stable:

Cases 1. and 2. For $\eta_n \approx \dfrac{|\Omega_i \mp \Omega_j|}{n}$, or $\eta_n = \eta_{0n} + \varepsilon\sigma_{1,2n}$, if the following conditions are satisfied:

$$\boxed{\begin{aligned}
&(3.38,\ 3.45),\ \Delta_1 > 0:\quad \Theta_{11} + \Theta_{22} > 0, \\
&(3.38,\ 3.45),\ \text{stable } \eta_{0,n}:\quad \Theta_{11}\Theta_{22} \pm \frac{Q_{12n}Q_{21n} + P_{12n}P_{21n}}{4\Omega_1\Omega_2} > \frac{\beta_n^2}{(\Theta_{11} + \Theta_{22})^2} \geq 0, \\
&(3.40,\ 3.46),\ \Delta_2 = 0:\quad \sigma_{1,n} = \sigma_{s,n} - \tilde{\sigma}_{12,n},\quad \sigma_{2,n} = \sigma_{s,n} + \tilde{\sigma}_{12,n},
\end{aligned}}$$

(3.58)

with the abbreviation

$$\beta_n = \left(\frac{Q_{12n}P_{21n} - P_{12n}Q_{21n}}{4\Omega_1\Omega_2}\right),$$

$$\sigma_{s,n} = \frac{(\Theta_{11} - \Theta_{22})\beta_n}{2\Theta_{11}\Theta_{22}\ 2n},\quad \tilde{\sigma}_{12,n} = \frac{(\Theta_{11} + \Theta_{22})}{2\Theta_{11}\Theta_{22}}\frac{\sqrt{d_n}}{n},$$

with $\quad d_n = -\Theta_{11}\Theta_{22}\left(\Theta_{11}\Theta_{22} \pm \dfrac{Q_{12n}Q_{21n} + P_{12n}P_{21n}}{4\Omega_1\Omega_2}\right) + \left(\dfrac{\beta_n}{2}\right)^2,$

or $\quad d_n = -\left(\Theta_{11}\Theta_{22} \pm \dfrac{Q_{12n}Q_{21n} + P_{12n}P_{21n}}{8\Omega_1\Omega_2}\right)^2 + \dfrac{(Q_{12n}^2 + P_{12n}^2)(Q_{21n}^2 + P_{21n}^2)}{64\Omega_1^2\Omega_2^2}.$

Note the different signs of the parametric excitation terms $(Q_{12n}Q_{21n} + P_{12n}P_{21n})$ in the case of $\eta_{0n} = |\Omega_1 - \Omega_2|/n$ and $\eta_{0n} = (\Omega_1 + \Omega_2)/n$, respectively. If we consider the case of stability for $\eta_{0n} = |\Omega_1 - \Omega_2|/n$ and for $(\eta_0 = \Omega_1 + \Omega_2)/n$ simultaneously, we obtain from above, (3.40) and (3.46), the following conditions for stability:

$$\left.\begin{aligned}
\eta_{0n} &= \frac{|\Omega_1 - \Omega_2|}{n}:\quad \Theta_{11}\Theta_{22} + \frac{Q_{12n}Q_{21n} + P_{12n}P_{21n}}{4\Omega_1\Omega_2} > \frac{\beta_n^2}{(\Theta_{11} + \Theta_{22})^2} \geq 0 \\
\eta_{0n} &= \frac{\Omega_1 + \Omega_2}{n}:\quad \Theta_{11}\Theta_{22} - \frac{Q_{12n}Q_{21n} + P_{12n}P_{21n}}{4\Omega_1\Omega_2} > \frac{\beta_n^2}{(\Theta_{11} + \Theta_{22})^2} \geq 0
\end{aligned}\right\} \Rightarrow 2\Theta_{11}\Theta_{22} > 0.$$

(3.59)

This means that in the case $\Theta_{11}\Theta_{22} < 0$ a simultaneously stable system at the parametric excitation frequency of the difference type $\eta_{0n} = |\Omega_1 - \Omega_2|/n$ and the summation type $\eta_{0n} = (\Omega_1 + \Omega_2)/n$ is *not* possible, the same result as in the case of a general harmonic stiffness variation in Section 3.2 and a single harmonic stiffness variation in Section 3.1! Additionally, if we ask for simultaneous stability at the frequency $\eta_0 = |\Omega_1 - \Omega_2|/n + \varepsilon\sigma_{s,n}$ and $\eta_0 = (\Omega_1 + \Omega_2)/n + \varepsilon\sigma_{s,n}$, respectively, we obtain the same result as in (3.59).

Furthermore, note that the shifts $\sigma_{s,n}$ are the same for a parametric combination frequency of the difference type $\eta_{0n} = |\Omega_1 - \Omega_2|/n$ and of the summation type $\eta_{0n} = (\Omega_1 + \Omega_2)/n$, respectively. The additional frequency shift $\sigma_{s,n}$ changes its sign at least if a modal damping term changes its sign, too. The critical case where one modal damping parameter is equal to zero, $\Theta_{11}\Theta_{22} = 0$, results in an infinite frequency shift $\sigma_{s,n}$. Hence, a change in the sign of $\sigma_{s,n}$ never occurs by crossing the frequency lines η_{0n}. This circumstance was pointed out in Tab. 3.3 on page 48.

If a general periodic stiffness variation collapses to a Fourier series of cosine functions, a Fourier cosine series, then the P_{ijn}-term vanish and the corresponding stability conditions from above simplifies to the conditions derived for the case of a single harmonic stiffness variation in Section 3.1 by performing the substitution $Q_{ij} \mapsto Q_{ijn}$.

3.3.3. Symmetry considerations

The sign of the parametric excitation term $Q_{12n}Q_{21n}$ is independent from the sign of the corresponding coefficients a_n, b_n of the Fourier series, because the Fourier coefficients occur squared:

$$\mathbf{Q}\mathbf{k}_c(t) + \mathbf{P}\mathbf{k}_s(t) = \mathbf{Q}\sum_{n=1}^{N} a_n \cos(n\omega t) + \mathbf{P}\sum_{n=1}^{N} b_n \sin(n\omega t),$$

$$Q_{12n}Q_{21n} = (Q_{12}a_n)(Q_{21}a_n) = a_n^2 Q_{12}Q_{21}, \qquad (3.60)$$

$$P_{12n}P_{21n} = (P_{12}b_n)(P_{21}b_n) = b_n^2 P_{12}P_{21}.$$

Hence, the same symmetry classifications can be made as in Tab. 3.5. In the case of Fourier cosine series, $P_{ijn} = 0$, the simpler classification in Tab. 3.2 is appropriate.

4. Time-periodic damping

The previous chapter theoretically revealed that in case of a time-periodic stiffness variation it is possible to stabilize a vibrating system, if certain conditions are satisfied. The equations investigated were a set of coupled Mathieu or Mathieu-like equations. In this chapter we analyze whether a periodic change of one or more damping parameters is also capable of generating additional positive damping in a system.

For the case of a single harmonic damping variation the mass and stiffness matrices are kept constant

$$\mathbf{M}(t) = \mathbf{M}_0, \quad \mathbf{K}(t) = \mathbf{K}_0,$$
$$\mathbf{C}(t) = \mathbf{C}_0 + \varepsilon \left\{ \mathbf{C}_c^s + \mathbf{C}_c^a \right\} \cos(\omega t),$$

and the general linear equations of motion (2.2) simplify to

$$\ddot{\mathbf{z}}(t) + \varepsilon \tilde{\mathbf{C}}_0 \dot{\mathbf{z}}(t) + \tilde{\mathbf{K}}_0 \mathbf{z}(t) = -\varepsilon \left\{ \tilde{\mathbf{C}}_c^s + \tilde{\mathbf{C}}_c^a \right\} \cos(\omega t) \mathbf{z}(t).$$

Note that, due to the fact that $\tilde{\mathbf{C}}_c$ is a matrix, more than one damping coefficient in the system can be varied. The restriction is that these variations have the same frequency ω and the same phase 0. These equations of motion in dimensional terms can be transformed ($\underline{z}(t) \longrightarrow \underline{z}(\tau)$) to the following non-dimensional normal form

$$\mathbf{z}''(\tau) + \varepsilon \mathbf{\Theta} \mathbf{z}'(\tau) + \mathbf{\Omega}^2 \mathbf{z}(\tau) = -\varepsilon \left(\mathbf{R}^s + \mathbf{R}^a \right) \cos(\eta \tau) \mathbf{z}'(\tau),$$
$$\text{where } \mathbf{R} = \mathbf{R}^s + \mathbf{R}^a = \mathbf{T}^{-1} \mathbf{M}_0^{-1} \mathbf{C}_c \mathbf{T},$$

as it was shown in Section 2.3. For a mechanical system with two degrees of freedom this yields

$$\mathbf{z}''(\tau) + \mathbf{\Omega}^2 \mathbf{z}(\tau) = -\varepsilon \left\{ \mathbf{\Theta} + \left(\mathbf{R}^s + \mathbf{R}^a \right) \cos(\eta \tau) \right\} \mathbf{z}'(\tau),$$

$$\mathbf{z}''(\tau) + \begin{bmatrix} \Omega_1^2 & 0 \\ 0 & \Omega_2^2 \end{bmatrix} \mathbf{z}(\tau) = -\varepsilon \left\{ \begin{bmatrix} \Theta_{11} & \Theta_{12} \\ \Theta_{21} & \Theta_{22} \end{bmatrix} + \begin{bmatrix} R_{11} & R_{12} \\ R_{21} & R_{22} \end{bmatrix} \cos(\eta \tau) \right\} \mathbf{z}'(\tau),$$

or formulated in the comprehensive index notation

$$z_i''(\tau) + \Omega_i^2 z_i(\tau) = -\varepsilon \left\{ \Theta_{ij} z_j'(\tau) + R_{ij} z_j'(\tau) \cos(\eta \tau) \right\}, \tag{4.1}$$

with $i, j = 1, 2$. The approach presented below follows the procedure in Section 3.1.

4.1. Transformation and averaging

Applying the procedure in Section 3.1, the time transformation (3.2) and the Taylor expansion (3.4), on (4.1) leads to

$$\ddot{z}_i + \varpi_i^2 z_i = -\frac{\varepsilon}{\eta_0^2}\left\{+\eta_0 \Theta_{ij}\dot{z}_j + \eta_0 R_{ij}\dot{z}_j \cos t - 2\Omega_i \varpi_i \sigma z_i\right\} + \mathcal{O}\left(\varepsilon^2\right), \qquad (4.2)$$

with the abbreviations $\varpi_i = \Omega_i/\eta_0$. Using the coordinate transformation in (3.6) the right hand sides of (4.2) become

$$F_i\left(\mathbf{u},t\right) = -\eta_0 \sum_j \left(\Theta_{ij} + R_{ij}\cos t\right)\left(-u_j \varpi_j s_j + v_j \varpi_j c_j\right) + 2\Omega_i \varpi_i \sigma \left(u_i c_i + v_i s_i\right) \qquad (4.3)$$

with the abbreviations $s_i = \sin \varpi_i t$ and $c_i = \cos \varpi_i t$ and the state vector $\mathbf{u} = \left(u_1, v_1, u_2, v_2\right)^T$. Comparing (4.3) with (3.8) points out the difference between a harmonic stiffness variation (Q_{ij}) and a harmonic damping variation (R_{ij}).

The final system equations for the slow amplitudes $\mathbf{u}$ are equal to (3.9), with quasi-periodic right hand sides. For this simple two mode system we obtain 12 different periods according to Appendix B by substituting $n = 1$ and performing $Q_{kl} \mapsto R_{kl}$. This is the same number of periods as for the single harmonic stiffness variation in (3.8). Averaging over these different periodic terms yields always zero, except for the case where a term becomes resonant:

Θ_{ii}, σ : no resonant terms, constant factor $\frac{1}{2}$,

Θ_{ij} : resonant terms for $\varpi_i \pm \varpi_j = 0$,

R_{ii} : resonant terms for $2\varpi_i \pm 1 = 0$,

R_{ij} : resonant terms for $\varpi_i \pm \varpi_j \pm 1 = 0$.

General case

If ϖ_1, ϖ_2 do not take on special values averaging over each trigonometric function gives zero and only the coefficients of Θ_{ii}, σ remain. We obtain from (4.3)

$$\dot{\hat{u}}_i = \frac{\varepsilon}{\eta_0^2 \varpi_i}\left\{-\frac{\eta_0}{2}\Theta_{ii}\varpi_i \hat{u}_i - \Omega_i \varpi_i \sigma \hat{v}_i\right\},$$
$$\dot{\hat{v}}_i = \frac{\varepsilon}{\eta_0^2 \varpi_i}\left\{-\frac{\eta_0}{2}\Theta_{ii}\varpi_i \hat{v}_i + \Omega_i \varpi_i \sigma \hat{u}_i\right\}. \qquad (4.4)$$

This system of equations is identical to the general case of a stiffness variation in (3.11) and yields for $\eta_0 > 0$

$$\boxed{\Theta_{11} > 0 \quad \text{and} \quad \Theta_{22} > 0,} \qquad (4.5)$$

according to (3.14). The critical values of σ are always purely imaginary. This means that in general a stability boundary does not exist. Globally the system is either stable or unstable, depending on the conditions in (4.5).

4 Time-periodic damping

Some resonant cases

If ϖ_1, ϖ_2 take on special values, some terms become resonant in the averaging process and (4.4) does not longer hold and the stability conditions in (4.5) have to be adapted. This will be shown in the following paragraphs. In physical meaningful systems the two natural frequencies Ω_1, Ω_2 are *positive*. The case where one frequency is equal to zero is again not considered for this 2dof system.

1. Simple parametric resonance frequencies of first kind

By defining

$$\boxed{\eta_0 = 2\Omega_k \quad \text{or} \quad 2\varpi_k = 1, \quad \text{with } k = 1 \text{ or } 2}$$

some terms become resonant and produce additional terms after averaging. Instead of (4.4) we obtain from (4.3)

$$\dot{\hat{u}}_i = \frac{\varepsilon}{\eta_0^2}\left\{\left(-\frac{\eta_0}{2}\Theta_{ii} + \frac{\eta_0 R_{ii}}{4\varpi_i}\delta_{ik}\right)\hat{u}_i - \Omega_i\sigma\hat{v}_i\right\},$$

$$\dot{\hat{v}}_i = \frac{\varepsilon}{\eta_0^2}\left\{\left(-\frac{\eta_0}{2}\Theta_{ii} - \frac{\eta_0 R_{ii}}{4\varpi_i}\delta_{ik}\right)\hat{v}_i + \Omega_i\sigma\hat{u}_i\right\},$$
(4.6)

where δ_{ik} is the Kronecker delta-function. Rewritten in matrix notation one gets equations of the form

$$\dot{\hat{\mathbf{u}}}(t) = \frac{\varepsilon}{\eta_0^2}\begin{bmatrix} \mathbf{A}_1 & 0 \\ 0 & \mathbf{A}_2 \end{bmatrix}\hat{\mathbf{u}}(t)$$

with $\hat{\mathbf{u}} = (\hat{u}_1, \hat{v}_1, \hat{u}_2, \hat{v}_2)^T$ and

$$\mathbf{A}_i = \begin{bmatrix} -\dfrac{\eta_0}{2}\Theta_{ii} + \dfrac{\eta_0 R_{ii}}{4}\delta_{ik} & -\Omega_i\sigma \\ \Omega_i\sigma & -\dfrac{\eta_0}{2}\Theta_{ii} - \dfrac{\eta_0 R_{ii}}{4}\delta_{ik} \end{bmatrix}.$$

Hence, the slow states $(\hat{u}_1, \hat{v}_1)$ and $(\hat{u}_2, \hat{v}_2)$ are decoupled. Comparing the coefficient matrix in (4.6) with the coefficient matrix of the single harmonic stiffness variation in (3.15) reveals the difference between these different types of variation: while for a stiffness variation the coefficients of its parametric excitation, Q_{ii}, are coefficients of $\hat{v}_i$, for a damping variation the coefficients of its parametric excitation term, R_{ii}, becomes coefficients of $\hat{u}_i$. But this difference is only formal, as the following paragraphs show.

After rescaling time by ε/η_0^2 we obtain the characteristic equation

$$\det(\lambda\mathbf{I}_4 - \mathbf{A}) = \det(\lambda\mathbf{I}_2 - \mathbf{A}_1)\det(\lambda\mathbf{I}_2 - \mathbf{A}_2) = 0.$$

At least one of the determinants has to be zero. Without restriction of generality we choose the first determinant to be zero, this means we are looking just for stable behavior of the states $(\hat{u}_1, \hat{v}_1)$

$$\lambda^2 + \eta_0\Theta_{11}\lambda + \frac{1}{4}\eta_0^2\Theta_{11}^2 + \Omega_1^2\sigma^2 - \frac{1}{4}\eta_0^2\frac{R_{11}^2}{4}\delta_{1k} = 0. \tag{4.7}$$

Although the coefficient matrices of the averaged equations in (4.6) and (3.15) are structurally different. This difference is only formal, because the resulting characteristic equations possess the same structure and we can proceed as in the previous chapter. Comparing the polynomials in (4.7) with the polynomials for a single harmonic stiffness variation in (3.16, 3.18) leads to the following correlations

$$Q_{ii}^2 \mapsto \Omega_i^2 R_{ii}^2. \tag{4.8}$$

Hence, we can conclude the stability conditions at $\eta_0 = 2\Omega_k$ from (3.17) for $\sigma = 0$, and from (3.19) for $\sigma \neq 0$. The interpretations of the resulting stability conditions are equivalent to the statements made for the single harmonic stiffness variation in Section 3.1. Recapitulating this special case of η_0, it is not possible to stabilize the system for a negative modal damping, as well as for a moderate modal positive damping that does not satisfy the conditions in (3.17), respecting the correlations (4.8).

2. Combination parametric resonance of difference type and first kind

By setting

$$\boxed{\eta_0 = |\Omega_1 - \Omega_2| > 0 \quad \text{or} \quad \varpi_1 - \varpi_2 = \mp 1}$$

some terms become resonant and produce additional terms after averaging. We get from (4.3)

$$\begin{aligned}
\dot{\hat{u}}_i &= \frac{\varepsilon}{\eta_0^2 \varpi_i} \left\{ -\frac{\eta_0}{2} \Theta_{ii} \varpi_i \hat{u}_i - \frac{\eta_0 \varpi_j}{4} R_{ij} \hat{u}_j - \Omega_i \varpi_i \sigma \hat{v}_i \right\} \\
\dot{\hat{v}}_i &= \frac{\varepsilon}{\eta_0^2 \varpi_i} \left\{ -\frac{\eta_0}{2} \Theta_{ii} \varpi_i \hat{v}_i - \frac{\eta_0 \varpi_j}{4} R_{ij} \hat{v}_j + \Omega_i \varpi_i \sigma \hat{u}_i \right\}
\end{aligned} \tag{4.9}$$

instead of (4.4). Rewritten in matrix notation leads to

$$\begin{pmatrix} \dot{\hat{u}}_1 \\ \dot{\hat{v}}_1 \\ \dot{\hat{u}}_2 \\ \dot{\hat{v}}_2 \end{pmatrix} = \frac{\varepsilon}{\eta_0^2} \begin{bmatrix} -\frac{\eta_0}{2}\Theta_{11} & -\Omega_1 \sigma & -\frac{\eta_0 \Omega_2}{4\Omega_1} R_{12} & 0 \\ \Omega_1 \sigma & -\frac{\eta_0}{2}\Theta_{11} & 0 & -\frac{\eta_0 \Omega_2}{4\Omega_1} R_{12} \\ -\frac{\eta_0 \Omega_1}{4\Omega_2} R_{21} & 0 & -\frac{\eta_0}{2}\Theta_{22} & -\Omega_2 \sigma \\ 0 & -\frac{\eta_0 \Omega_1}{4\Omega_2} R_{21} & \Omega_2 \sigma & -\frac{\eta_0}{2}\Theta_{22} \end{bmatrix} \begin{pmatrix} \hat{u}_1 \\ \hat{v}_1 \\ \hat{u}_2 \\ \hat{v}_2 \end{pmatrix} = \frac{\varepsilon}{\eta_0^2} \mathbf{A}_4 \hat{\mathbf{u}}.$$

After rescaling time by ε/η_0^2 we obtain the characteristic equation

$$\det(\lambda \mathbf{I}_4 - \mathbf{A}_4) = 0 = \lambda^4 + b_1 \lambda^3 + b_2 \lambda^2 + b_3 \lambda + b_4,$$

a polynomial of order four. Using the abbreviations

$$\hat{w}_i = \hat{u}_i + j\hat{v}_i,$$

where $j = \sqrt{-1}$ is the complex unit, (4.9) is equivalent to

$$\dot{\hat{w}}_i = \frac{\varepsilon}{\eta_0^2} \left\{ \left(-\frac{\eta_0}{2} \Theta_{ii} + j\Omega_i \sigma \right) \hat{w}_i - \frac{\eta_0 \Omega_k}{4\Omega_i} R_{ik} \hat{w}_k \right\} \quad k \neq i.$$

4 Time-periodic damping

Rewritten in matrix notation gives equations of the form

$$\dot{\hat{\mathbf{w}}}(t) = \frac{\varepsilon}{\eta_0^2} \begin{bmatrix} -\frac{\eta_0}{2}\Theta_{11} + j\Omega_1\sigma & -\frac{\eta_0\Omega_2}{4\Omega_1}R_{12} \\ -\frac{\eta_0\Omega_1}{4\Omega_2}R_{21} & -\frac{\eta_0}{2}\Theta_{22} + j\Omega_2\sigma \end{bmatrix} \hat{\mathbf{w}}(t) = \frac{\varepsilon}{\eta_0^2} \mathbf{A}_2 \hat{\mathbf{w}}(t),$$

with the complex state vector $\hat{\mathbf{w}} = (\hat{w}_1, \hat{w}_2)^T$. After rescaling time by ε/η_0^2 the characteristic equation of the coefficient matrix is reduced from a real polynomial of order four to a complex polynomial of order two

$$\det(\lambda \mathbf{I}_2 - \mathbf{A}_2) = 0,$$

$$\lambda^2 + \left(\tfrac{1}{2}\eta_0(\Theta_{11} + \Theta_{22}) - j(\Omega_1 + \Omega_2)\sigma\right)\lambda + \left(\tfrac{1}{2}\eta_0\Theta_{11} - j\Omega_1\sigma\right)\left(\tfrac{1}{2}\eta_0\Theta_{22} - j\Omega_2\sigma\right) - \frac{\eta_0^2}{16}R_{12}R_{21} = 0. \quad (4.10)$$

Comparing this polynomial with the polynomial obtained in (3.22) for a single harmonic stiffness variation if the frequency η_0 is equal to parametric combination frequency of *difference* type $|\Omega_1 - \Omega_2|$ gives the following mapping

$$Q_{12}Q_{21} \mapsto -\Omega_1\Omega_2 R_{12}R_{21}. \quad (4.11)$$

On the other hand, comparing the characteristic equation (4.10) with the polynomial in (3.37) for a general harmonic stiffness variation with the parametric combination frequency of *difference* type $|\Omega_1 - \Omega_2|$ leads to the following mapping

$$\begin{aligned} Q_{12} &\mapsto 0, & Q_{21} &\mapsto 0, \\ P_{12} &\mapsto R_{12}, & P_{21} &\mapsto -R_{21}. \end{aligned} \quad (4.12)$$

Hence, the results from (3.23, 3.24) or (3.38, 3.40) can be adopted by applying the substitutions (4.11) or (4.12).

First analyzing the case for $\sigma = 0$, checking whether the system is stable at the frequency $\eta_0 = |\Omega_1 - \Omega_2|$ or not, the necessary and sufficient conditions are

$$\begin{aligned} \Theta_{11} + \Theta_{22} &> 0, \\ \Theta_{11}\Theta_{22} - \frac{1}{4}R_{12}R_{21} &> 0, \end{aligned} \quad (4.13)$$

in analogy to (3.23) or (3.38). For the case $\sigma \neq 0$, the following stability conditions

$$\Delta_1 : \tfrac{1}{2}\eta_0 \left(\Theta_{11} + \Theta_{22}\right) > 0,$$

$$\Delta_2 : \begin{cases} \tfrac{1}{16}\eta_0^4 \left(\Theta_{11} + \Theta_{22}\right)^2 \left(\Theta_{11}\Theta_{22} - \tfrac{1}{4}R_{12}R_{21} - \tfrac{4}{\eta_0^2}\Omega_1\Omega_2\sigma^2\right) + \\ + \tfrac{1}{4}\eta_0^2 \left(\Theta_{11}\Omega_2 + \Theta_{22}\Omega_1\right)\left(\Theta_{11}\Omega_1 + \Theta_{22}\Omega_2\right)\sigma^2 > 0, \\ a_0\sigma^2 + a_2 > 0, \\ \text{critical values:} \\ \sigma_{1,2} = \pm \dfrac{\eta_0 \left(\Theta_{11} + \Theta_{22}\right)}{2\left(\Omega_1 - \Omega_2\right)} \sqrt{-\dfrac{\Theta_{11}\Theta_{22} - \tfrac{1}{4}R_{12}R_{21}}{\Theta_{11}\Theta_{22}}}, \end{cases} \quad (4.14)$$

are obtained from (3.24) or (3.40). Note the change in the sign of the parametric excitation term compared to the corresponding case for a single harmonic stiffness variation at $\eta_0 = |\Omega_1 - \Omega_2|$ in (3.24). Therefore a harmonic damping variation at $|\Omega_1 - \Omega_2|$ resembles the harmonic stiffness variation at $\Omega_1 + \Omega_2$ in (3.29), although the derivation of the stability conditions (4.13, 4.14) is based on (3.23, 3.24). The critical values $\sigma_{1,2}$ represent the boundary between stable and unstable parameter domains. In combination with (4.13) a classification can be made for distinct modal damping parameters, similar to the case of a harmonic stiffness variation at $\eta_0 = \Omega_1 + \Omega_2$ in the Section 3.1. Furthermore, note that again the critical case $\sigma_1 = 0$ in (4.14) is an equivalent formulation of the critical case of the second condition in (4.13). This result is summarized in Tab. 4.1.

From the above conditions it can be easily concluded, that a parametric anti-resonance at $\eta_0 = |\Omega_1 - \Omega_2|$ can only occur if the term $R_{12}R_{21}$ is *negative*. A more detailed study on the symmetry property of the parametric excitation term $R_{12}R_{21}$ is investigated in Section 4.3.

3. Combination parametric resonance of summation type and first kind

By defining

$$\boxed{\eta_0 = \Omega_i + \Omega_j \quad \text{or} \quad \varpi_i + \varpi_j = 1}$$

some terms become resonant and produce additional terms after averaging. Instead of (4.4) we obtain from (4.3)

$$\begin{aligned}
\dot{\hat{u}}_i &= \frac{\varepsilon}{\eta_0^2 \varpi_i}\left\{-\frac{\eta_0}{2}\Theta_{ii}\varpi_i\hat{u}_i + \frac{\eta_0 \varpi_j}{4}R_{ij}\hat{u}_j - \Omega_i\varpi_i\sigma\hat{v}_i\right\}, \\
\dot{\hat{v}}_i &= \frac{\varepsilon}{\eta_0^2 \varpi_i}\left\{-\frac{\eta_0}{2}\Theta_{ii}\varpi_i\hat{v}_i + \frac{\eta_0 \varpi_j}{4}R_{ij}\hat{v}_j + \Omega_i\varpi_i\sigma\hat{u}_i\right\}.
\end{aligned} \quad (4.15)$$

Comparing with (4.9) reveals that the signs of all parametric excitation terms R_{ij} change. Using now the abbreviations

$$\hat{w}_i = \hat{u}_i + j\left(-1\right)^{i-1}\hat{v}_i,$$

4 Time-periodic damping

where $j = \sqrt{-1}$ is the complex unit, (4.15) is equivalent to

$$\dot{\hat{w}}_i = \frac{\varepsilon}{\eta_0^2}\left\{\left(-\frac{\eta_0}{2}\Theta_{ii} + j(-1)^{i-1}\Omega_i\sigma\right)\hat{w}_i + \frac{\eta_0\Omega_k}{4\Omega_i}R_{ik}\hat{w}_k\right\} \quad k \neq i.$$

Rewritten in matrix notation we get equations of the form

$$\dot{\hat{\mathbf{w}}}(t) = \frac{\varepsilon}{\eta_0^2}\begin{bmatrix} -\frac{\eta_0}{2}\Theta_{11} + j\Omega_1\sigma & \frac{\eta_0\Omega_2}{4\Omega_1}R_{12} \\ \frac{\eta_0\Omega_1}{4\Omega_2}R_{21} & -\frac{\eta_0}{2}\Theta_{22} - j\Omega_2\sigma \end{bmatrix}\hat{\mathbf{w}}(t) = \frac{\varepsilon}{\eta_0^2}\mathbf{A}\hat{\mathbf{w}}(t),$$

with the complex state vector $\hat{\mathbf{w}} = (\hat{w}_1, \hat{w}_2)^T$. After rescaling time by ε/η_0^2 the characteristic equation of the coefficient matrix is a polynomial of order two

$$\det(\lambda\mathbf{I}_2 - \mathbf{A}) = 0,$$

$$\lambda^2 + \left(\tfrac{1}{2}\eta_0(\Theta_{11} + \Theta_{22}) - j(\Omega_1 - \Omega_2)\sigma\right)\lambda$$
$$+ \left(\tfrac{1}{2}\eta_0\Theta_{11} - j\Omega_1\sigma\right)\left(\tfrac{1}{2}\eta_0\Theta_{22} + j\Omega_2\sigma\right) - \frac{\eta_0^2}{16}R_{12}R_{21} = 0.$$

Comparing with the averaging process for the general case in (4.4) the determinant is extended by $R_{12}R_{21}$ while the trace remains unchanged. Note the different sign of terms containing Ω_2 compared to the previous section, where $\eta_0 = |\Omega_1 - \Omega_2|$. Hence, we can conclude the stability conditions for $\eta_0 = \Omega_1 + \Omega_2$ by performing the substitution $\Omega_2 \mapsto -\Omega_2$ in the resultant stability conditions of $\eta_0 = |\Omega_1 - \Omega_2|$.

First, analyzing the case of $\sigma = 0$ the necessary and sufficient stability conditions are

$$\Theta_{11} + \Theta_{22} > 0,$$
$$\Theta_{11}\Theta_{22} - \frac{1}{4}R_{12}R_{21} > 0, \quad (4.16)$$

according to (4.13) or by substituting (4.11) in (3.23). The conditions (4.16) and (4.13) are equivalent because the parametric excitation term does not contain the frequency Ω_2 and therefore is not affected by performing the substitution $\Omega_2 \mapsto -\Omega_2$.

For the case $\sigma \neq 0$ stability conditions

$$\Delta_1 : \tfrac{1}{2}\eta_0(\Theta_{11} + \Theta_{22}) > 0,$$

$$\Delta_2 : \begin{cases} \tfrac{1}{16}\eta_0^4(\Theta_{11}+\Theta_{22})^2\left(\Theta_{11}\Theta_{22} - \dfrac{1}{4}R_{12}R_{21} + \dfrac{4}{\eta_0^2}\Omega_1\Omega_2\sigma^2\right) - \\ -\tfrac{1}{4}\eta_0^2(\Theta_{11}\Omega_2 - \Theta_{22}\Omega_1)(\Theta_{11}\Omega_1 - \Theta_{22}\Omega_2)\sigma^2 > 0, \\ a_0\sigma^2 + a_2 > 0, \\ \text{critical values:} \\ \sigma_{1,2} = \pm\dfrac{\eta_0(\Theta_{11}+\Theta_{22})}{2(\Omega_1+\Omega_2)}\sqrt{-\dfrac{\Theta_{11}\Theta_{22} - \tfrac{1}{4}R_{12}R_{21}}{\Theta_{11}\Theta_{22}}}, \end{cases} \quad (4.17)$$

are obtained from (4.14). Because the square root terms does not contain the frequency Ω_2, the stability conditions remain unchanged after performing the substitution $\Omega_2 \mapsto -\Omega_2$. This results in identical stability conditions (4.14) for $\eta_0 = |\Omega_1 - \Omega_2|$ and (4.17) for $\eta_0 = \Omega_1 + \Omega_2$, respectively, and the same statements as before can be made. This result is summarized in Tab. 4.1. A more detailed study on the symmetry property of the parametric excitation term $R_{12}R_{21}$ is presented in Section 4.3.

4.2. Summary for small single harmonic damping variations

Whether the system in (4.1) is stable or not depends mostly on the frequency η of its damping variation. Globally the system is stable for arbitrary values η iff

$$\Theta_{11} > 0 \quad \text{and} \quad \Theta_{22} > 0,$$

according to (4.5). If the frequency η is close to a certain combination frequency the system might be locally stable:

Case 1. For $\eta \approx 2\Omega_k$, if the conditions in (4.13) are satisfied:

$$\Theta_{ii} > \frac{|R_{ii}|}{2}\delta_{ik}.$$

In this case, however, the critical values $\sigma_{1,2}$ in (4.14) are purely imaginary and no stability boundary occurs.

Cases 2. and 3. For $\eta \approx \Omega_i - \Omega_j$ and $\eta \approx \Omega_i + \Omega_j$, respectively, if the following conditions are satisfied:

(4.13, 4.16), $\Delta_1 = 0$:	$\Theta_{11} + \Theta_{22} > 0,$
(4.13, 4.16), stable η_0 :	$\Theta_{11}\Theta_{22} - \frac{1}{4}R_{12}R_{21} > 0,$
(4.14, 4.17), $\Delta_2 = 0$:	$\sigma_{1,2} = \pm\dfrac{\Theta_{11} + \Theta_{22}}{2}\sqrt{-\dfrac{\Theta_{11}\Theta_{22} - \frac{1}{4}R_{12}R_{21}}{\Theta_{11}\Theta_{22}}},$
(4.14), $\Delta_2 > 0$:	$\Theta_{11}\Theta_{22} < 0$ \qquad\qquad $\Theta_{11}\Theta_{22} > 0$
	$\sigma_{1,2}$ is real-valued \quad $\sigma_{1,2}$ is purely imaginary
	$-|\sigma_1| < \sigma < |\sigma_1|$ \quad no boundary

Other than for a single harmonic stiffness variation (3.23, 3.24, 3.28, 3.29), for a single harmonic damping variation the sign of the parametric excitation term does *not* change in the resulting stability conditions (4.13, 4.14, 4.16, 4.17). Thus, both stability conditions coincide for the case of single harmonic damping the stability areas for the summation and difference combination frequency. Recall that we obtain two different stability conditions in case of a single harmonic stiffness variation.

4 Time-periodic damping

If we ask for stability at $\eta_0 = |\Omega_1 - \Omega_2|$ and at $\eta_0 = \Omega_1 + \Omega_2$ *simultaneously*, we obtain from above, (4.17) and (4.14), the following conditions for stability

$$\left.\begin{array}{l} \eta_0 = |\Omega_1 - \Omega_2| : \\ \eta_0 = \Omega_1 + \Omega_2 : \end{array}\right\} \Rightarrow \Theta_{11}\Theta_{22} - \frac{1}{4}R_{12}R_{21} > 0.$$

This means, that in the case where one modal damping becomes negative, $\Theta_{11}\Theta_{22} < 0$, it may

Table 4.1.: Summary for single harmonic damping variation at parametric combination resonance frequencies.

$\eta_0 = |\Omega_1 \mp \Omega_2|$	unstable η_0	stable η_0
	$\Theta_{11} + \Theta_{22} > 0$ $\Theta_{11}\Theta_{22} - \frac{1}{4}R_{12}R_{21} < 0$	$\Theta_{11} + \Theta_{22} > 0$ $\Theta_{11}\Theta_{22} - \frac{1}{4}R_{12}R_{21} > 0$
$\Theta_{11}\Theta_{22} > 0$	σ_1 is real-valued $-|\sigma_1| < \sigma < |\sigma_1|$ **parametric resonance**	σ_1 is purely imaginary no boundary **stable**
$\Theta_{11}\Theta_{22} < 0$	σ_1 is purely imaginary no boundary **unstable**	σ_1 is real-valued $-|\sigma_1| < \sigma < |\sigma_1|$ **parametric anti-resonance**
critical value (switch):	$\sigma_1 = \dfrac{\Theta_{11} + \Theta_{22}}{2}\sqrt{-\dfrac{\Theta_{11}\Theta_{22} - \frac{1}{4}R_{12}R_{21}}{\Theta_{11}\Theta_{22}}}$	

be possible to stabilize the system at the parametric excitation frequency of the difference type $\eta_0 = |\Omega_1 - \Omega_2|$ and the summation type $\eta_0 = \Omega_1 + \Omega_2$ simultaneously, which is different to the results for a single harmonic stiffness variation! On one hand, a single harmonic damping variation is capable of stabilizing a system at both parametric resonance frequencies, but on the other hand, for a single harmonic damping variation with a negative modal damping the parametric excitation term has to be *negative*, $R_{12}R_{21} < 0$. Hence, a parametric anti-resonance can only be achieved by a skew-symmetric parametric excitation matrix. These results are summarized in Tab. 4.1. More detailed explanations are found in the description of Tab. 3.1 on page 36.

The case where $\Theta_{11} + \Theta_{22} < 0$ is valid, leads always to an unstable system. In this case the stability map for a certain system parameter looks like in the sketch in the left bottom corner in Tab. 4.1. A detailed listing is therefore omitted. These results are valid on the time scale $1/\varepsilon$.

4 Time-periodic damping

4.3. Symmetry considerations

The sign of the parametric excitation term $R_{12}R_{21}$ is crucial for the position of the parametric anti-resonance frequency, whether the anti-resonance is located at $\eta_0 = |\Omega_1 - \Omega_2|$ or $\eta_0 = \Omega_1 + \Omega_2$. Due to the fact that the stability conditions coincide at both frequencies, a classification of the symmetry properties of the parametric excitation matrix, and their consequences on the stability conditions are much more simple, than in case of a single harmonic stiffness variation as shown in Tab. 3.2. The results are presented in Tab. 4.2.

Table 4.2.: Symmetry considerations of parametric excitation matrix at frequencies $\eta_0 = |\Omega_1 \mp \Omega_2|$ for single harmonic damping variation.

$\eta_0 = \|\Omega_1 \mp \Omega_2\|$	(4.5) is fulfilled: $\Theta_{11} + \Theta_{22} > 0$ $\Theta_{11}\Theta_{22} > 0$	(4.5) is violated: $\Theta_{11} + \Theta_{22} > 0$ $\Theta_{11}\Theta_{22} < 0$
$R_{12}R_{21} < 0$ (skew-sym.)	(4.13, 4.16) are fulfilled: **stable** σ_1 is purely imaginary no boundary	additional restriction: **parametric anti-resonance** for $0 > \Theta_{11}\Theta_{22} > \frac{1}{4}R_{12}R_{21}$ σ_1 is real-valued $-\|\sigma_1\| < \sigma < \|\sigma_1\|$
$R_{12}R_{21} > 0$ (sym.)	additional restriction: **parametric resonance** for $0 < \Theta_{11}\Theta_{22} < \frac{1}{4}R_{12}R_{21}$ σ_1 is real-valued $-\|\sigma_1\| < \sigma < \|\sigma_1\|$	(4.13, 4.16) are violated: **unstable** σ_1 is purely imaginary no boundary

5. Synchronous time-periodic damping and stiffness

In the previous chapters we investigated the stability conditions for systems where either the stiffness coefficients or the damping coefficients were varied periodically with time. In some cases it might happen that both physical parameters are varied simultaneously. In this chapter we analyze the case, where the stiffness and the damping coefficients are varied with the same phase, synchronously. An easy to imagine physical example is a beam with a time varying length. Varying the length periodically, causes a periodical change of the bending stiffness of the beam. Simultaneously, the structural damping is varied periodically, because the part of the beam that experiences vibrations changes periodically, too. Both variations proceed synchronous.

For the case of a synchronous damping and stiffness variation the inertia matrix is kept constant

$$\mathbf{M}(t) = \mathbf{M}_0,$$
$$\mathbf{K}(t) = \mathbf{K}_0 + \varepsilon \left\{ \mathbf{K}_c^s + \mathbf{K}_c^a \right\} \cos(\omega t),$$
$$\mathbf{C}(t) = \mathbf{C}_0 + \varepsilon \left\{ \mathbf{C}_c^s + \mathbf{C}_c^a \right\} \cos(\omega t),$$

and the general linear equations of motion (2.2) simplify to

$$\ddot{\mathbf{z}}(t) + \varepsilon \tilde{\mathbf{C}}_0 \dot{\mathbf{z}}(t) + \tilde{\mathbf{K}}_0 \mathbf{z}(t) = -\varepsilon \left\{ \left(\tilde{\mathbf{C}}_c^s + \tilde{\mathbf{C}}_c^a \right) \dot{\mathbf{z}}(t) + \left(\tilde{\mathbf{K}}_c^s + \tilde{\mathbf{K}}_c^a \right) \mathbf{z}(t) \right\} \cos(\omega t).$$

These equations of motion in dimensional terms can be transformed ($\mathbf{z}(t) \longrightarrow \mathbf{z}(\tau)$) into the following non-dimensional normal form

$$\mathbf{z}''(\tau) + \varepsilon \boldsymbol{\Theta} \mathbf{z}'(\tau) + \boldsymbol{\Omega}^2 \mathbf{z}(\tau) = -\varepsilon \left\{ \left(\mathbf{R}^s + \mathbf{R}^a \right) \mathbf{z}'(\tau) + \left(\mathbf{Q}^s + \mathbf{Q}^a \right) \mathbf{z}(\tau) \right\} \cos(\eta \tau),$$

where $\mathbf{Q} = \mathbf{Q}^s + \mathbf{Q}^a = \mathbf{T}^{-1} \mathbf{M}_0^{-1} \mathbf{K}_c \mathbf{T},$

$$\mathbf{R} = \mathbf{R}^s + \mathbf{R}^a = \mathbf{T}^{-1} \mathbf{M}_0^{-1} \mathbf{C}_c \mathbf{T},$$

as it was shown in Section 2.3 or rewritten in the comprehensive index notation for a system with two degrees of freedom

$$z_i''(\tau) + \Omega_i^2 z_i(\tau) = -\varepsilon \left\{ \Theta_{ij} z_j'(\tau) + \left\{ R_{ij} z_j'(\tau) + Q_{ij} z_j(\tau) \right\} \cos(\eta \tau) \right\}, \qquad (5.1)$$

with $i, j = 1, 2$. The approach presented below follows the procedure presented in Sections 3.1 and 3.2.

5.1. Transformation and averaging

Applying the procedure in Section 3.1, the time transformation (3.2) and the Taylor expansion (3.4), on (5.1) yields

$$\ddot{z}_i + \frac{\Omega_i^2}{\eta^2} z_i = -\frac{\varepsilon}{\eta} \Theta_{ij} \dot{z}_j - \frac{\varepsilon}{\eta^2} \left(Q_{ij} z_j + R_{ij} \dot{z}_j \right) \cos t, \qquad \text{with } i, j = 1, 2, \qquad (5.2)$$

with the abbreviations $\varpi_i = \Omega_i / \eta_0$. Using the coordinate transformation in (3.6) the right hand sides of (5.2) become

$$F_i(\mathbf{u}, t) = -\eta_0 \sum_j \left(\Theta_{ij} + R_{ij} \cos t \right) \left(-u_j \varpi_j s_j + v_j \varpi_j c_j \right) -$$

$$- \sum_j Q_{ij} \cos t \left(u_j c_j + v_j s_j \right) + 2\Omega_i \varpi_i \sigma \left(u_i c_i + v_i s_i \right), \qquad (5.3)$$

with the abbreviations $s_i = \sin \varpi_i t$ and $c_i = \cos \varpi_i t$ and the state vector $\mathbf{u} = (u_1, v_1, u_2, v_2)^T$. Comparing (5.3) with (3.8) and (4.3), points out the difference between a single harmonic stiffness variation (Q_{ij}), a single harmonic damping variation (R_{ij}) and a synchronous single harmonic stiffness and damping variation (Q_{ij}, R_{ij}).

The final system equations for the slow amplitudes $\mathbf{u}$ are equal to (3.9), with quasi-periodic right hand sides. For this simple two mode system we obtain again 12 different periods, according to Appendix B by substituting $n = 1$, and performing $R_{kl} \mapsto Q_{kl}$. This is the same number of periods as for the single harmonic stiffness variation in (3.8) or the single harmonic damping variation in (4.3). Averaging over these different periodic terms yields always zero, except for the case where a term becomes resonant:

$\Theta_{ii}, \sigma:$ no resonant terms, constant factor $\frac{1}{2}$,

$\Theta_{ij}:$ resonant terms for $\varpi_i \pm \varpi_j = 0$,

$Q_{ii}, R_{ii}:$ resonant terms for $2\varpi_i \pm 1 = 0$,

$Q_{ij}, R_{ij}:$ resonant terms for $\varpi_i \pm \varpi_j \pm 1 = 0$.

General case

If ϖ_1, ϖ_2 do not take on special values averaging over each trigonometric function gives zero, only the terms Θ_{ii}, σ remain and we obtain from (4.3) the same equations as in (3.11). The stability conditions yields for $\eta_0 > 0$

$$\boxed{\Theta_{11} > 0 \quad \text{and} \quad \Theta_{22} > 0,} \qquad (5.4)$$

according to (3.14), and the critical values of σ are always purely imaginary. Globally the system is either stable or unstable depending on whether the conditions in (5.4) are satisfied or not.

5 Synchronous time-periodic damping and stiffness 77

Some resonant cases

1. Combination parametric resonance of difference type and first kind

By setting
$$\boxed{\eta_0 = \Omega_2 - \Omega_1 > 0 \quad \text{or} \quad \varpi_1 - \varpi_2 = -1}$$
some terms become resonant and produce additional terms after averaging and we obtain from (4.3)

$$\begin{aligned}
\dot{\hat{u}}_i &= \frac{\varepsilon}{\eta_0^2 \varpi_i} \left\{ -\frac{\eta_0}{2} \Theta_{ii} \varpi_i \hat{u}_i + \frac{Q_{ij}}{4} \hat{v}_j - \eta_0 \frac{R_{ij}}{4} \hat{u}_j - \Omega_i \varpi_i \sigma \hat{v}_i \right\}, \\
\dot{\hat{v}}_i &= \frac{\varepsilon}{\eta_0^2 \varpi_i} \left\{ -\frac{\eta_0}{2} \Theta_{ii} \varpi_i \hat{v}_i - \frac{Q_{ij}}{4} \hat{u}_j - \eta_0 \frac{R_{ij}}{4} \hat{v}_j + \Omega_i \varpi_i \sigma \hat{u}_i \right\}.
\end{aligned} \quad (5.5)$$

Rewritten in matrix notation yields

$$\begin{pmatrix} \dot{\hat{u}}_1 \\ \dot{\hat{v}}_1 \\ \dot{\hat{u}}_2 \\ \dot{\hat{v}}_2 \end{pmatrix} = \frac{\varepsilon}{\eta_0^2} \begin{bmatrix} -\dfrac{\eta_0}{2}\Theta_{11} & -\Omega_1\sigma & -\dfrac{\eta_0\Omega_2}{4\Omega_1}R_{12} & \dfrac{\eta_0}{4\Omega_1}Q_{12} \\ \Omega_1\sigma & -\dfrac{\eta_0}{2}\Theta_{11} & -\dfrac{\eta_0}{4\Omega_1}Q_{12} & -\dfrac{\eta_0\Omega_2}{4\Omega_1}R_{12} \\ -\dfrac{\eta_0\Omega_1}{4\Omega_2}R_{21} & \dfrac{\eta_0}{4\Omega_2}Q_{21} & -\dfrac{\eta_0}{2}\Theta_{22} & -\Omega_2\sigma \\ -\dfrac{\eta_0}{4\Omega_2}Q_{21} & -\dfrac{\eta_0\Omega_1}{4\Omega_2}R_{21} & \Omega_2\sigma & -\dfrac{\eta_0}{2}\Theta_{22} \end{bmatrix} \begin{pmatrix} \hat{u}_1 \\ \hat{v}_1 \\ \hat{u}_2 \\ \hat{v}_2 \end{pmatrix}.$$

Comparing (5.5) with (3.35) gives the following mapping rules
$$P_{12} \mapsto \Omega_2 R_{12}, \qquad P_{21} \mapsto -\Omega_1 R_{21}. \quad (5.6)$$
Performing these substitutions the stability conditions for $\sigma = 0$ result in

$$\Theta_{11} + \Theta_{22} > 0,$$

$$\Theta_{11}\Theta_{22} + \frac{Q_{12}Q_{21} - \Omega_1\Omega_2 R_{12}R_{21}}{4\Omega_1\Omega_2} > \left(\frac{-\Omega_1 Q_{12}R_{21} - \Omega_2 R_{12}Q_{21}}{4\Omega_1\Omega_2(\Theta_{11}+\Theta_{22})} \right)^2 \geq 0, \quad (5.7)$$

according to (3.38) and the stability conditions for $\sigma \neq 0$ in

$\Delta_1 : \quad \frac{1}{2}\eta_0 (\Theta_{11} + \Theta_{22}) > 0,$

$$\Delta_2 : \begin{cases}
a_0 \sigma^2 + a_1(Q_{ij}, R_{ij})\sigma + a_2(Q_{ij}, R_{ij}) > 0, \\
\text{critical values:} \\
\sigma_{1,2} = \sigma_s \pm \tilde{\sigma}_{1,2}, \\
\sigma_s = \dfrac{\eta_0(\Theta_{11} - \Theta_{22})}{2\Theta_{11}\Theta_{22}(\Omega_1 - \Omega_2)} \dfrac{\beta}{2}, \\
\tilde{\sigma}_{1,2} = \dfrac{\eta_0(\Theta_{11}+\Theta_{22})}{2\Theta_{11}\Theta_{22}(\Omega_1-\Omega_2)}\sqrt{d}, \\
\text{with} \quad d = -\Theta_{11}\Theta_{22}\left(\Theta_{11}\Theta_{22} + \dfrac{Q_{12}Q_{21} - \Omega_1\Omega_2 R_{12}R_{21}}{4\Omega_1\Omega_2}\right) + \left(\dfrac{\beta}{2}\right)^2, \\
\text{and} \quad \beta = \dfrac{-\Omega_1 Q_{12}R_{21} - \Omega_2 R_{12}Q_{21}}{4\Omega_1\Omega_2},
\end{cases} \quad (5.8)$$

according to (3.40).

2. Combination parametric resonance of summation type and first kind
By setting
$$\boxed{\eta_0 = \Omega_1 + \Omega_2 \quad \text{or} \quad \varpi_1 + \varpi_2 = 1}$$
some terms become resonant and produce additional terms after averaging and we obtain from (4.3)
$$\dot{\hat{u}}_i = \frac{\varepsilon}{\eta_0^2 \varpi_i} \left\{ -\frac{\eta_0}{2} \Theta_{ii} \varpi_i \hat{u}_i - \frac{Q_{ij}}{4} \hat{v}_j + \eta_0 \frac{R_{ij}}{4} \hat{u}_j - \Omega_i \varpi_i \sigma \hat{v}_i \right\},$$
$$\dot{\hat{v}}_i = \frac{\varepsilon}{\eta_0^2 \varpi_i} \left\{ -\frac{\eta_0}{2} \Theta_{ii} \varpi_i \hat{v}_i - \frac{Q_{ij}}{4} \hat{u}_j - \eta_0 \frac{R_{ij}}{4} \hat{v}_j + \Omega_i \varpi_i \sigma \hat{u}_i \right\}.$$
(5.9)

Comparing (5.9) with (3.43) the following correlations can be identified
$$P_{12} \mapsto \Omega_2 R_{12}, \qquad P_{21} \mapsto \Omega_1 R_{21}. \tag{5.10}$$

Note the different sign of the P_{21} term compared to (5.9). Performing these substitutions the stability conditions for $\sigma = 0$ result in
$$\Theta_{11} + \Theta_{22} > 0,$$
$$\Theta_{11}\Theta_{22} - \frac{Q_{12}Q_{21} + \Omega_1\Omega_2 R_{12} R_{21}}{4\Omega_1\Omega_2} > \left(\frac{\Omega_1 Q_{12} R_{21} - \Omega_2 R_{12} Q_{21}}{4\Omega_1\Omega_2 (\Theta_{11} + \Theta_{22})} \right)^2 \geq 0, \tag{5.11}$$
according to (3.45) and the stability conditions for $\sigma \neq 0$ in

$\Delta_1 : \quad \frac{1}{2}\eta_0 (\Theta_{11} + \Theta_{22}) > 0,$

$\Delta_2 :$
$$\begin{cases} \tilde{a}_0 \sigma^2 + \tilde{a}_1 (Q_{ij}, R_{ij}) \sigma + \tilde{a}_2 (Q_{ij}, R_{ij}) > 0, \\ \text{critical values:} \\ \sigma_{1,2} = \sigma_s \pm \tilde{\sigma}_{1,2}, \\ \sigma_s = \frac{\eta_0 (\Theta_{11} - \Theta_{22})}{2\Theta_{11}\Theta_{22}(\Omega_1 + \Omega_2)} \frac{\beta}{2}, \\ \tilde{\sigma}_{1,2} = \frac{\eta_0 (\Theta_{11} + \Theta_{22})}{2\Theta_{11}\Theta_{22}(\Omega_1 + \Omega_2)} \sqrt{d}, \\ \text{with} \quad d = -\Theta_{11}\Theta_{22}\left(\Theta_{11}\Theta_{22} - \frac{Q_{12}Q_{21} + \Omega_1\Omega_2 R_{12} R_{21}}{4\Omega_1\Omega_2}\right) + \left(\frac{\beta}{2}\right)^2, \\ \text{and} \quad \beta = \frac{\Omega_1 Q_{12} R_{21} - \Omega_2 R_{12} Q_{21}}{4\Omega_1\Omega_2}. \end{cases}$$
(5.12)

Note the change in the sign, which not only occurs for the parametric excitation term but also in the denominator. Furthermore, note that the condition $\sigma_1 = 0$ in (5.12) is *not* an equivalent formulation for the critical case of the second condition in (5.11), as in the case of a harmonic stiffness variation in (3.46, 3.45) in Chapter 3.2.

5.2. Summary for small variations

Whether a system is stable or not depends mainly on the frequency η of the parametric excitation terms. Globally the system is stable for arbitrary frequency values η iff

$$\boxed{\Theta_{11} > 0 \quad \text{and} \quad \Theta_{22} > 0,}$$

according to (5.4). If the frequency η is close to a certain combination frequency the system might be locally stable:

Cases 1. and 2. For $\eta \approx |\Omega_1 \mp \Omega_2| = \eta_0$ if the following conditions are satisfied:

$$\boxed{\begin{aligned} &(5.7,\ 5.11),\ \Delta_1 > 0:\ \Theta_{11} + \Theta_{22} > 0, \\ &(5.7,\ 5.11),\ \text{stable}\ \eta_0:\ \Theta_{11}\Theta_{22} \pm \frac{Q_{12}Q_{21}}{4\Omega_1\Omega_2} - \frac{R_{12}R_{21}}{4} > \frac{\beta^2}{(\Theta_{11} + \Theta_{22})^2} \geq 0, \\ &(5.8,\ 5.12),\ \Delta_2 = 0:\ \sigma_{1,2} = \sigma_s \pm \tilde{\sigma}_{1,2}, \end{aligned}}$$

with the abbreviation

$$\beta^{(\mp)} = \left(\mp\frac{Q_{12}R_{21}}{4\Omega_2} - \frac{R_{12}Q_{21}}{4\Omega_1}\right),$$

$$\sigma_s^{(\mp)} = \frac{(\Theta_{11} - \Theta_{22})}{2\Theta_{11}\Theta_{22}}\frac{\beta^{(\mp)}}{2},\quad \tilde{\sigma}_{1,2}^{(\mp)} = \frac{(\Theta_{11} + \Theta_{22})}{2\Theta_{11}\Theta_{22}}\sqrt{d^{(\mp)}},$$

with $\quad d^{(\mp)} = -\Theta_{11}\Theta_{22}\left(\Theta_{11}\Theta_{22} \pm \frac{Q_{12}Q_{21}}{4\Omega_1\Omega_2} - \frac{R_{12}R_{21}}{4}\right) + \left(\frac{\beta^{(\mp)}}{2}\right)^2.$

Note that contrary to the case of a general harmonic stiffness variation in Section 3.2, now, depending whether the frequency of the stiffness variation is near to a parametric combination frequency of the difference type $\eta_0 = |\Omega_1 - \Omega_2|$ or of the summation type $\eta_0 = \Omega_1 + \Omega_2$, the frequency shift σ_s is different. This results from the expression $\beta^{(\mp)}$, which does not just change its sign, if we switch from $\eta_0 = |\Omega_1 - \Omega_2|$ to $\eta_0 = \Omega_1 + \Omega_2$. The resulting value of $\beta^{(\mp)}$ is completely different, because the sign of the second expression in β, $R_{12}R_{21}$, is conserved, while the sign of the first expression, $Q_{12}Q_{21}$, alters.

If one demands a design with stability at $\eta_0 = |\Omega_1 - \Omega_2|$ and *simultaneously* at $\eta_0 = \Omega_1 + \Omega_2$, we obtain from above, (5.7) and (5.11), the following conditions for stability

$$\left.\begin{aligned} \eta_0 = |\Omega_1 - \Omega_2|:\ &\Theta_{11}\Theta_{22} + \frac{Q_{12}Q_{21}}{4\Omega_1\Omega_2} - \frac{R_{12}R_{21}}{4} > \frac{\beta^{2(-)}}{(\Theta_{11} + \Theta_{22})^2} \geq 0 \\ \eta_0 = \Omega_1 + \Omega_2:\ &\Theta_{11}\Theta_{22} - \frac{Q_{12}Q_{21}}{4\Omega_1\Omega_2} - \frac{R_{12}R_{21}}{4} > \frac{\beta^{2(+)}}{(\Theta_{11} + \Theta_{22})^2} \geq 0 \end{aligned}\right\}$$

$$\Rightarrow 2\Theta_{11}\Theta_{22} - \frac{R_{12}R_{21}}{2} > 0.$$

This means that in the case $\Theta_{11}\Theta_{22} < 0$ it may be possible to stabilize a system at the parametric excitation frequency of the difference type $\eta_0 = |\Omega_1 - \Omega_2|$ and the summation type $\eta_0 = \Omega_1 + \Omega_2$ simultaneously, if at least (necessary condition)

$$0 > \Theta_{11}\Theta_{22} > \frac{R_{12}R_{21}}{4}. \tag{5.13}$$

This requires the single harmonic damping matrix to be skew-symmetric! Note that (5.13) is a necessary, but not sufficient conditions for simultaneous stability. Respecting the terms $\beta^{2(\mp)}$ yields the stricter necessary condition

$$0 > \Theta_{11}\Theta_{22} > \frac{R_{12}R_{21}}{4} + \frac{1}{16(\Theta_{11}+\Theta_{22})^2}\left(\left(\frac{Q_{12}R_{21}}{\Omega_2}\right)^2 + \left(\frac{R_{12}Q_{21}}{\Omega_1}\right)^2\right). \tag{5.14}$$

The frequency shift σ_s changes its sign if a modal damping term changes its sign, too. The critical case where one modal damping parameter is equal to zero results in an infinite frequency shift σ_s. Hence, a change in the sign of σ_s never occurs by crossing the frequency line η_0. If additionally $\beta^{(\mp)} > 0$ is satisfied, then the same classification of the adapted skeleton line can be carried out as in the case of a general harmonic stiffness variation in Tab. 3.3 on page 48 by performing the substitution

$$(Q_{12}P_{21} - P_{12}Q_{21}) \mapsto (\mp \Omega_1 Q_{12}R_{21} - \Omega_2 R_{12}Q_{21}).$$

The upper sign corresponds to the frequency $\eta_0 = |\Omega_1 - \Omega_2|$ and the lower sign to the frequency $\eta_0 = \Omega_1 + \Omega_2$. The fact that the first expression alters its sign, when we change the frequency η_0, while the second expressions is conserved, leads to an effect that differs from the interpretation of a harmonic stiffness variation in Tab. 3.3. Contrary to Section 3.2, now for a fixed system parameter p the size of the frequency shift σ_s is *not* the same at the frequency $\eta_0 = |\Omega_1 - \Omega_2|$ and $\eta_0 = \Omega_1 + \Omega_2$, respectively. Furthermore, if the condition

$$\Omega_1 Q_{12} R_{21} < \Omega_2 R_{12} Q_{21} \tag{5.15}$$

is fulfilled the resulting sign of σ_s differs, too. This consequence is sketched for $A = -\Omega_1 Q_{12} R_{21}$, $B = -\Omega_2 R_{12} Q_{21}$ and $A > B$ in Fig. 5.1. To correlate this result with Tab. 3.3, we have to set $B = 0$ and $A = (Q_{12}P_{21} - P_{12}Q_{21})$.

In analogy to the case of a harmonic stiffness variation in (3.49), the adequate critical condition for a non-vanishing interaction term in case of a synchronous damping and stiffness variation follows from a stability analysis of the original characteristic polynomial in (3.37), respecting the mappings (5.6) or (5.10), by setting $\sigma = \sigma_s$. The shifted frequency line $\eta_0 + \varepsilon\sigma_s$ is stable iff

$$\Delta_1 : \tfrac{1}{2}\eta_0(\Theta_{11} + \Theta_{22}) > 0,$$

$$\Delta_2 : \begin{aligned}\tfrac{1}{16}\eta_0^4(\Theta_{11}+\Theta_{22})^2\left(\Theta_{11}\Theta_{22} \pm \frac{Q_{12}Q_{21}}{4\Omega_1\Omega_2} - \frac{R_{12}R_{21}}{4} - \right.\\ \left. -\frac{1}{\Theta_{11}\Theta_{22}}\left(\frac{\mp\Omega_1 Q_{12}R_{21} - \Omega_2 R_{12}Q_{21}}{8\Omega_1\Omega_2}\right)^2\right) > 0,\end{aligned}$$

5 Synchronous time-periodic damping and stiffness

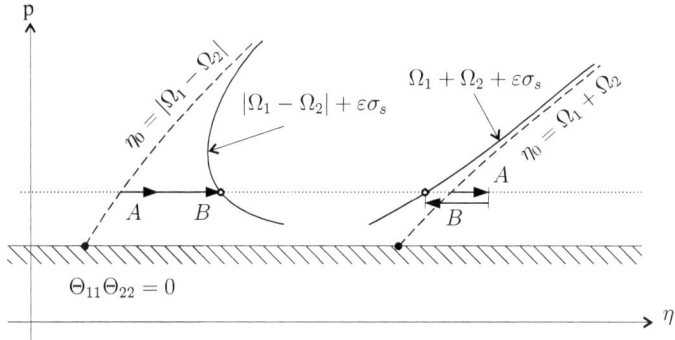

Figure 5.1.: Shift of skeleton lines if condition (5.15) is fulfilled.

are fulfilled, according to (A.2). These conditions can be rewritten for $\eta_0 > 0$ as

$$\Theta_{11} + \Theta_{22} > 0,$$
$$-\frac{d}{\Theta_{11}\Theta_{22}} > 0, \qquad (5.16)$$

which are equivalent to the conditions found in (3.49). Using the inequalities in (5.16) enables a classification of the conditions (5.8, 5.12) for a synchronous damping and stiffness variation that is equivalent to Tab. 3.4 on page 52. A discussion of the results is therefore omitted.

The presented results in Tab. 3.4 hold as long as the interaction term does not vanish, $(\mp\Omega_1 Q_{12}R_{21} - \Omega_2 R_{12}Q_{21}) \neq 0$. If the interaction term disappears, then the qualitative behavior is equivalent to the Tab. 3.1 on page 36. In this case the frequency shift σ_s vanishes and the critical values $\tilde{\sigma}_{1,2}$ can become also purely imaginary.

If we ask now for simultaneous stability at $\eta_0 = |\Omega_1 - \Omega_2|$ and $\eta_0 = \Omega_1 + \Omega_2$, then the necessary condition from (5.14) can be rearranged as

$$0 > \Theta_{11}\Theta_{22} > \frac{R_{12}R_{21}}{4} + \frac{\Omega_1^2 Q_{12}^2 R_{21}^2 + \Omega_2^2 R_{12}^2 Q_{21}^2}{\Theta_{11}\Theta_{22}},$$

in analogy to (3.49).

5.3. Symmetry considerations

The sign of the parametric excitation term $(\pm Q_{12}Q_{21} - \Omega_1\Omega_2 R_{12}R_{21})$ decides the position of the parametric anti-resonance frequency, whether the anti-resonance is situated at $\eta_0 = |\Omega_1 - \Omega_2|$ or at $\eta_0 = \Omega_1 + \Omega_2$.

As it was shown in the previous section the conditions for a system with negative modal damping, $\Theta_{11}\Theta_{22} < 0$, to be stable at $\eta_0 = \Omega_1 + \Omega_2$ are

$$\Theta_{11} + \Theta_{22} > 0,$$
$$0 > \alpha > \frac{Q_{12}Q_{21}}{4\Omega_1\Omega_2} + \frac{R_{12}R_{21}}{4},$$

with the abbreviation

$$\alpha = \Theta_{11}\Theta_{22} - \frac{\beta^2}{(\Theta_{11} + \Theta_{22})^2}.$$

If the parametric excitation matrices

$$\mathbf{Q} = \begin{bmatrix} Q_{11} & Q_{12} \\ Q_{21} & Q_{22} \end{bmatrix}, \quad \mathbf{R} = \begin{bmatrix} R_{11} & R_{12} \\ R_{21} & R_{22} \end{bmatrix}$$

are symmetric matrices, then the product $(Q_{12}Q_{21} + \Omega_1\Omega_2 R_{12}R_{21})$ is positive and the conditions above are satisfiable. On the other hand, if the parametric excitation matrices are skew-symmetric, the product $(Q_{12}Q_{21} + \Omega_1\Omega_2 R_{12}R_{21})$ is negative. For $(Q_{12}Q_{21} + \Omega_1\Omega_2 R_{12}R_{21}) < 0$ the second inequality would require

$$0 > \alpha > 0.$$

In this case it is impossible to fulfill the second inequality for a non-vanishing modal damping coefficients.

Contrary to a single harmonic variation in the previous paragraph where $\eta_0 = \Omega_1 + \Omega_2$, if both parametric excitation matrices $\mathbf{Q}$ and $\mathbf{R}$ possess the same symmetry, a classification cannot be made for a stability desired at the frequency $\eta_0 = |\Omega_1 - \Omega_2|$. Here the conditions

$$\Theta_{11} + \Theta_{22} > 0,$$
$$0 > \alpha > -\frac{Q_{12}Q_{21}}{4\Omega_1\Omega_2} + \frac{R_{12}R_{21}}{4},$$

have to be fulfilled. Now if the parametric excitation matrices are symmetric the sign of the product $(Q_{12}Q_{21} - \Omega_1\Omega_2 R_{12}R_{21}) > 0$ cannot be concluded without further restrictions. In this case we have to consider a skew-symmetric matrix $\mathbf{Q}$ and a symmetric matrix $\mathbf{R}$ which results again in

$$0 > \alpha > 0.$$

Hence, the second inequality is unsatisfiable. Note that a *simultaneous* parametric anti-resonance at $\eta_0 = |\Omega_1 - \Omega_2|$ and $\eta_0 = \Omega_1 + \Omega_2$ in the case of a single negative modal damping, $\Theta_{11}\Theta_{22} < 0$, is only possible for a skew-symmetric matrix $\mathbf{R}$, $R_{12}R_{21} < 0$. This result was already obtained in the summary in Section 3.2.2, and is confirmed for the present synchronous variation. The results are summarized in Tab. 5.1. A classification that is equivalent to Tab. 3.5 in Section 3.2.3 is presented in Tab. 5.2.

5 Synchronous time-periodic damping and stiffness

Table 5.1.: Overview on symmetry considerations of parametric excitation matrices at frequencies $\eta_0 = |\Omega_1 \mp \Omega_2|$ for synchronous stiffness and damping variations.

$\Theta_{11}\Theta_{22} \lessgtr 0$	$R_{12}R_{21} < 0$ (skew-sym.)	$R_{12}R_{21} > 0$ (sym.)
$Q_{12}Q_{21} < 0$ (skew-sym.)	$\boxed{0 < \gamma_{(+)} > \gamma_{(-)}}$ always stable if: $(0>)\,\alpha > -\dfrac{\gamma_{(+)}}{4\Omega_1\Omega_2}$, stable at $\eta_0 = |\Omega_1 - \Omega_2|$ if: $(0>)\,-\dfrac{\gamma_{(+)}}{4\Omega_1\Omega_2} > \alpha > -\dfrac{\gamma_{(-)}}{4\Omega_1\Omega_2}$	$\boxed{\gamma_{(-)} < 0}$ unstable at $\eta_0 = |\Omega_1 - \Omega_2|$, stable at $\eta_0 = \Omega_1 + \Omega_2$ if: $(0<)\,-\dfrac{\gamma_{(-)}}{4\Omega_1\Omega_2} > \alpha > -\dfrac{\gamma_{(+)}}{4\Omega_1\Omega_2}$
$Q_{12}Q_{21} > 0$ (sym.)	$\boxed{0 < \gamma_{(-)} > \gamma_{(+)}}$ always stable if: $(0>)\,\alpha > -\dfrac{\gamma_{(-)}}{4\Omega_1\Omega_2}$, stable at $\eta_0 = \Omega_1 + \Omega_2$ if: $(0>)\,-\dfrac{\gamma_{(-)}}{4\Omega_1\Omega_2} > \alpha > -\dfrac{\gamma_{(+)}}{4\Omega_1\Omega_2}$	$\boxed{\gamma_{(+)} < 0}$ unstable at $\eta_0 = \Omega_1 + \Omega_2$, stable at $\eta_0 = |\Omega_1 - \Omega_2|$ if: $(0<)\,-\dfrac{\gamma_{(+)}}{4\Omega_1\Omega_2} > \alpha > -\dfrac{\gamma_{(-)}}{4\Omega_1\Omega_2}$

$$\eta_0 = |\Omega_1 - \Omega_2| : \quad \gamma_{(-)} = +Q_{12}Q_{21} - \Omega_1\Omega_2 R_{12}R_{21},$$
$$\eta_0 = \Omega_1 + \Omega_2 : \quad \gamma_{(+)} = -Q_{12}Q_{21} - \Omega_1\Omega_2 R_{12}R_{21}$$

$$\alpha = \Theta_{11}\Theta_{22} - \frac{\beta^2}{(\Theta_{11} + \Theta_{22})^2}$$

Table 5.2.: Symmetry considerations of parametric excitation matrices at frequencies $\eta_0 = |\Omega_1 \mp \Omega_2|$ for synchronous stiffness and damping variations.

$\eta_0 = |\Omega_1 - \Omega_2|$	(5.4) is fulfilled: $\Theta_{11} + \Theta_{22} > 0$ $\alpha > 0$	(5.4) is violated: $\Theta_{11} + \Theta_{22} > 0$ $\alpha < 0$
$-\gamma_{(-)} < 0$	additional restriction: **parametric resonance** for $0 < \alpha < -\dfrac{\gamma_{(-)}}{4\Omega_1\Omega_2}$, $\tilde{\sigma}_{1,2}$ is real-valued: $\sigma_s - |\tilde{\sigma}_{1,2}| < \sigma < \sigma_s + |\tilde{\sigma}_{1,2}|$	(5.7) is violated: **unstable** no boundary
$-\gamma_{(-)} > 0$	(5.7) is fulfilled: **stable** no boundary	additional restriction: **parametric anti-resonance** for $0 > \alpha > -\dfrac{\gamma_{(-)}}{4\Omega_1\Omega_2}$, $\tilde{\sigma}_{1,2}$ is real-valued: $-|\tilde{\sigma}_{1,2}| < \sigma < |\tilde{\sigma}_{1,2}|$
$\eta_0 = \Omega_1 + \Omega_2$	(5.4) is fulfilled: $\Theta_{11} + \Theta_{22} > 0$ $\alpha > 0$	(5.4) is violated: $\Theta_{11} + \Theta_{22} > 0$ $\alpha < 0$
$-\gamma_{(+)} < 0$	(5.11) is fulfilled: **stable** no boundary	additional restriction: **parametric anti-resonance** for $0 > \alpha > -\dfrac{\gamma_{(+)}}{4\Omega_1\Omega_2}$, $\tilde{\sigma}_{1,2}$ is real-valued: $\sigma_s - |\tilde{\sigma}_{1,2}| < \sigma < \sigma_s + |\tilde{\sigma}_{1,2}|$
$-\gamma_{(+)} > 0$	additional restriction: **parametric resonance** for $0 < \alpha < -\dfrac{\gamma_{(+)}}{4\Omega_1\Omega_2}$, $\tilde{\sigma}_{1,2}$ is real-valued: $\sigma_s - |\tilde{\sigma}_{1,2}| < \sigma < \sigma_s + |\tilde{\sigma}_{1,2}|$	(5.11) is violated: **unstable** no boundary

$$\eta_0 = |\Omega_1 - \Omega_2| : \quad \gamma_{(-)} = +Q_{12}Q_{21} - \Omega_1\Omega_2 R_{12}R_{21},$$

$$\eta_0 = \Omega_1 + \Omega_2 : \quad \gamma_{(+)} = -Q_{12}Q_{21} - \Omega_1\Omega_2 R_{12}R_{21}$$

$$\alpha = \Theta_{11}\Theta_{22} - \frac{\beta^2}{(\Theta_{11} + \Theta_{22})^2}$$

6. Time-periodic inertia

For the case of a single harmonic inertia variation the inertia matrix is varied periodically in time, while the stiffness and damping matrices are kept constant

$$\mathbf{M}(t) = \mathbf{M}_0 + \varepsilon \mathbf{M}_c^s \cos(\omega t),$$
$$\mathbf{K}(t) = \mathbf{K}_0, \ \mathbf{C}(t) = \mathbf{C}_0.$$

Contrary to stiffness and damping matrices the inertia matrix is always symmetric and the general linear equations of motion (2.2) simplify to

$$\mathbf{M}_0 \ddot{\mathbf{z}}(t) + \varepsilon \mathbf{C}_0 \dot{\mathbf{z}}(t) + \mathbf{K}_0 \mathbf{z}(t) = -\varepsilon \mathbf{M}_c^s \ddot{\mathbf{z}}(t) \cos(\omega t). \tag{6.1}$$

This equation cannot be treated by the same way as the Averaging Method was used in the previous chapters about harmonic stiffness or damping variations. The variable coefficient of the acceleration vector $\ddot{\mathbf{z}}$ leads to a singular denominator in the expression of the right hand side (3.9, 4.3, 3.32, 5.3). After dividing (6.1) by the time-dependent part and expanding the coefficients for a small amplification factor ε into Taylor series

$$\frac{1}{1 + \varepsilon \cos \eta \tau} = 1 - \varepsilon \cos \eta \tau + \varepsilon^2 \cos^2 \eta \tau + \mathcal{O}(\varepsilon^3),$$
$$\frac{\varepsilon \eta \sin \eta \tau}{1 + \varepsilon \cos \eta \tau} = \eta \left[\varepsilon \sin \eta \tau - \varepsilon^2 \sin^2 \eta \tau + \mathcal{O}(\varepsilon^3) \right],$$

and by respecting terms up to the first order we can simplify (6.1) to

$$\ddot{\mathbf{z}}(t) + \varepsilon \tilde{\mathbf{C}}_0 \dot{\mathbf{z}}(t) + \tilde{\mathbf{K}}_0 \mathbf{z}(t) = \varepsilon \left\{ \tilde{\mathbf{K}}_0 \mathbf{z}(t) \cos(\omega t) + \omega \tilde{\mathbf{M}}_c^s \dot{\mathbf{z}}(t) \sin(\omega t) \right\} + \mathcal{O}(\varepsilon^2).$$

Now the coefficient of the acceleration vector $\ddot{\mathbf{z}}$ is constant and we can follow the approach presented in Sections 3.1 and 3.2. The equations of motion in dimensional terms can be transformed ($\mathbf{z}(t) \longrightarrow \mathbf{z}(\tau)$) to non-dimensional normal form

$$\mathbf{z}''(\tau) + \varepsilon \Theta \mathbf{z}'(\tau) + \Omega^2 \mathbf{z}(\tau) = \varepsilon \left\{ \eta \left(\mathbf{S}^s + \mathbf{S}^a \right) \mathbf{z}'(\tau) \sin(\eta \tau) + \left(\mathbf{Q}^s + \mathbf{Q}^a \right) \mathbf{z}(\tau) \cos(\eta \tau) \right\} + \mathcal{O}(\varepsilon^2),$$

$$\text{where } \mathbf{Q} = \mathbf{Q}^s + \mathbf{Q}^a = \mathbf{T}^{-1} \mathbf{M}_0^{-1} \mathbf{K}_0 \mathbf{T},$$
$$\mathbf{S} = \mathbf{S}^s + \mathbf{S}^a = \mathbf{T}^{-1} \mathbf{M}_0^{-1} \mathbf{M}_c^s \mathbf{T},$$

where

$$\mathbf{Q} = \mathbf{Q}^s + \mathbf{Q}^a = \mathbf{T}^{-1} \mathbf{M}_0^{-1} \mathbf{K}_0 \mathbf{T},$$
$$\mathbf{S} = \mathbf{S}^s + \mathbf{S}^a = \mathbf{T}^{-1} \mathbf{M}_0^{-1} \mathbf{M}_c^s \mathbf{T},$$

as shown in Section 2.3 or rewritten in the comprehensive index notation for a system with two degrees of freedom

$$z_i''(\tau) + \Omega_i^2 z_i(\tau) = \varepsilon\left\{-\Theta_{ij} z_j'(\tau) + \eta S_{ij} z_j'(\tau)\sin(\eta\tau) + Q_{ij} z_j(\tau)\cos(\eta\tau)\right\} + \mathcal{O}\left(\varepsilon^2\right), \quad (6.2)$$

with $i, j = 1, 2$.

Comparing (6.2) with (5.1) reveals, that for a first order approximation an inertia variation is equivalent to a stiffness and simultaneous damping variation with a phase shift of $\pi/2$ – an asynchronous stiffness and damping variation! Hence, contrary to a stiffness variation only, an inertia variation should be capable of stabilizing vibrations both in the vicinity of the frequency $\eta_0 = |\Omega_1 - \Omega_2|$ and the frequency $\eta_0 = \Omega_1 + \Omega_2$. This will become more clear in the next sections.

6.1. Transformation and averaging

Applying the procedure as used in Section 3.1, the time transformation (3.2) and the Taylor expansion (3.4), to (5.1) yields

$$\ddot{z}_i + \frac{\Omega_i^2}{\eta^2} z_i = -\frac{\varepsilon}{\eta}\Theta_{ij}\dot{z}_j - \varepsilon S_{ij}\dot{z}_j \sin t - \frac{\varepsilon}{\eta^2}Q_{ij} z_j \cos t + \mathcal{O}\left(\varepsilon^2\right), \quad \text{with } i,j = 1,2, \quad (6.3)$$

with the abbreviations $\varpi_i = \Omega_i/\eta_0$. Using the coordinate transformation in (3.6) the right hand sides of (6.3) becomes

$$F_i(\mathbf{u}, t) = -\eta_0 \sum_j (\Theta_{ij} + \eta_0 S_{ij} \sin t)(-u_j \varpi_j s_j + v_j \varpi_j c_j) -$$
$$- \sum_j Q_{ij} \cos t (u_j c_j + v_j s_j) + 2\Omega_i \varpi_i \sigma (u_i c_i + v_i s_i) \Bigg\}, \quad (6.4)$$

with the abbreviations $s_i = \sin \varpi_i t$ and $c_i = \cos \varpi_i t$ and the state vector $\mathbf{u} = (u_1, v_1, u_2, v_2)^T$. Comparing (6.4) with (5.3) points out the difference between a synchronous stiffness and damping variation, (Q_{ij}, R_{ij}), and a single harmonic inertia variation (Q_{ij}, S_{ij}). The structure of (6.4) and (5.3) is the same, but compared to (5.3), in this chapter the damping variation is phase shifted by $\pi/2$. Hence, for a first order approximation, we can interpret a single harmonic inertia variation as an asynchronous stiffness and damping variation.

The final first order system equations for the slow amplitudes $\mathbf{u}$ are the same as (3.9), with quasi-periodic right hand sides. For this simple two mode system we obtain again 12 different periods according to Appendix B by substituting $n = 1$ and performing $S_{kl} \mapsto Q_{kl}$. This is the same number of periods as for the single harmonic stiffness variation in (3.8) or the single harmonic damping variation in (4.3). Averaging over these different periodic terms yields always

6 Time-periodic inertia

zero, except for the case where a term becomes resonant:

Θ_{ii}, σ : no resonant terms, constant factor $\frac{1}{2}$,

Θ_{ij} : resonant terms for $\varpi_i \pm \varpi_j = 0$,

Q_{ii}, S_{ii} : resonant terms for $2\varpi_i \pm 1 = 0$,

Q_{ij}, S_{ij} : resonant terms for $\varpi_i \pm \varpi_j \pm 1 = 0$.

General Case

If ϖ_1, ϖ_2 do not take on special values averaging over each trigonometric function gives zero, only the terms Θ_{ii}, σ remain and we obtain from (6.4) the same equations as in (3.11). The stability conditions yield for $\eta_0 > 0$

$$\boxed{\Theta_{11} > 0 \quad \text{and} \quad \Theta_{22} > 0.} \tag{6.5}$$

according to (3.14), and the critical values of σ are always purely imaginary. Globally the system is either stable or unstable depending on whether the conditions in (6.5) are satisfied or not.

Some resonant cases

1. Parametric combination resonance of first kind

By defining

$$\boxed{\eta_0 = 2\Omega_k \quad \text{or} \quad 2\varpi_k = 1, \quad \text{with } k = 1 \text{ or } 2}$$

some terms become resonant and produce additional terms after averaging. Instead of (3.11) we obtain from (6.4)

$$\begin{aligned}
\dot{\hat{u}}_i &= \frac{\varepsilon}{\eta_0^2} \left\{ -\frac{\eta_0}{2}\Theta_{ii}\hat{u}_i - \left(\frac{Q_{ii}}{4\varpi_i}\delta_{ik} - \eta_0^2\frac{S_{ii}}{4}\delta_{ik} + \Omega_i\sigma\right)\hat{v}_i \right\}, \\
\dot{\hat{v}}_i &= \frac{\varepsilon}{\eta_0^2} \left\{ -\frac{\eta_0}{2}\Theta_{ii}\hat{v}_i - \left(\frac{Q_{ii}}{4\varpi_i}\delta_{ik} - \eta_0^2\frac{S_{ii}}{4}\delta_{ik} - \Omega_i\sigma\right)\hat{u}_i \right\},
\end{aligned} \tag{6.6}$$

where δ_{ik} is the Kronecker delta-function. Comparing (6.6) with (3.15) gives the following mapping rule

$$Q_{ii} - \eta_0^2 \varpi_i S_{ii} = Q_{ii} - \eta_0 \Omega_i S_{ii} \mapsto Q_{ii}.$$

Thus for stability at $\eta_0 = 2\Omega_k$ the conditions

$$\boxed{\Theta_{ii} > \frac{|Q_{ii} - \eta_0 \Omega_i S_{ii}|}{2\Omega_i}\delta_{ik} = \left|\frac{Q_{ii}}{2\Omega_i} - \Omega_i S_{ii}\right|\delta_{ik} \geq 0} \tag{6.7}$$

have to be satisfied, according to (3.17). For the case $\sigma \neq 0$ we obtain

$$\boxed{\Theta_{ii} > 0, \qquad \sigma_{1,2} = \pm \frac{\eta_0}{2\Omega_i} \sqrt{-\Theta_{ii}^2 + \frac{(Q_{ii} - \eta_0 \Omega_i S_{ii})^2}{4\Omega_i^2} \delta_{ik}}}, \qquad (6.8)$$

according to (3.19).

2. Combination parametric resonance of difference type and first kind

By setting

$$\boxed{\eta_0 = |\Omega_1 - \Omega_2| > 0 \qquad \text{or} \qquad \varpi_1 - \varpi_2 = \pm 1}$$

some terms become resonant and produce additional terms after averaging and we obtain from (6.4)

$$\dot{\hat{u}}_i = \frac{\varepsilon}{\eta_0^2 \varpi_i} \left\{ -\frac{\eta_0}{2} \Theta_{ii} \varpi_i \hat{u}_i + \left(\frac{Q_{ij}}{4} + (-1)^{i+1} \eta_0^2 \varpi_j \frac{S_{ij}}{4} \right) \hat{v}_j - \Omega_i \varpi_i \sigma \hat{v}_i \right\},$$

$$\dot{\hat{v}}_i = \frac{\varepsilon}{\eta_0^2 \varpi_i} \left\{ -\frac{\eta_0}{2} \Theta_{ii} \varpi_i \hat{v}_i - \left(\frac{Q_{ij}}{4} + (-1)^{i+1} \eta_0^2 \varpi_j \frac{S_{ij}}{4} \right) \hat{u}_j + \Omega_i \varpi_i \sigma \hat{u}_i \right\},$$
(6.9)

instead of (3.11). Rewritten in matrix notation yields

$$\begin{pmatrix} \dot{\hat{u}}_1 \\ \dot{\hat{v}}_1 \\ \dot{\hat{u}}_2 \\ \dot{\hat{v}}_2 \end{pmatrix} = \frac{\varepsilon}{\eta_0^2} \begin{bmatrix} -\frac{\eta_0}{2}\Theta_{11} & -\Omega_1 \sigma & 0 & \frac{\eta_0}{4\Omega_1}\widehat{Q}_{12}^{(-)} \\ \Omega_1 \sigma & -\frac{\eta_0}{2}\Theta_{11} & -\frac{\eta_0}{4\Omega_1}\widehat{Q}_{12}^{(-)} & 0 \\ 0 & \frac{\eta_0}{4\Omega_2}\widehat{Q}_{21}^{(-)} & -\frac{\eta_0}{2}\Theta_{22} & -\Omega_2 \sigma \\ -\frac{\eta_0}{4\Omega_2}\widehat{Q}_{21}^{(-)} & 0 & \Omega_2 \sigma & -\frac{\eta_0}{2}\Theta_{22} \end{bmatrix} \begin{pmatrix} \hat{u}_1 \\ \hat{v}_1 \\ \hat{u}_2 \\ \hat{v}_2 \end{pmatrix},$$

with the abbreviations

$$\widehat{Q}_{12}^{(-)} = Q_{12} \pm \eta_0 \Omega_2 S_{12},$$
$$\widehat{Q}_{21}^{(-)} = Q_{21} \mp \eta_0 \Omega_1 S_{21}.$$
(6.10)

Comparing (6.9) with (3.20) gives now the following mapping relations

$$\widehat{Q}_{12}^{(-)} \mapsto Q_{12}, \qquad \widehat{Q}_{21}^{(-)} \mapsto Q_{21}.$$

Performing these substitutions the stability conditions for $\sigma = 0$ results in

$$\Theta_{11} + \Theta_{22} > 0,$$
$$\Theta_{11}\Theta_{22} + \frac{1}{4\Omega_1 \Omega_2} \widehat{Q}_{12}^{(-)} \widehat{Q}_{21}^{(-)} > 0,$$
(6.11)

according to (3.23) and the stability conditions for $\sigma \neq 0$ in

$$\Delta_1: \quad \tfrac{1}{2}\eta_0 \left(\Theta_{11} + \Theta_{22} \right) > 0,$$

$$\Delta_2: \quad \begin{cases} \text{critical values:} \\ \sigma_{1,2} = \pm \dfrac{\eta_0 \left(\Theta_{11} + \Theta_{22} \right)}{2 \left(\Omega_1 - \Omega_2 \right)} \sqrt{-\dfrac{\Theta_{11}\Theta_{22} + \dfrac{1}{4\Omega_1\Omega_2}\widehat{Q}_{12}^{(-)}\widehat{Q}_{21}^{(-)}}{\Theta_{11}\Theta_{22}}}, \end{cases} \qquad (6.12)$$

6 Time-periodic inertia

according to (3.24). In general the parametric excitation term $\hat{Q}_{12}^{(-)}\hat{Q}_{21}^{(-)}$ changes depending on whether the upper or the lower signs in $\eta_0 = \pm\Omega_1 \mp \Omega_2$ are applied. This result is somehow surprising, because this statement is not true for a harmonic stiffness variation in Section 3.1 as well as for a harmonic damping variation in Chapter 4, even for a synchronous stiffness and damping variation in Chapter 5. Only in the present case, where the stiffness and damping coefficients are varied asynchronous, it is relevant whether $\Omega_1 > \Omega_2$ or not! Furthermore, note that similar to the case of a single harmonic stiffness or damping variation, the condition $\sigma_1 = 0$ in (6.12) is an equivalent formulation for the critical case of the second condition in (6.11).

3. Combination parametric resonance of summation type and first kind

By setting

$$\boxed{\eta_0 = \Omega_1 + \Omega_2 \quad \text{or} \quad \varpi_1 + \varpi_2 = 1}$$

some terms become resonant and produce additional terms after averaging and we obtain from (6.4)

$$\begin{aligned}
\dot{\hat{u}}_i &= \frac{\varepsilon}{\eta_0^2 \varpi_i}\left\{-\frac{\eta_0}{2}\Theta_{ii}\varpi_i\hat{u}_i - \left(\frac{Q_{ij}}{4} - (-1)^{i+1}\eta_0^2\varpi_j\frac{S_{ij}}{4}\right)\hat{v}_j - \Omega_i\varpi_i\sigma\hat{v}_i\right\}, \\
\dot{\hat{v}}_i &= \frac{\varepsilon}{\eta_0^2 \varpi_i}\left\{-\frac{\eta_0}{2}\Theta_{ii}\varpi_i\hat{v}_i - \left(\frac{Q_{ij}}{4} - (-1)^{i+1}\eta_0^2\varpi_j\frac{S_{ij}}{4}\right)\hat{u}_j + \Omega_i\varpi_i\sigma\hat{u}_i\right\}.
\end{aligned} \quad (6.13)$$

Rewritten in matrix notation yields

$$\begin{pmatrix} \dot{\hat{u}}_1 \\ \dot{\hat{v}}_1 \\ \dot{\hat{u}}_2 \\ \dot{\hat{v}}_2 \end{pmatrix} = \frac{\varepsilon}{\eta_0^2}\begin{bmatrix} -\frac{\eta_0}{2}\Theta_{11} & -\Omega_1\sigma & 0 & -\frac{\eta_0}{4\Omega_1}\hat{Q}_{12}^{(+)} \\ \Omega_1\sigma & -\frac{\eta_0}{2}\Theta_{11} & -\frac{\eta_0}{4\Omega_1}\hat{Q}_{12}^{(+)} & 0 \\ 0 & -\frac{\eta_0}{4\Omega_2}\hat{Q}_{21}^{(+)} & -\frac{\eta_0}{2}\Theta_{22} & -\Omega_2\sigma \\ -\frac{\eta_0}{4\Omega_2}\hat{Q}_{21}^{(+)} & 0 & \Omega_2\sigma & -\frac{\eta_0}{2}\Theta_{22} \end{bmatrix}\begin{pmatrix} \hat{u}_1 \\ \hat{v}_1 \\ \hat{u}_2 \\ \hat{v}_2 \end{pmatrix},$$

with the abbreviations

$$\begin{aligned}
\hat{Q}_{12}^{(+)} &= Q_{12} - \eta_0\Omega_2 S_{12}, \\
\hat{Q}_{21}^{(+)} &= Q_{21} - \eta_0\Omega_1 S_{21}.
\end{aligned} \quad (6.14)$$

Comparing (6.13) with (3.25) gives the following relations

$$\hat{Q}_{12}^{(+)} \mapsto Q_{12}, \quad \hat{Q}_{21}^{(+)} \mapsto Q_{21}.$$

Performing these substitutions the stability conditions for $\sigma = 0$ results in

$$\begin{aligned}
\Theta_{11} + \Theta_{22} &> 0, \\
\Theta_{11}\Theta_{22} + \frac{1}{4\Omega_1\Omega_2}\hat{Q}_{12}^{(+)}\hat{Q}_{21}^{(+)} &> 0,
\end{aligned} \quad (6.15)$$

according to (3.28) and the stability conditions for $\sigma \neq 0$ in

$$\Delta_1 : \tfrac{1}{2}\eta_0\left(\Theta_{11}+\Theta_{22}\right)>0,$$

$$\Delta_2 : \begin{cases} \text{critical values:} \\ \sigma_{1,2}=\pm\dfrac{\eta_0\left(\Theta_{11}+\Theta_{22}\right)}{2\left(\Omega_1+\Omega_2\right)}\sqrt{-\dfrac{\Theta_{11}\Theta_{22}-\dfrac{1}{4\Omega_1\Omega_2}\widehat{Q}_{12}^{(+)}\widehat{Q}_{21}^{(+)}}{\Theta_{11}\Theta_{22}}}, \end{cases} \quad (6.16)$$

according to (3.29). Note the change in the sign not only for the parametric excitation term but also in the denominator. Further note that the condition $\sigma_1 = 0$ in (6.16) is an equivalent formulation for the critical case of the second condition in (6.15), as in the case of a single harmonic stiffness variation in the Section 3.1.

6.2. Summary for small single harmonic inertia variations

Whether a system is stable or not depends mostly on the frequency η of its inertia variation. Globally the system is stable for arbitrary frequency values η iff

$$\boxed{\Theta_{11}>0 \quad \text{and} \quad \Theta_{22}>0,}$$

according to (6.5). If the frequency η is close to a certain combination frequency the system might be locally stable:

Case 1. For $\eta \approx 2\Omega_k = \eta_0$ if the conditions in (6.7) are satisfied

$$\boxed{\Theta_{ii}>\dfrac{|Q_{ii}-\eta_0\Omega_i S_{ii}|}{2\Omega_i}\delta_{ik}.}$$

In this case the critical values $\sigma_{1,2}$ in (6.8) are purely imaginary and no stability boundary occurs.

Cases 2. and 3. For $\eta \approx |\Omega_1 \mp \Omega_2| = \eta_0$ if the following conditions are satisfied:

$$\boxed{\begin{array}{ll} (6.11,\,6.15),\,\Delta_1>0: & \Theta_{11}+\Theta_{22}>0, \\[4pt] (6.11,\,6.15),\,\text{stable }\eta_0: & \Theta_{11}\Theta_{22}\pm\dfrac{\widehat{Q}_{12}^{(\mp)}\widehat{Q}_{21}^{(\mp)}}{4\Omega_1\Omega_2}>0, \\[4pt] (6.12,\,6.16),\,\Delta_2=0: & \sigma_{1,2}=\dfrac{\eta_0\left(\Theta_{11}+\Theta_{22}\right)}{2\left(\Omega_1\mp\Omega_2\right)}\sqrt{-\dfrac{\Theta_{11}\Theta_{22}\pm\dfrac{1}{4\Omega_1\Omega_2}\widehat{Q}_{12}^{(\mp)}\widehat{Q}_{21}^{(\mp)}}{\Theta_{11}\Theta_{22}}}, \end{array}}$$

with the abbreviations in (6.10) and (6.14), respectively:

$$\begin{array}{ll} \eta_0=\pm\Omega_1\mp\Omega_2: & \eta_0=\Omega_1+\Omega_2: \\ \widehat{Q}_{12}^{(-)}=Q_{12}\pm\eta_0\Omega_2 S_{12}, & \widehat{Q}_{12}^{(+)}=Q_{12}-\eta_0\Omega_2 S_{12}, \\ \widehat{Q}_{21}^{(-)}=Q_{21}\mp\eta_0\Omega_1 S_{21}, & \widehat{Q}_{21}^{(+)}=Q_{21}-\eta_0\Omega_1 S_{21}. \end{array} \quad \text{and}$$

6 Time-periodic inertia

Note the different signs of the parametric excitation term $\widehat{Q}_{12}^{(\mp)}\widehat{Q}_{21}^{(\mp)}$ in the case of $\eta_0 = |\Omega_1 - \Omega_2|$ and $\eta_0 = \Omega_1 + \Omega_2$.

If we want the system to be stable at $\eta_0 = |\Omega_1 - \Omega_2|$ and at $\eta_0 = \Omega_1 + \Omega_2$ *simultaneously*, then the conditions (6.11) and (6.15) have to be satisfied:

$$\eta_0 = \pm\Omega_1 \mp \Omega_2 : \quad \Theta_{11}\Theta_{22} + \frac{1}{4\Omega_1\Omega_2}\widehat{Q}_{12}^{(-)}\widehat{Q}_{21}^{(-)} > 0,$$

$$\eta_0 = \Omega_1 + \Omega_2 : \quad \Theta_{11}\Theta_{22} - \frac{1}{4\Omega_1\Omega_2}\widehat{Q}_{12}^{(+)}\widehat{Q}_{21}^{(+)} > 0.$$

which yields for $\Omega_2 > \Omega_1$,

$$\Theta_{11}\Theta_{22} - \frac{\eta_0^2 S_{21} S_{12}}{4} + \frac{\eta_0}{4\Omega_2}Q_{12}S_{21} > 0 \qquad (6.17)$$

and for $\Omega_1 > \Omega_2$

$$\Theta_{11}\Theta_{22} - \frac{\eta_0^2 S_{21} S_{12}}{4} + \frac{\eta_0}{4\Omega_1}Q_{21}S_{12} > 0. \qquad (6.18)$$

This means that, in the case where one modal damping parameter becomes negative, $\Theta_{11}\Theta_{22} < 0$, it is possible to stabilize the system at the parametric excitation frequency of the difference type $\eta_0 = |\Omega_1 - \Omega_2|$ and the summation type $\eta_0 = \Omega_1 + \Omega_2$ simultaneously, as in the case of a single harmonic damping variation in Chapter 4! Note that (6.17) and (6.18) are necessary, but not sufficient conditions for simultaneous stability. The relation with Chapter 4 becomes evident by mapping $Q_{kl} \mapsto 0$ and $\eta_0 S_{kl} \mapsto R_{kl}$.

Using the substitutions $\widehat{Q}_{kl} \mapsto Q_{kl}$ and the mapping rules (6.10) for $\eta_0 = |\Omega_1 - \Omega_2|$ and (6.14) for $\eta_0 = \Omega_1 + \Omega_2$ reveals, that a single harmonic inertia variation at these special frequencies η_0 is equivalent to a single harmonic stiffness variation as investigated in Section 3.1. Therefore a detailed discussion of the results is omitted here and can be found in Section 3.1.2 and summarized in Tab. 3.1 on page 36. These results are valid on the time scale $1/\varepsilon$.

6.3. Symmetry considerations

The sign of the parametric excitation term $\widehat{Q}_{12}\widehat{Q}_{21}$ determines the position of the parametric anti-resonance frequency, whether the anti-resonance is located at $\eta_0 = |\Omega_1 - \Omega_2|$ or at $\eta_0 = \Omega_1 + \Omega_2$.

Using the substitutions $\widehat{Q}_{kl} \mapsto Q_{kl}$ and the mapping rules (6.10) for $\eta_0 = |\Omega_1 - \Omega_2|$ and (6.14) for $\eta_0 = \Omega_1 + \Omega_2$ the same inferences can be made than for a single harmonic stiffness variation investigated in Section 3.1.3 and summarized in Tab. 3.2. A detailed discussion of the results is omitted here.

7. Simultaneous time-periodic inertia, damping and stiffness

With the results from the previous chapters we can deduct stability formulaes for the most general linear system with two degrees of freedoms and with time-harmonic coefficients with a single frequency as introduced in (2.2) and explicitly written in (2.6)

$$\mathbf{M}_0\ddot{\mathbf{x}}(t) + \varepsilon\mathbf{C}_0\dot{\mathbf{x}}(t) + \mathbf{K}_0\mathbf{x}(t) = \\ -\varepsilon\mathbf{M}_s^s\ddot{\mathbf{x}}(t)\sin(\omega t) - \varepsilon\mathbf{M}_c^s\ddot{\mathbf{x}}(t)\cos(\omega t) \\ -\varepsilon\left\{\mathbf{C}_s^s + \mathbf{C}_s^a\right\}\dot{\mathbf{x}}(t)\sin(\omega t) - \varepsilon\left\{\mathbf{C}_c^s + \mathbf{C}_c^a\right\}\dot{\mathbf{x}}(t)\cos(\omega t) \\ -\varepsilon\left\{\mathbf{K}_s^s + \mathbf{K}_s^a\right\}\mathbf{x}(t)\sin(\omega t) - \varepsilon\left\{\mathbf{K}_c^s + \mathbf{K}_c^a\right\}\mathbf{x}(t)\cos(\omega t), \tag{7.1}$$

where ε is a small amplification factor. Because the parametric excitation terms on the right hand side are additively associated and the time-variation is single frequency, the averaged equations can be derived from superimposing the separate results obtained in the previous chapters. The equations of motion in dimensional terms can be transformed ($\mathbf{z}(t) \longrightarrow \mathbf{z}(\tau)$) to the following non-dimensional normal form

$$\mathbf{z}''(\tau) + \varepsilon\boldsymbol{\Theta}\mathbf{z}'(\tau) + \boldsymbol{\Omega}^2\mathbf{z}(\tau) = -\varepsilon\left\{(\mathbf{R}^s + \mathbf{R}^a)\mathbf{z}'(\tau) + (\mathbf{Q}^s + \mathbf{Q}^a)\mathbf{z}(\tau)\right\}\cos(\eta\tau) \\ -\varepsilon\left\{(\mathbf{S}^s + \mathbf{S}^a)\mathbf{z}'(\tau) + (\mathbf{P}^s + \mathbf{P}^a)\mathbf{z}(\tau)\right\}\sin(\eta\tau) + \mathcal{O}\left(\varepsilon^2\right), \tag{7.2}$$

where

$$\mathbf{Q} = \mathbf{Q}^s + \mathbf{Q}^a = \mathbf{T}^{-1}\mathbf{M}_0^{-1}\left(\mathbf{K}_c + \mathbf{K}_0\right)\mathbf{T},$$
$$\mathbf{P} = \mathbf{P}^s + \mathbf{P}^a = \mathbf{T}^{-1}\mathbf{M}_0^{-1}\left(\mathbf{K}_s - \mathbf{K}_0\right)\mathbf{T},$$
$$\mathbf{R} = \mathbf{R}^s + \mathbf{R}^a = \mathbf{T}^{-1}\mathbf{M}_0^{-1}\left(\mathbf{C}_c + \eta\mathbf{M}_c^s\right)\mathbf{T},$$
$$\mathbf{S} = \mathbf{S}^s + \mathbf{S}^a = \mathbf{T}^{-1}\mathbf{M}_0^{-1}\left(\mathbf{C}_s + \eta\mathbf{M}_s^s\right)\mathbf{T},$$

as it was shown in Section 2.3. Note that the frequency η is associated with the inertia matrices $\mathbf{M}_{c/o}^s$ that arise from the Taylor expansion. Rewritten in the comprehensive index notation for a system with two degrees of freedom yields

$$z_i''(\tau) + \Omega_i^2 z_i(\tau) = -\varepsilon\left\{\Theta_{ij}z_j'(\tau) + \left\{R_{ij}z_j'(\tau) + Q_{ij}z_j(\tau)\right\}\cos(\eta\tau)\right\} \\ -\varepsilon\left\{S_{ij}z_j'(\tau) + P_{ij}z_j(\tau)\right\}\sin(\eta\tau) + \mathcal{O}(\varepsilon^2), \tag{7.3}$$

with $i, j = 1, 2$. The higher order terms in ε results from the Taylor expansion of the inertia variation in (6.2).

7.1. Transformation and averaging

The following stability conditions are obtained by comparing (7.3) with the general harmonic stiffness variation (Q_{kl}, P_{kl}) in (3.31) in Section 3.2, the synchronous stiffness and damping variation (Q_{kl}, R_{kl}) in (5.1) in Chapter 5 and the asynchronous stiffness and damping variation (Q_{kl}, S_{kl}) in (6.2) in Chapter 6.

Superimposing the right hand sides F_i from (3.32), (5.3) and (6.4) leads to

$$F_i(\mathbf{u}, t) = -\eta_0 \sum_j (\Theta_{ij} + R_{ij} \cos t + S_{ij} \sin t)(-u_j \varpi_j s_j + v_j \varpi_j c_j) -$$
$$- \sum_j (Q_{ij} \cos t + P_{ij} \sin t)(u_j c_j + v_j s_j) + 2\Omega_i \varpi_i \sigma (u_i c_i + v_i s_i), \qquad (7.4)$$

with the abbreviations $s_i = \sin \varpi_i t$ and $c_i = \cos \varpi_i t$ and the state vector $\mathbf{u} = (u_1, v_1, u_2, v_2)^T$. The final system equations for the slow amplitudes $\mathbf{u}$ are equal to (3.9), with quasi-periodic right hand sides. As in the previous chapters we obtain 12 different periods, for this simple two degrees of freedom system, according to according to Appendix B. Averaging over a basic trigonometric term yields always zero, except for the case where a term becomes resonant:

Θ_{ii}, σ : no resonant terms, constant factor $\frac{1}{2}$,

Θ_{ij} : resonant terms for $\varpi_i \pm \varpi_j = 0$,

$Q_{ii}, P_{ii}, R_{ii}, S_{ii}$: resonant terms for $2\varpi_i \pm 1 = 0$,

$Q_{ij}, P_{ij}, R_{ij}, S_{ij}$: resonant terms for $\varpi_i \pm \varpi_j \pm 1 = 0$.

General Case

If ϖ_1, ϖ_2 do not take on special values averaging over each trigonometric function gives zero and only the terms Θ_{ii}, σ remain. In this case the system is stable for $\eta_0 > 0$ just if

$$\boxed{\Theta_{11} > 0 \quad \text{and} \quad \Theta_{22} > 0,} \qquad (7.5)$$

are fulfilled, according to (3.14). σ is always purely imaginary. Globally the system is either stable or unstable depending on whether the conditions in (7.5) are satisfied or not.

Some resonant cases

1. Combination parametric resonance of difference type and first kind
By setting

$$\boxed{\eta_0 = |\Omega_1 - \Omega_2| > 0 \quad \text{or} \quad \varpi_1 - \varpi_2 = \pm 1}$$

7 Simultaneous time-periodic inertia, damping and stiffness

some terms become resonant and produce additional terms after averaging (7.3). Instead of (3.11) we obtain from (3.35), (3.42), (5.5) and (6.9)

$$
\begin{pmatrix} \dot{\hat{u}}_1 \\ \dot{\hat{v}}_1 \\ \dot{\hat{u}}_2 \\ \dot{\hat{v}}_2 \end{pmatrix} = \frac{\varepsilon}{\eta_0^2} \begin{bmatrix} -\frac{\eta_0}{2}\Theta_{11} & -\Omega_1\sigma & -\frac{\eta_0}{4\Omega_1}\widehat{P}_{12}^{(-)} & \frac{\eta_0}{4\Omega_1}\widehat{Q}_{12}^{(-)} \\ \Omega_1\sigma & -\frac{\eta_0}{2}\Theta_{11} & -\frac{\eta_0}{4\Omega_1}\widehat{Q}_{12}^{(-)} & -\frac{\eta_0}{4\Omega_1}\widehat{P}_{12}^{(-)} \\ \frac{\eta_0}{4\Omega_2}\widehat{P}_{21}^{(-)} & \frac{\eta_0}{4\Omega_2}\widehat{Q}_{21}^{(-)} & -\frac{\eta_0}{2}\Theta_{22} & -\Omega_2\sigma \\ -\frac{\eta_0}{4\Omega_2}\widehat{Q}_{21}^{(-)} & \frac{\eta_0}{4\Omega_2}\widehat{P}_{21}^{(-)} & \Omega_2\sigma & -\frac{\eta_0}{2}\Theta_{22} \end{bmatrix} \begin{pmatrix} \hat{u}_1 \\ \hat{v}_1 \\ \hat{u}_2 \\ \hat{v}_2 \end{pmatrix}, \quad (7.6)
$$

with the abbreviations

$$
\widehat{Q}_{12}^{(-)} = Q_{12} \pm \Omega_2 S_{12}, \qquad \widehat{P}_{12}^{(-)} = \mp P_{12} + \Omega_2 R_{12},
$$
$$
\widehat{Q}_{21}^{(-)} = Q_{21} \mp \Omega_1 S_{21}, \qquad \widehat{P}_{21}^{(-)} = \mp P_{21} - \Omega_1 R_{21}.
$$
(7.7)

Apply the upper signs for $\eta_0 = \Omega_1 - \Omega_2$ and the lower signs for $\eta_0 = \Omega_2 - \Omega_1$. Compared to a single harmonic inertia variation in (6.10) the prefactor η_0 of S_{kl} is missing because this factor is already respected in (7.3). Comparing (7.6) with (3.35) gives the following mapping rules

$$\widehat{Q}_{ij}^{(-)} \mapsto Q_{ij}, \qquad \widehat{P}_{ij}^{(-)} \mapsto P_{ij}.$$

Performing these substitutions the stability conditions for $\sigma = 0$ results in

$$\Theta_{11} + \Theta_{22} > 0,$$
$$\Theta_{11}\Theta_{22} + \frac{\widehat{Q}_{12}^{(-)}\widehat{Q}_{21}^{(-)} + \widehat{P}_{12}^{(-)}\widehat{P}_{21}^{(-)}}{4\Omega_1\Omega_2} > \left(\frac{\widehat{Q}_{12}^{(-)}\widehat{P}_{21}^{(-)} - \widehat{P}_{12}^{(-)}\widehat{Q}_{21}^{(-)}}{4\Omega_1\Omega_2(\Theta_{11} + \Theta_{22})}\right)^2 \geq 0,$$
(7.8)

according to (3.38) and the stability conditions for $\sigma \neq 0$ in

$\Delta_1: \quad \frac{1}{2}\eta_0(\Theta_{11} + \Theta_{22}) > 0,$

$\Delta_2: \begin{cases} \text{critical values:} \\ \sigma_{1,2} = \sigma_s \pm \tilde{\sigma}_{1,2}, \\ \sigma_s = \dfrac{\eta_0(\Theta_{11} - \Theta_{22})}{2\Theta_{11}\Theta_{22}(\Omega_1 - \Omega_2)}\left(\dfrac{\widehat{Q}_{12}^{(-)}\widehat{P}_{21}^{(-)} - \widehat{P}_{12}^{(-)}\widehat{Q}_{21}^{(-)}}{8\Omega_1\Omega_2}\right), \\ \tilde{\sigma}_{1,2} = \dfrac{\eta_0(\Theta_{11} + \Theta_{22})}{2\Theta_{11}\Theta_{22}(\Omega_1 - \Omega_2)}\sqrt{d}, \\ \text{with } d = -\Theta_{11}\Theta_{22}\left(\Theta_{11}\Theta_{22} + \dfrac{\widehat{Q}_{12}^{(-)}\widehat{Q}_{21}^{(-)} + \widehat{P}_{12}^{(-)}\widehat{P}_{21}^{(-)}}{4\Omega_1\Omega_2}\right) + \left(\dfrac{\widehat{Q}_{12}^{(-)}\widehat{P}_{21}^{(-)} - \widehat{P}_{12}^{(-)}\widehat{Q}_{21}^{(-)}}{8\Omega_1\Omega_2}\right)^2, \end{cases}$
(7.9)

according to (3.40). In general the parametric excitation term $\widehat{Q}_{12}^{(-)}\widehat{Q}_{21}^{(-)} + \widehat{P}_{12}^{(-)}\widehat{P}_{21}^{(-)}$ as well as the interaction term $\widehat{Q}_{12}^{(-)}\widehat{P}_{21}^{(-)} - \widehat{P}_{12}^{(-)}\widehat{Q}_{21}^{(-)}$ change depending on whether the upper or the lower signs in $\eta_0 = \pm\Omega_1 \mp \Omega_2$ are applied. Note that, contrary to the case of a single harmonic stiffness or damping variation in Section 3.1 or Chapter 4, the condition $\sigma_1 = 0$ in (7.9) is *not* an equivalent formulation for the critical case of the second condition in (7.8). This will become more clear in Section 7.2.

2. Combination parametric resonance of summation type and first kind

By setting

$$\boxed{\eta_0 = \Omega_1 + \Omega_2 \quad \text{or} \quad \varpi_1 + \varpi_2 = 1}$$

some terms become resonant and produce additional terms after averaging (7.3) and we obtain from (3.25), (3.43) and (6.13)

$$\begin{pmatrix} \dot{\hat{u}}_1 \\ \dot{\hat{v}}_1 \\ \dot{\hat{u}}_2 \\ \dot{\hat{v}}_2 \end{pmatrix} = \frac{\varepsilon}{\eta_0^2} \begin{bmatrix} -\frac{\eta_0}{2}\Theta_{11} & -\Omega_1\sigma & \frac{\eta_0}{4\Omega_1}\widehat{P}_{12}^{(+)} & -\frac{\eta_0}{4\Omega_1}\widehat{Q}_{12}^{(+)} \\ \Omega_1\sigma & -\frac{\eta_0}{2}\Theta_{11} & -\frac{\eta_0}{4\Omega_1}\widehat{Q}_{12}^{(+)} & -\frac{\eta_0}{4\Omega_1}\widehat{P}_{12}^{(+)} \\ \frac{\eta_0}{4\Omega_2}\widehat{P}_{21}^{(+)} & -\frac{\eta_0}{4\Omega_2}\widehat{Q}_{21}^{(+)} & -\frac{\eta_0}{2}\Theta_{22} & -\Omega_2\sigma \\ -\frac{\eta_0}{4\Omega_2}\widehat{Q}_{21}^{(+)} & -\frac{\eta_0}{4\Omega_2}\widehat{P}_{21}^{(+)} & \Omega_2\sigma & -\frac{\eta_0}{2}\Theta_{22} \end{bmatrix} \begin{pmatrix} \hat{u}_1 \\ \hat{v}_1 \\ \hat{u}_2 \\ \hat{v}_2 \end{pmatrix}, \quad (7.10)$$

with the abbreviations

$$\begin{aligned} \widehat{Q}_{12}^{(+)} &= Q_{12} - \Omega_2 S_{12}, & \widehat{P}_{12}^{(+)} &= P_{12} + \Omega_2 R_{12}, \\ \widehat{Q}_{21}^{(+)} &= Q_{21} - \Omega_1 S_{21}, & \widehat{P}_{21}^{(+)} &= P_{21} + \Omega_1 R_{21}. \end{aligned} \quad (7.11)$$

Compared to a single harmonic inertia variation in (6.14) the prefactor η_0 of S_{kl} is missing because this factor is already respected in (7.3). Comparing (6.13) with (3.43) gives the following relations

$$\widehat{Q}_{ij}^{(+)} \mapsto Q_{ij}, \qquad \widehat{P}_{ij}^{(+)} \mapsto P_{ij}.$$

Performing these substitutions the stability conditions for $\sigma = 0$ results in

$$\Theta_{11} + \Theta_{22} > 0,$$

$$\Theta_{11}\Theta_{22} - \frac{\widehat{Q}_{12}^{(+)}\widehat{Q}_{21}^{(+)} + \widehat{P}_{12}^{(+)}\widehat{P}_{21}^{(+)}}{4\Omega_1\Omega_2} > \left(\frac{\widehat{Q}_{12}^{(+)}\widehat{P}_{21}^{(+)} - \widehat{P}_{12}^{(+)}\widehat{Q}_{21}^{(+)}}{4\Omega_1\Omega_2(\Theta_{11} + \Theta_{22})}\right)^2 \geq 0, \quad (7.12)$$

7 Simultaneous time-periodic inertia, damping and stiffness

according to (3.45) and the stability conditions for $\sigma \neq 0$ in

$\Delta_1: \quad \frac{1}{2}\eta_0 \left(\Theta_{11} + \Theta_{22}\right) > 0,$

$\Delta_2: \quad \begin{cases} \text{critical values:} \\ \sigma_{1,2} = \sigma_s \pm \tilde{\sigma}_{1,2}, \\ \sigma_s = \dfrac{\eta_0 \left(\Theta_{11} - \Theta_{22}\right)}{2\Theta_{11}\Theta_{22}\left(\Omega_1 + \Omega_2\right)} \left(\dfrac{\widehat{Q}_{12}^{(+)} \widehat{P}_{21}^{(+)} - \widehat{P}_{12}^{(+)} \widehat{Q}_{21}^{(+)}}{8\Omega_1\Omega_2} \right), \\ \tilde{\sigma}_{1,2} = \dfrac{\eta_0 \left(\Theta_{11} + \Theta_{22}\right)}{2\Theta_{11}\Theta_{22}\left(\Omega_1 + \Omega_2\right)} \sqrt{d}, \\ \text{with} \quad d = -\Theta_{11}\Theta_{22}\left(\Theta_{11}\Theta_{22} - \dfrac{\widehat{Q}_{12}^{(+)} \widehat{Q}_{21} + \widehat{P}_{12}^{(+)} \widehat{P}_{21}^{(+)}}{4\Omega_1\Omega_2}\right) + \left(\dfrac{\widehat{Q}_{12}^{(+)} \widehat{P}_{21}^{(+)} - \widehat{P}_{12}^{(+)} \widehat{Q}_{21}^{(+)}}{8\Omega_1\Omega_2}\right)^2, \end{cases}$
(7.13)

according to (3.46). Note the change in the sign not only for the parametric excitation term but also in the denominator. Further note that, contrary to the case of a single harmonic stiffness or damping variation in Section 3.1 or Chapter 4, the condition $\sigma_1 = 0$ in (7.13) is *not* an equivalent formulation for the critical case of the second condition in (7.12). This will become more clear in the next section.

7.2. Summary for small variations

Whether a system is stable or not depends mainly on the frequency η of its stiffness variation. Globally the system is stable for arbitrary frequency values η iff

$$\boxed{\Theta_{11} > 0 \quad \text{and} \quad \Theta_{22} > 0,}$$

according to (7.5). If the frequency η is close to a certain combination frequency the system might be locally stable:

Cases 1. and 2. For $\eta \approx |\Omega_1 \mp \Omega_2| = \eta_0$ if the following conditions are satisfied:

$$\boxed{\begin{aligned}
&(7.8, 7.12), \Delta_1 > 0: \quad \Theta_{11} + \Theta_{22} > 0, \\
&(7.8, 7.12), \text{stable } \eta_0: \quad \Theta_{11}\Theta_{22} \pm \frac{\widehat{Q}_{12}^{(\mp)}\widehat{Q}_{21}^{(\mp)} + \widehat{P}_{12}^{(\mp)}\widehat{P}_{21}^{(\mp)}}{4\Omega_1\Omega_2} > \frac{\beta^{2(\mp)}}{(\Theta_{11} + \Theta_{22})^2} \geq 0, \\
&(7.9, 7.13), \Delta_2 = 0: \quad \sigma_{1,2}^{(\mp)} = \sigma_s^{(\mp)} \pm \tilde{\sigma}_{1,2}^{(\mp)},
\end{aligned}}$$

with the abbreviation

$$\beta^{(\mp)} = \left(\frac{\widehat{Q}_{12}^{(\mp)}\widehat{P}_{21}^{(\mp)} - \widehat{P}_{12}^{(\mp)}\widehat{Q}_{21}^{(\mp)}}{4\Omega_1\Omega_2}\right),$$

$$\sigma_s^{(\mp)} = \frac{(\Theta_{11} - \Theta_{22})}{2\Theta_{11}\Theta_{22}}\frac{\beta^{(\mp)}}{2}, \quad \tilde{\sigma}_{1,2}^{(\mp)} = \frac{(\Theta_{11} + \Theta_{22})}{2\Theta_{11}\Theta_{22}}\sqrt{d^{(\mp)}},$$

$$d^{(\mp)} = -\Theta_{11}\Theta_{22}\left(\Theta_{11}\Theta_{22} \pm \frac{\widehat{Q}_{12}^{(\mp)}\widehat{Q}_{21}^{(\mp)} + \widehat{P}_{12}^{(\mp)}\widehat{P}_{21}^{(\mp)}}{4\Omega_1\Omega_2}\right) + \left(\frac{\beta^{(\mp)}}{2}\right)^2,$$

and the mapping relations (7.7) and (7.11).

Note that, contrary to the case of a general harmonic stiffness variation in Chapter 3.2, but in confirmation with the case of a synchronous stiffness and damping variation in Chapter 5, the frequency shift σ_s is different whether the frequency of the stiffness variation is near to a parametric combination frequency of the difference type $\eta_0 = |\Omega_1 - \Omega_2|$ or of the summation type $\eta_0 = \Omega_1 + \Omega_2$.

If one considers the case with stability at $\eta_0 = |\Omega_1 - \Omega_2|$ and *simultaneously* at $\eta_0 = \Omega_1 + \Omega_2$ we obtain from above, (7.8) and (7.12), the following conditions for stability

$$\eta_0 = |\Omega_1 - \Omega_2|: \quad \Theta_{11}\Theta_{22} + \frac{\widehat{Q}_{12}^{(-)}\widehat{Q}_{21}^{(-)} + \widehat{P}_{12}^{(-)}\widehat{P}_{21}^{(-)}}{4\Omega_1\Omega_2} > \frac{\beta^{2(-)}}{(\Theta_{11} + \Theta_{22})^2} \geq 0,$$

$$\eta_0 = \Omega_1 + \Omega_2: \quad \Theta_{11}\Theta_{22} - \frac{\widehat{Q}_{12}^{(+)}\widehat{Q}_{21}^{(+)} + \widehat{P}_{12}^{(+)}\widehat{P}_{21}^{(+)}}{4\Omega_1\Omega_2} > \frac{\beta^{2(+)}}{(\Theta_{11} + \Theta_{22})^2} \geq 0,$$

For simultaneous stability at least

$$\frac{\widehat{Q}_{12}^{(-)}\widehat{Q}_{21}^{(-)} + \widehat{P}_{12}^{(-)}\widehat{P}_{21}^{(-)}}{4\Omega_1\Omega_2} - \frac{\widehat{Q}_{12}^{(+)}\widehat{Q}_{21}^{(+)} + \widehat{P}_{12}^{(+)}\widehat{P}_{21}^{(+)}}{4\Omega_1\Omega_2} \geq 0$$

7 Simultaneous time-periodic inertia, damping and stiffness   99

has to be satisfied. This means that in the case $\Theta_{11}\Theta_{22} < 0$ it may be possible that the system is stable at the parametric excitation frequency of the difference type $\eta_0 = |\Omega_1 - \Omega_2|$ and the summation type $\eta_0 = \Omega_1 + \Omega_2$ simultaneously, at least for $\Omega_1 > \Omega_2$ if (necessary condition)

$$0 > \Theta_{11}\Theta_{22} > -\frac{S_{12}S_{21}}{4} + \frac{R_{12}R_{21}}{4} + \frac{S_{12}Q_{21}}{4\Omega_1} - \frac{P_{12}R_{21}}{4\Omega_2} \quad (7.14)$$

and for $\Omega_2 > \Omega_1$ if

$$0 > \Theta_{11}\Theta_{22} > -\frac{S_{12}S_{21}}{4} + \frac{R_{12}R_{21}}{4} + \frac{Q_{12}S_{21}}{4\Omega_2} - \frac{P_{12}R_{21}}{4\Omega_1} \quad . \quad (7.15)$$

is satisfied. While the first two terms on the right hand side are equal, the last two alters their indices depending on whether $\Omega_2 > \Omega_1$ or not. Using the mapping rules $R_{kl} \mapsto 0$, $P_{kl} \mapsto 0$ and $S_{kl} \mapsto \eta_0 S_{kl}$ these conditions simplify to (6.18) and (6.17) for a single harmonic inertia variation. Using the mapping rules $P_{kl} \mapsto 0$, $S_{kl} \mapsto 0$ and $\eta_0 R_{kl} \mapsto R_{kl}$, these conditions simplify to only one condition (5.13) for a synchronous stiffness and damping variation. Note that the conditions (7.14) and (7.15) are necessary, but not sufficient conditions for simultaneous stability.

In Chapter 3.2 about a harmonic stiffness variation we found that the adequate critical condition (3.49) on page 51 for non-vanishing interaction term follows from a stability analysis by setting $\sigma = \sigma_s$. The shifted frequency line $\eta_0 + \varepsilon\sigma_s$ is stable iff

$\Delta_1 : \frac{1}{2}\eta_0 (\Theta_{11} + \Theta_{22}) > 0,$

$\Delta_2 : \frac{1}{16}\eta_0^4 (\Theta_{11} + \Theta_{22})^2$

$$\times \left(\Theta_{11}\Theta_{22} \pm \frac{\widehat{Q}_{12}^{(\mp)}\widehat{Q}_{21}^{(\mp)} + \widehat{P}_{12}^{(\mp)}\widehat{P}_{21}^{(\mp)}}{4\Omega_1\Omega_2} - \frac{1}{\Theta_{11}\Theta_{22}} \left(\frac{\widehat{Q}_{12}^{(\mp)}\widehat{P}_{21}^{(\mp)} - \widehat{P}_{12}^{(\mp)}\widehat{Q}_{21}^{(\mp)}}{8\Omega_1\Omega_2} \right)^2 \right) > 0,$$

are fulfilled. These conditions can be rewritten for $\eta_0 > 0$ as

$$\begin{aligned}\Theta_{11} + \Theta_{22} &> 0, \\ -\frac{d^{(\mp)}}{\Theta_{11}\Theta_{22}} &> 0.\end{aligned} \quad (7.16)$$

Using the inequalities in (7.16) enables a classification of the conditions (7.9, 7.13) for the most general harmonic variation that is similar to the simple case of a single harmonic stiffness variation in Tab. 3.1, even if the additional frequency shift σ_s of the skeleton line η_0 occurs. By using the substitutions $\widehat{Q}_{kl}^{(\mp)} \mapsto Q_{kl}$ and $\widehat{P}_{kl}^{(\mp)} \mapsto P_{kl}$ for corresponding η_0 this result is equivalent to Tab. 3.4 on page 52. A detailed discussion is therefore omitted here.

Additionally, if we ask for simultaneous stability at the frequency $\eta_0 = |\Omega_1 - \Omega_2| + \varepsilon\sigma_s$ and $\eta_0 = \Omega_1 + \Omega_2 + \varepsilon\sigma_s$, respectively, we obtain from

$$-\frac{d^{(-)}}{\Theta_{11}\Theta_{22}} = \Theta_{11}\Theta_{22} + \frac{\widehat{Q}_{12}^{(-)}\widehat{Q}_{21}^{(-)} + \widehat{P}_{12}^{(-)}\widehat{P}_{21}^{(-)}}{4\Omega_1\Omega_2} - \frac{1}{\Theta_{11}\Theta_{22}}\left(\frac{\beta^{(-)}}{2}\right)^2 > 0,$$

$$-\frac{d^{(+)}}{\Theta_{11}\Theta_{22}} = \Theta_{11}\Theta_{22} - \frac{\widehat{Q}_{12}^{(+)}\widehat{Q}_{21}^{(+)} + \widehat{P}_{12}^{(+)}\widehat{P}_{21}^{(+)}}{4\Omega_1\Omega_2} - \frac{1}{\Theta_{11}\Theta_{22}}\left(\frac{\beta^{(+)}}{2}\right)^2 > 0,$$

the stricter necessary conditions for $\Omega_1 > \Omega_2$

$$0 > \Theta_{11}\Theta_{22} > \Theta_{11}\Theta_{22} > -\frac{S_{12}S_{21}}{4} + \frac{R_{12}R_{21}}{4} + \frac{S_{12}Q_{21}}{4\Omega_1} - \frac{P_{12}R_{21}}{4\Omega_2}$$

$$+\frac{1}{\Theta_{11}\Theta_{22}}\left(\left(\left(Q_{12} + \Omega_2 S_{12}\right)\left(-P_{21} - \Omega_1 R_{21}\right) - \left(P_{12} + \Omega_2 R_{12}\right)\left(Q_{21} - \Omega_1 S_{21}\right)\right)^2\right.$$

$$\left.\left(\left(Q_{12} - \Omega_2 S_{12}\right)\left(P_{21} + \Omega_1 R_{21}\right) - \left(P_{12} + \Omega_2 R_{12}\right)\left(Q_{21} - \Omega_1 S_{21}\right)\right)^2\right),$$

and for $\Omega_2 > \Omega_1$

$$0 > \Theta_{11}\Theta_{22} > \Theta_{11}\Theta_{22} > -\frac{S_{12}S_{21}}{4} + \frac{R_{12}R_{21}}{4} + \frac{Q_{12}S_{21}}{4\Omega_2} - \frac{P_{12}R_{21}}{4\Omega_1}$$

$$+\frac{1}{\Theta_{11}\Theta_{22}}\left(\left(\left(Q_{12} - \Omega_2 S_{12}\right)\left(P_{21} - \Omega_1 R_{21}\right) - \left(P_{12} + \Omega_2 R_{12}\right)\left(Q_{21} + \Omega_1 S_{21}\right)\right)^2\right.$$

$$\left.\left(\left(Q_{12} - \Omega_2 S_{12}\right)\left(P_{21} + \Omega_1 R_{21}\right) - \left(P_{12} + \Omega_2 R_{12}\right)\left(Q_{21} - \Omega_1 S_{21}\right)\right)^2\right),$$

similar to the results in case of a synchronous stiffness and damping variation in (5.14).

Part II.

Interpretations and optimization

8. Equivalent damping

In this chapter we develop analytical approximations for the amount of damping that is gained by introducing a proper parametric excitation in a system. The amount of damping is called equivalent damping, because this damping mechanism is based on additional motion of the system, which is quite opposite to the operation of a classical damper. First, we derive explicit expressions of equivalent damping coefficients for the equations of motion in normal form, based on the procedure presented in the previous chapters. Without loss of generality only the case of a time-harmonic stiffness variation is considered.

8.1. Single harmonic stiffness variation

In the subsequent paragraphs we investigate the following equations of motion in normal form

$$\begin{pmatrix}\ddot{z}_1\\\ddot{z}_2\end{pmatrix}+\begin{bmatrix}\Omega_1^2 & 0\\0 & \Omega_2^2\end{bmatrix}\begin{pmatrix}z_1\\z_2\end{pmatrix}=-\varepsilon\begin{bmatrix}\Theta_{11} & \Theta_{12}\\\Theta_{21} & \Theta_{22}\end{bmatrix}\begin{pmatrix}\dot{z}_1\\\dot{z}_2\end{pmatrix}-\varepsilon\begin{bmatrix}Q_{11} & Q_{12}\\Q_{21} & Q_{22}\end{bmatrix}\begin{pmatrix}z_1\\z_2\end{pmatrix}\cos\eta\tau, \qquad (8.1)$$

as introduced in (3.1). Considerable vibration suppression occurs only at a parametric anti-resonance that is located either at $\eta \approx |\Omega_1 - \Omega_2|$ or at $\eta \approx \Omega_1 + \Omega_2$, as shown in Tab. 3.1 on page 36. As described in Chapter 3.1, we expand the frequency η of the parametric excitation into Taylor series according to (3.4),

$$\eta = \eta_0 + \varepsilon\sigma,$$

then introduce the coordinate transformation as in (3.6) and apply the averaging process respecting only terms up to the first order of ε. For $\eta_0 = |\Omega_1 \mp \Omega_2|$ the coefficient matrix of the complex state vector $\hat{\mathbf{w}} = (\hat{w}_1, \hat{w}_2)^T$ yields

$$\mathbf{A}_2 = \frac{\varepsilon}{\eta_0^2}\begin{bmatrix}-\frac{\eta_0}{2}\Theta_{11}+j\Omega_1\sigma & -j\frac{\eta_0}{4\Omega_1}Q_{12}\\\mp j\frac{\eta_0}{4\Omega_2}Q_{21} & -\frac{\eta_0}{2}\Theta_{22}\pm j\Omega_2\sigma\end{bmatrix}, \qquad (8.2)$$

according to (3.21) and (3.26). The eigenvalues of this complex 2×2 coefficient matrix can be calculated explicitly and the characteristic polynomial yields

$$\lambda^2 + \left(\tfrac{1}{2}\eta_0\left(\Theta_{11}+\Theta_{22}\right)-j\left(\Omega_1\pm\Omega_2\right)\sigma\right)\lambda$$
$$+\left(\frac{\eta_0}{2}\Theta_{11}-j\Omega_1\sigma\right)\left(\frac{\eta_0}{2}\Theta_{22}\mp j\Omega_2\sigma\right)\pm\frac{\eta_0^2}{16\Omega_1\Omega_2}Q_{12}Q_{21}=0, \qquad (8.3)$$

as derived in (3.22) and (3.27). The upper sign corresponds to $\eta_0 = |\Omega_1 - \Omega_2|$ and the lower sign to $\eta_0 = \Omega_1 + \Omega_2$. The hat accent indicates averaged variables. The complex states $\hat{w}_i$ consist of a linear combination of the averaged states $\hat{u}_i, \hat{v}_i$. Due to the transformation (3.6) and the averaging process, these states describe the slow motion of the vibration. The slow motion represents the envelope of the time evolution of the vibrations. Hence, the slopes of the envelopes are good approximations for the overall damping present in the system. We focus the following analysis on the case of $\eta_0 = |\Omega_1 - \Omega_2|$, which corresponds to the upper signs in (8.2) and (8.3), but the results for the case of $\eta_0 = \Omega_1 + \Omega_2$ can be concluded easily.

First, we analyze the system dynamics of the equations of motion (8.1) without parametric excitation, $Q_{ij} = 0$, and therefore constant system parameters. The eigenvalues of the coefficient matrix (8.2) are

$$\lambda_{1,2}^0 = -\tfrac{1}{4}\eta_0 \left(\Theta_{11} + \Theta_{22}\right) \pm \tfrac{1}{4}\eta_0 \left(\Theta_{11} - \Theta_{22}\right),$$
$$\lambda_1^0 = -\tfrac{1}{2}\eta_0 \Theta_{11} \quad \text{and} \quad \lambda_2^0 = -\tfrac{1}{2}\eta_0 \Theta_{22}, \tag{8.4}$$

according to (8.3). The mean value for both eigenvalues becomes relevant in the following analysis and is equal to

$$\lambda_m^0 = \tfrac{1}{2}\left(\lambda_1^0 + \lambda_2^0\right) = -\tfrac{1}{4}\eta_0 \left(\Theta_{11} + \Theta_{22}\right). \tag{8.5}$$

Note that λ_m^0 is proportional to the expression in the main Routh-Hurwitz-condition $(\Theta_{11} + \Theta_{22}) > 0$, as stated for example in Tab. 3.1. Hence, if vibration suppression, and consequently a parametric anti-resonance, shall occur, the main Routh-Hurwitz condition demands that λ_m^0 is negative. Both eigenvalues λ_1^0 and λ_2^0 are located symmetrically to the mean value at the distance of

$$\Delta\lambda^0 = \tfrac{1}{2}\left(\lambda_2^0 - \lambda_1^0\right) = \tfrac{1}{4}\eta_0 \left(\Theta_{11} - \Theta_{22}\right). \tag{8.6}$$

An example for a distribution of the eigenvalues from (8.4) with negative mean value is shown in Fig. 8.1, where we arbitrarily choose the modal damping Θ_{11} to be positive and Θ_{22} to be negative. The symmetry in terms of λ_m^0 is indicated by a circle with the radius $\Delta\lambda^0$. Due to the fact, that the eigenvalue λ_2^0 lies in the right half plane of the complex domain, the system without parametric excitation is dynamically unstable.

Introducing a parametric excitation that satisfies the second Routh-Hurwitz condition as presented in Tab. 3.1, is capable of improving the transient behavior by a parametric anti-resonance if, and only if, the mean value λ_m^0 is negative. Hence, the presence of a negative mean value is the main requirement to exploit the method of damping by parametric excitation. We analyze the polynomial in (8.3) for $\sigma = 0$, where the parametric excitation is activated and its frequency is exactly equal to the parametric combination resonance $\eta_0 = |\Omega_1 - \Omega_2|$. In this case the eigenvalues of the coefficient matrix (8.4) are equal to

$$\lambda_{1,2} = -\frac{\eta_0}{4}\left(\Theta_{11} + \Theta_{22}\right) \pm \frac{\eta_0}{4}\sqrt{\left(\Theta_{11} - \Theta_{22}\right)^2 - \frac{Q_{12}Q_{21}}{\Omega_1\Omega_2}} = \lambda_m^0 \pm \Delta\lambda. \tag{8.7}$$

8 Single harmonic stiffness variation

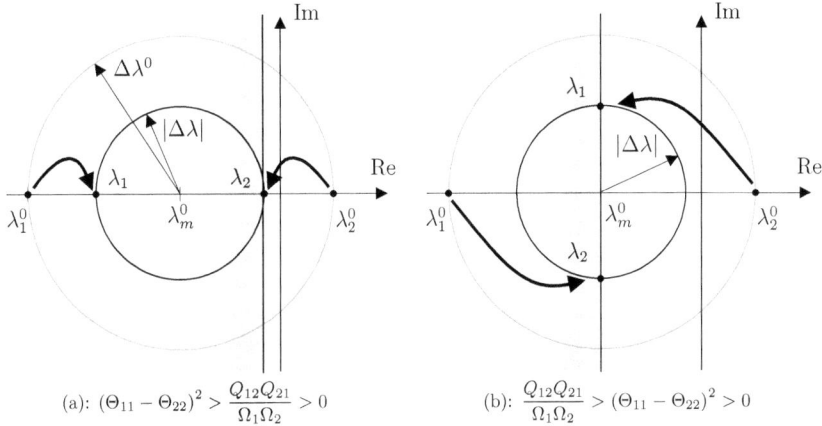

(a): $(\Theta_{11} - \Theta_{22})^2 > \dfrac{Q_{12}Q_{21}}{\Omega_1 \Omega_2} > 0$ 　　　　(b): $\dfrac{Q_{12}Q_{21}}{\Omega_1 \Omega_2} > (\Theta_{11} - \Theta_{22})^2 > 0$

Figure 8.1.: Stable eigenvalues in case of a single harmonic stiffness variation at $\eta = \eta_0 = |\Omega_1 - \Omega_2|$ for $\Theta_{11} > 0$, $\Theta_{22} < 0$ and $Q_{12}Q_{21} > 0$.

The new eigenvalues $\lambda_{1,2}$ are also located symmetrically to the mean value λ_m^0 of the eigenvalues λ_1^0, λ_2^0 as in the case of a system without parametric excitation in (8.5), because the trace of matrix $\mathbf{A}_2$ in (8.2) is not affected by the parametric excitation. The difference now is that the distance $|\Delta\lambda|$ from the symmetry point becomes a function of the parametric excitation term $Q_{12}Q_{21}$. While $\Delta\lambda^0$ in (8.6) is always real-valued, $\Delta\lambda$ in (8.7) can be real-valued or purely imaginary, depending on the size and the sign of the parametric excitation term:

$$\Delta\lambda \text{ is real-valued for } (\Theta_{11} - \Theta_{22})^2 > \frac{Q_{12}Q_{21}}{\Omega_1\Omega_2}, \tag{8.8}$$

$$\text{and } \Delta\lambda \text{ purely imaginary for } (\Theta_{11} - \Theta_{22})^2 < \frac{Q_{12}Q_{21}}{\Omega_1\Omega_2}. \tag{8.9}$$

Finally, if the root in (8.7) vanishes, $\Delta\lambda = 0$, then both eigenvalues λ_1, λ_2 coincide and are equal to λ_m^0. Both conditions above can be rewritten by using the expressions gained from the Routh-Hurwitz conditions in (3.23) as

$$\frac{1}{4}\left(\Theta_{11}^2 + \Theta_{22}^2\right) \gtreqless \Theta_{11}\Theta_{22} + \frac{Q_{12}Q_{21}}{4\Omega_1\Omega_2}. \tag{8.10}$$

For a vanishing parametric excitation term $Q_{12}Q_{21} = 0$, the eigenvalues $\lambda_{1,2}$ from (8.7) are equivalent to the eigenvalues $\lambda_{1,2}^0$ in (8.4). They are located on the real axis, symmetric to the origin λ_m^0, indicated by the circle with the radius $\Delta\lambda^0$, see Fig. 8.1a. Introducing a parametric excitation with a small, but positive parametric excitation term, $Q_{12}Q_{21} > 0$, leads to a new real-valued radius $\Delta\lambda$ that is smaller than the original radius $\Delta\lambda^0$, $\Delta\lambda < \Delta\lambda^0$. A smaller radius $\Delta\lambda$ means, that the eigenvalues $\lambda_{1,2}$ are shifted towards the mean value λ_m^0. Hence, the lower eigenvalue λ_1^0 is shifted to the right on the real axis, while the larger eigenvalue λ_2^0 is shifted to the left, but still remain in the unstable right half plane. Further increase of the initially small parametric excitation term shifts the larger eigenvalue λ_2^0 further to the left, until the condition $\lambda_2^0 = 0$ is reached. Now, the radius $\Delta\lambda$ on the circle is equal to $|\lambda_m^0|$ and the imaginary axis is a tangent of the circle. Further increasing of $Q_{12}Q_{21}$ shifts the larger eigenvalue into the left half plane, see Fig. 8.1a. The shaded area indicates the boundary of the largest eigenvalue. All eigenvalues lie to the left of or on this boundary. This is the first time when the system becomes dynamically stable – a parametric anti-resonance occurs. Until now, the inequality (8.8) is valid and the term $\Delta\lambda$ is always real-valued. Further increasing of $Q_{12}Q_{21}$ decreases the radius $\Delta\lambda$ until the size of $Q_{12}Q_{21}$ is just of the same size as $(\Theta_{11} - \Theta_{22})^2$ and the root in (8.7) vanishes. A vanishing root is equivalent to a vanishing radius, $\Delta\lambda = 0$. The eigenvalues $\lambda_{1,2}$ degenerate and collapse into λ_m^0.

By a further increase of $Q_{12}Q_{21}$ the inequality (8.9) is satisfied. Now, the term $\Delta\lambda$ becomes purely imaginary and the radius of the circle is $|\Delta\lambda|$, as shown in Fig. 8.1b. Hence, increasing the parametric excitation term $Q_{12}Q_{21}$ for (8.9) results in an increase of the radius, too. While the eigenvalues $\lambda_{1,2}$ were real-valued if the condition in (8.8) is satisfied, now they are complex. The real part of this complex eigenvalues is equal to λ_m^0, and the imaginary part is $|\Delta\lambda|$. Note that in this case the parametric excitation affects only the imaginary part of the eigenvalues and the real part is independent of the parametric excitation. From this fact it follows the important conclusion, that for the normal form in (8.1) the equivalent maximum damping possible is just λ_m^0. This equivalent damping is achieved if the parametric excitation term satisfies the condition

$$\frac{Q_{12}Q_{21}}{\Omega_1\Omega_2} \geq (\Theta_{11} - \Theta_{22})^2. \tag{8.11}$$

A further increase of the parametric excitation term introduces a modulation of the slow motion with a frequency $|\Delta\lambda|$, that envelopes the original time evolutions z_1, z_2, but the equivalent damping, which corresponds to the real part of the complex eigenvalues in (8.7) remain constant.

Schematic time plots are shown in Fig. 8.2. The envelopes of the time evolutions described by the calculated eigenvalues in (8.7) are plotted thick-lined. The imaginary part of these eigenvalues leads to a modulation as shown in Fig. 8.2b. Nevertheless, the dashed envelope corresponds to the damping of this transient evolution and is proportional to the real part only.

Evaluating (8.7) for the critical case $\lambda_2^0 = 0$, where the larger eigenvalue lies on the origin in the complex domain, while the smaller one is located on the left half plane, gives

$$\lambda_2^0 = 0: \quad (\Theta_{11} + \Theta_{22}) = \sqrt{(\Theta_{11} - \Theta_{22})^2 - \frac{Q_{12}Q_{21}}{\Omega_1\Omega_2}}.$$

This relation is equivalent to the critical case of the second Routh-Hurwitz condition in (3.23):

$$\Theta_{11}\Theta_{22} + \frac{Q_{12}Q_{21}}{4\Omega_1\Omega_2} = 0. \tag{8.12}$$

Note that for a purely imaginary $\Delta\lambda$, and therefore complex eigenvalues $\lambda_{1,2}$, the only crucial condition for stability is the first Hurwitz condition in (3.23). The second condition is trivially

8 Single harmonic stiffness variation

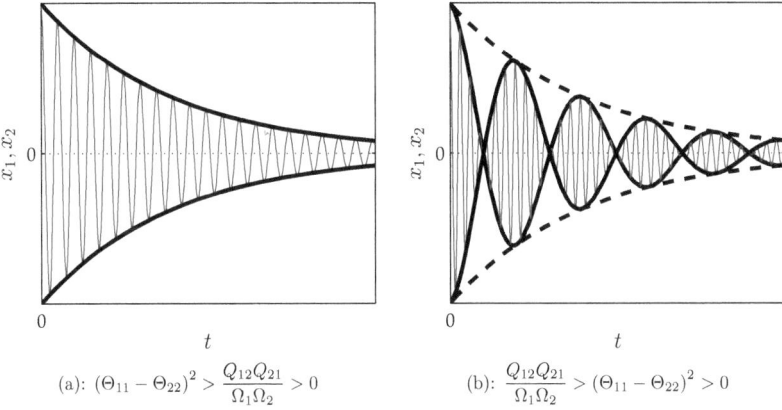

(a): $(\Theta_{11} - \Theta_{22})^2 > \dfrac{Q_{12}Q_{21}}{\Omega_1\Omega_2} > 0$ \qquad (b): $\dfrac{Q_{12}Q_{21}}{\Omega_1\Omega_2} > (\Theta_{11} - \Theta_{22})^2 > 0$

Figure 8.2.: Exemplary time series of x_1, x_2 for stable eigenvalues corresponding to Fig. 8.1.

satisfied, see also Fig. 8.1b. On the other hand, for a real-valued $\Delta\lambda$, and therefore real-valued eigenvalues $\lambda_{1,2}$, both conditions are important. The first condition guarantees that the center of the circle, λ_m, is located on the left half plane, while the second condition ensures that the larger eigenvalue λ_2^0 lies in the left half plane, too, see Fig. 8.1a.

Calculating the eigenvalues for the case $\eta_0 = \Omega_1 + \Omega_2$ in analogy to (8.7) results in

$$\lambda_{1,2} = -\frac{\eta_0}{4}(\Theta_{11} + \Theta_{22}) \pm \frac{\eta_0}{4}\sqrt{(\Theta_{11} - \Theta_{22})^2 + \frac{Q_{12}Q_{21}}{\Omega_1\Omega_2}} = \lambda_m^0 \pm \Delta\lambda. \qquad (8.13)$$

Now, it follows trivially, that a parametric excitation at the frequency $\Omega_1+\Omega_2$ always destabilizes the system in case of $Q_{12}Q_{21} > 0$, because in this case $|\Delta\lambda|$ is always larger than the radius $\Delta\lambda^0$ of the original system without parametric excitation. Therefore, the larger eigenvalue λ_2^0, which already lies in the right half plane, is shifted further to the right and the system remains unstable, as predicted in Tab. 3.2 on page 38.

Analyzing the full characteristic polynomial from (8.3) for $\eta_0 = |\Omega_1 - \Omega_2|$ and $\sigma \neq 0$ yields, instead of (8.7), the following explicit expressions for the eigenvalues

$$\lambda_{1,2} = -\frac{\eta_0}{4}(\Theta_{11} + \Theta_{22}) + j\frac{1}{2}(\Omega_1 + \Omega_2)\sigma \pm \sqrt{a + jb} = \lambda_m^0 \pm \Delta\lambda, \qquad (8.14)$$

where

$$a = \frac{\eta_0^2}{16}\left((\Theta_{11} - \Theta_{22})^2 - \frac{Q_{12}Q_{21}}{\Omega_1\Omega_2}\right) + \frac{\eta_0^2}{4}\left((\Theta_{11} + \Theta_{22})\sigma - \left(1 - \frac{4\Omega_1\Omega_2}{\eta_0^2}\right)\sigma^2\right),$$
$$b = \frac{\eta_0}{2}(\Theta_{11}\Omega_2 - \Theta_{22}\Omega_1)\sigma. \qquad (8.15)$$

The square root of the complex expression in (8.14) can be easily calculated by rewriting the complex variable $a + jb$ using the Euler formula

$$\Delta\lambda = \sqrt{a+jb} = \sqrt{r}e^{j\frac{1}{2}(\alpha+2n\pi)} \quad \text{with} \quad n = 0, 1,$$

where

$$r = \sqrt{a^2 + b^2} \quad \text{and} \quad \alpha = \arctan\frac{b}{a}. \tag{8.16}$$

The square root leads to the two solutions

$$\Delta\lambda = \pm\sqrt{r}e^{j\frac{\alpha}{2}} \quad (= \Delta\lambda_{1,2}), \tag{8.17}$$

and the radius of the circle becomes $|\Delta\lambda| = \sqrt{r}$, as indicated in Fig. 8.3a.

For a system without parametric excitation, $Q_{12}Q_{21} = 0$, the variable a in (8.15) simplifies to a^0 and the eigenvalues from (8.14) results in

$$\lambda_{1,2}^0 = \lambda_m^0 \pm \Delta\lambda^0, \quad \text{with} \quad \Delta\lambda^0 = \sqrt{a^0 + jb}. \tag{8.18}$$

The following relation holds

$$\text{Re}\left\{\lambda_m^0 \text{ in } (8.14)\right\} = \lambda_m^0 \text{ in } (8.5). \tag{8.19}$$

Again the center of the circle with the radius $|\Delta\lambda^0|$ is the mean value of the eigenvalues $\lambda_{1,2}$ for the case without parametric excitation.

Consequently, the eigenvalues $\lambda_{1,2}$ from (8.14) lie on a circle with the radius $\Delta\lambda$, see Fig 8.3b. For the case of $\sigma = 0$, as explained in Fig. 8.1, the center of the circle is located on the real axis with the distance λ_m^0 from the origin of the complex domain. Now, for the case of $\sigma \neq 0$, the center of the circle is shifted from the real axis proportional to σ, according to (8.14). The real part of the center of the circle remains at the same position λ_m^0, see Fig 8.3b.

For a parametric anti-resonance the eigenvalues $\lambda_{1,2}$ are shifted towards the center of the circle, as in the case of $\sigma = 0$ in Fig. 8.1. The main difference between the cases $\sigma = 0$ and $\sigma \neq 0$ is, that now the eigenvalues possess both a real and an imaginary part, respectively, see Fig. 8.3b. Hence, a modulation as shown in Fig. 8.2b occurs always for $\sigma \neq 0$, where the parametric excitation frequency does not exactly match the parametric anti-resonance frequency at $|\Omega_1 - \Omega_2|$ or $\Omega_1 + \Omega_2$.

According to (8.15) the parametric excitation term $Q_{12}Q_{21}$ affects just the variable a, the real part of $\Delta\lambda$ in (8.14). Increasing $Q_{12}Q_{21}$ for $a, b > 0$ starting with small values, a is decreased and, as shown in Fig. 8.3a, the angle α is increased. Thus, increasing the parametric excitation term for $a > 0$ leads to a smaller real part of $\Delta\lambda$, but the imaginary part of $\Delta\lambda$ is increased. A larger imaginary part of $\Delta\lambda$ leads to a larger distance parallel to the imaginary axis. For the example considered, the eigenvalues $\lambda_{1,2}^0$ of the system without parametric excitation are shifted. The real parts of $\lambda_{1,2}^0$ are shifted toward the origin λ_m^0 of the circle, while the imaginary parts of $\lambda_{1,2}^0$ are shifted from the origin, as it is indicated in Fig. 8.3b.

8 General harmonic stiffness variation   109

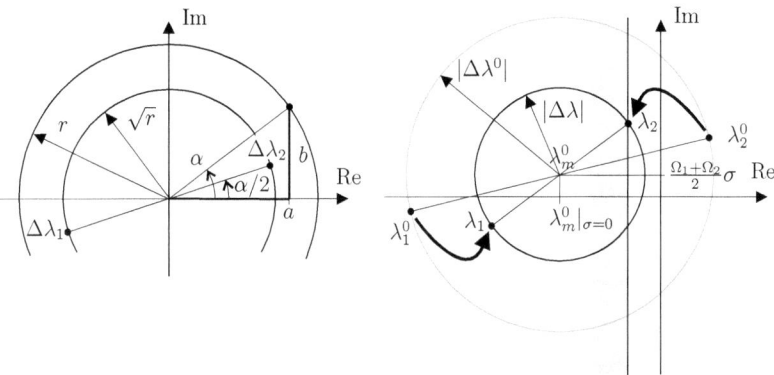

(a): Square root of the complex variable $a + jb$, with $a, b > 0$.

(b): Considering the branch $\Delta\lambda = \Delta\lambda_2$, $a, b > 0$ and small σ.

Figure 8.3.: Stable eigenvalues in case of a single harmonic stiffness variation at $\eta = \eta_0 + \varepsilon\sigma$ for $\eta_0 = |\Omega_1 - \Omega_2|$, $\Theta_{11} > 0$ and $\Theta_{22} < 0$.

Again, the second Routh-Hurwitz condition for the general case of $\sigma \neq 0$ in Tab. 3.1

$$\sigma < \frac{\Theta_{11} + \Theta_{22}}{2} \sqrt{-\frac{1}{\Theta_{11}\Theta_{22}}\left(\Theta_{11}\Theta_{22} + \frac{Q_{12}Q_{21}}{4\Omega_1\Omega_2}\right)} \qquad (8.20)$$

guarantees, that the larger eigenvalue λ_2 lies in the left half plane. The equivalent damping corresponds to the larger real part of the complex eigenvalues $\lambda_{1,2}$ and is equal to

$$\max\{\operatorname{Re}\{\lambda_{1,2}\}\} = \max\left\{\left.\lambda_m^0\right|_{\sigma=0} \pm \sqrt[4]{a^2 + b^2}\cos\frac{\alpha}{2}\right\}. \qquad (8.21)$$

8.2. General harmonic stiffness variation

In the previous section we considered the most simple case of a stiffness variation with a cosine function. Now we expand the parametric excitation considered in the equations of motion (8.1) to a general harmonic variation as introduced in (3.31) in Chapter 3.2. For this general harmonic stiffness variation we obtain the coefficient matrix

$$\mathbf{A}_2 = \frac{\varepsilon}{\eta_0^2}\begin{bmatrix} -\frac{\eta_0}{2}\Theta_{11} + j\Omega_1\sigma & \frac{\eta_0}{4\Omega_1}(P_{12} - jQ_{12}) \\ \frac{\eta_0}{4\Omega_2}(P_{21} + jQ_{21}) & -\frac{\eta_0}{2}\Theta_{22} - j\Omega_2\sigma \end{bmatrix}, \qquad (8.22)$$

instead of (8.2) according to (3.36) and (3.44). To keep expressions simple we restrict our study to the case $\sigma = 0$, where the frequency η of the parametric excitation exactly meets a parametric combination frequency $|\Omega_1 - \Omega_2|$ or $\Omega_1 + \Omega_2$. The characteristic polynomial of (8.22) for $\sigma = 0$ is

$$\lambda^2 + \frac{\eta_0}{2}(\Theta_{11} + \Theta_{22})\lambda + \frac{\eta_0^2}{4}\Theta_{11}\Theta_{22} - \frac{\eta_0^2}{16\Omega_1\Omega_2}(P_{12} - jQ_{12})(P_{21} + jQ_{21}) = 0. \qquad (8.23)$$

8 General harmonic stiffness variation

The solutions of this polynomial can be expressed explicitly and yield

$$\lambda_{1,2} = -\frac{\eta_0}{4}(\Theta_{11} + \Theta_{22}) \pm \frac{\eta_0}{4}\sqrt{(\Theta_{11} - \Theta_{22})^2 - \frac{P_{12}P_{21} + Q_{12}Q_{21}}{\Omega_1\Omega_2} + j\frac{P_{12}Q_{21} - P_{21}Q_{12}}{\Omega_1\Omega_2}}$$
$$= \lambda_m^0 \pm \Delta\lambda. \tag{8.24}$$

The occurrence of the parametric excitation coefficients P_{ij} leads to a complex-valued $\Delta\lambda$, while the mean value λ_m^0 remains unchanged and is equal to (8.5). The eigenvalues $\lambda_{1,2}$ lie on the circle with the radius

$$|\Delta\lambda| = \frac{\eta_0}{4}\sqrt{\left((\Theta_{11} - \Theta_{22})^2 - \frac{P_{12}P_{21} + Q_{12}Q_{21}}{\Omega_1\Omega_2}\right)^2 + \left(\frac{P_{12}Q_{21} - P_{21}Q_{12}}{\Omega_1\Omega_2}\right)^2}, \tag{8.25}$$

located opposite to each other in relation to the center λ_m^0 of the circle, see Fig. 8.4. For a vanishing interaction term $P_{12}Q_{21} - P_{21}Q_{12}$ the eigenvalues $\lambda_{1,2}$ lie on a circle with the radius $|\Delta\lambda|$ and the center λ_m^0, as indicated by the pictures presented in Fig. 8.1. In this case the variable $\Delta\lambda$ is either real-valued or purely imaginary, depending on the ratio of $(\Theta_{11} - \Theta_{22})^2$ and $(P_{12}P_{21} + Q_{12}Q_{21})/(\Omega_1\Omega_2)$. For real-valued $\Delta\lambda$ the eigenvalues $\lambda_{1,2}$ are located at the horizontal outer positions, see Fig. 8.1a, while for purely imaginary $\Delta\lambda$ the eigenvalues $\lambda_{1,2}$ lie at the vertical outer positions. Contrary to this result, for non-vanishing interaction term, $P_{12}Q_{21} - P_{21}Q_{12} \neq 0$, the variable $\Delta\lambda$ becomes complex-valued and the eigenvalues can be found anywhere on the circle with the radius $|\Delta\lambda|$ from (8.25), as shown in Fig. 8.4.

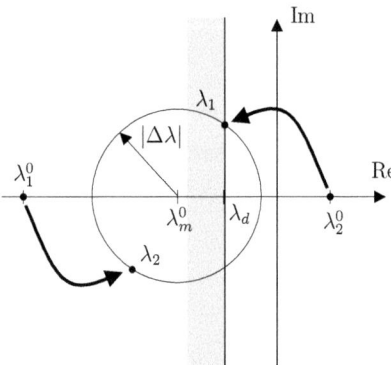

Figure 8.4.: Stable eigenvalues in case of a general harmonic stiffness variation at $\eta = \eta_0 = |\Omega_1 - \Omega_2|$ for $\Theta_{11} > 0$, $\Theta_{22} < 0$ and $Q_{12}Q_{21} > 0$.

8.3. Adopting stability formulae

In the previous chapters and sections we approximated the stability border in the parameter space by analytical conditions as summarized in the Tables 3.1, 3.4 and 4.1 and Sections 5.2, 6.2 and 7.2. We can adopt these results to derive conditions for the system parameters to achieve a desired equivalent damping. Note that although the eigenvalues λ_1, λ_2 of the averaged system may be real-valued and lie on the real axis as shown in Fig. 8.1a, the eigenvalues of the non-averaged system possess an additionally imaginary part, as indicated by the thin-lined curves in Fig. 8.2.

For example, for a system with a time-harmonic stiffness variation only, as investigated in (8.1), the conditions for stability are derived from the Routh-Hurwitz conditions as stated in Table 3.1. On the other hand, according to (8.7), the analytically approximated eigenvalues of the slow motion are

$$\lambda_{1,2} = -\frac{\eta_0}{4}\left(\Theta_{11} + \Theta_{22}\right) \pm \frac{\eta_0}{4}\sqrt{\left(\Theta_{11} - \Theta_{22}\right)^2 - \frac{Q_{12}Q_{21}}{\Omega_1\Omega_2}} = \lambda_m^0 \pm \Delta\lambda. \qquad (8.26)$$

Hence, for a parameter set for which the system operates at its stability border the following relation holds

$$\max\left\{\operatorname{Re}\left\{\lambda_{1,2}\right\}\right\} = 0. \qquad (8.27)$$

This condition is an equivalent formulation of the formulae in Tab. 3.1. The stability border represents a system where the modes coupled by a parametric excitation neither decay nor increase. Such modes have an equivalent damping of zero.

The approximated eigenvalues in (8.26) can be used to conclude conditions to achieve a desired equivalent damping which corresponds to the eigenvalue λ_d which lies in the left half plane, see the border of the shaded are in Fig. 8.4. Introducing λ_d is equivalent to a shift of the polynomial in (8.3) by a negative value, $\lambda_d < 0$, parallel to the real axis

$$\lambda \mapsto \lambda + (\lambda_d + j0). \qquad (8.28)$$

This transformation is equivalent to a shift of the imaginary axis to the right and the condition in (8.27) is adopted to

$$\lambda_d = \max\left\{\operatorname{Re}\left\{\lambda_{1,2}\right\}\right\}. \qquad (8.29)$$

For a system at the stability border we get $\lambda_d = 0$.

The polynomial in (8.3) corresponds to the matrix presented in (8.2) and leads to the eigenvalues in (8.26), but in fact the coefficient matrix (8.2) is multiplied by a factor, too. Hence, the equivalent damping is just proportional to λ_d and is equal to

$$\frac{\varepsilon}{\eta_0^2}\lambda_d. \qquad (8.30)$$

8.4. Systems with more degrees of freedom

In this section we consider a multi-degrees of freedom system with a single frequency parametric excitation. We assume that the system possesses only separate natural frequencies Ω_i, the case of degenerated eigenvalues is not considered. The normal form of the equations of motion for a system with n degrees of freedom with a stiffness variation and a right hand side multiplied by a small parameter can be written in analogy to (3.1) as

$$\ddot{\mathbf{z}} + \mathbf{\Omega}^2 \mathbf{z} = -\varepsilon \mathbf{\Theta} \dot{\mathbf{z}} - \varepsilon \mathbf{Q} \mathbf{z} \cos \eta \tau, \qquad (8.31)$$

with the position vector $\mathbf{z} = (z_1, z_2, \ldots, z_n)^T$, a diagonal matrix $\mathbf{\Omega}^2$ of natural frequencies

$$\Omega_1 < \Omega_2 < \cdots < \Omega_n,$$

and, in general, fully occupied damping matrix $\mathbf{\Theta}$ and parametric excitation matrix $\mathbf{Q}$.

We introduce a parametric excitation with a frequency η that is equal to a parametric combination frequency $|\Omega_k - \Omega_l|$ or $\Omega_k + \Omega_l$. Applying a procedure similar to the one presented in Chapter 3.1, which is based on [1], leads to the following averaged equations of motion

$$\begin{pmatrix} \ddot{\hat{z}}_k \\ \ddot{\hat{z}}_l \end{pmatrix} + \begin{bmatrix} \Omega_k^2 & 0 \\ 0 & \Omega_l^2 \end{bmatrix} \begin{pmatrix} \hat{z}_k \\ \hat{z}_l \end{pmatrix} = -\varepsilon \begin{bmatrix} \Theta_{kk} & 0 \\ 0 & \Theta_{ll} \end{bmatrix} \begin{pmatrix} \dot{\hat{z}}_k \\ \dot{\hat{z}}_l \end{pmatrix} - \varepsilon \begin{bmatrix} 0 & Q_{kl} \\ Q_{lk} & 0 \end{bmatrix} \begin{pmatrix} \hat{z}_k \\ \hat{z}_l \end{pmatrix} \cos \eta \tau, \qquad (8.32)$$

$$\ddot{\hat{z}}_i + \Omega_i^2 \hat{z}_i = -\varepsilon \Theta_{ii} \dot{\hat{z}}_i, \qquad i \neq k, l, \qquad (8.33)$$

where $i = 1, 2, \ldots, n$, as can be concluded from [42]. Note that the result above is only valid if the following two main restrictions are satisfied. First, the analysis performed is a local analysis near the special frequency $|\Omega_k - \Omega_l|$ or $\Omega_k + \Omega_l$, where we demand that the remaining combination frequencies

$$\frac{\Omega_i \pm \Omega_j}{k}, \quad \text{where} \quad i, j, k = 1, 2, \ldots, n, \qquad (8.34)$$

are sufficiently separated from these frequencies. Second, by introducing the small parameter ε in (8.31) we assume the case of small entries in the damping matrix and in the parametric excitation matrix, respectively. For a stronger damped system, where the values of the damping coefficients are of the same size as the remaining system coefficients, also the non-diagonal damping coefficients become important. In this case the approximations in (8.32) and (8.33) are no longer valid.

As can be concluded from (8.32) and (8.33), if we assume small damping coefficients only the diagonal matrix elements Θ_{ii} are crucial for stable transient behavior as can be seen in (8.32) and (8.33). Furthermore, from the fully occupied parametric excitation matrix of size $n \times n$ in

8 Systems with more degrees of freedom

(8.31)

$$\mathbf{Q} = \begin{bmatrix} Q_{11} & Q_{12} & \cdots & Q_{1k} & \cdots & Q_{1l} & \cdots & Q_{1n} \\ Q_{21} & Q_{22} & \cdots & Q_{2k} & \cdots & Q_{2l} & \cdots & Q_{2n} \\ \vdots & \vdots & \ddots & \vdots & & \vdots & & \vdots \\ Q_{k1} & Q_{k2} & \cdots & Q_{kk} & \cdots & \boxed{Q_{kl}} & \cdots & Q_{kn} \\ \vdots & \vdots & & \vdots & \ddots & \vdots & & \vdots \\ Q_{l1} & Q_{l2} & \cdots & \boxed{Q_{lk}} & \cdots & Q_{ll} & \cdots & Q_{ln} \\ \vdots & \vdots & & \vdots & & \vdots & \ddots & \vdots \\ Q_{n1} & Q_{n2} & \cdots & Q_{nk} & \cdots & Q_{nl} & \cdots & Q_{nn} \end{bmatrix}, \qquad (8.35)$$

only the two cross-coupling terms Q_{kl} and Q_{lk} that correspond to the frequencies Ω_k and Ω_l have an effect on the transient dynamics of the system. This means that only two of the n modes are coupled. Mode coupling occurs between the mode k and the mode l, which requires for stable transient motion that at least one of these modes is stable. This fact is expressed already in the conclusion of (8.5). The main consequence of (8.35) is that introducing a parametric anti-resonance in a dynamical system is capable of stabilizing just one mode. Choosing the frequencies Ω_k and Ω_l for the parametric anti-resonance determines the effective matrix elements in (8.35). Hence, we can interpret a parametric excitation at the parametric anti-resonance as a device that selectively distributes the kinetic energy between two modes, while the other modes are not affected – selective energy drain. A detailed analysis of the energetic flow in a parametrically excited system will be performed in Chapter 9.

Note that if the main parametric anti-resonance frequency η_0, which is equal to $|\Omega_k - \Omega_l|$ or $|\Omega_k + \Omega_l|$, is near to any primary resonance frequency $2\Omega_i/n$,

$$\eta_0 \approx \frac{2\Omega_i}{n}, \qquad \text{with } n \in \mathbb{N}, \qquad (8.36)$$

the relations in (8.33) have to be adapted to

$$\ddot{\hat{z}}_i + \Omega_i^2 \hat{z}_i = -\varepsilon \Theta_{ii} \dot{\hat{z}}_i - \varepsilon Q_{ii} \hat{z}_i \cos(\eta\tau), \qquad i \neq k, l. \qquad (8.37)$$

In this case, a parametric excitation with η_0 leads to a stability gain as well, but the resulting stability region is overlapped by regions of instability corresponding to the primary resonance frequencies $2\Omega_i/n$.

9. Energy flow

In this chapter we investigate the energy flow of a system with two degrees of freedom and a time-periodic variation of one or more system parameters. The corresponding equations of motion yield a linear time-periodic homogenous system, as stated in (2.6).

For a configuration with isolated system borders the energy conservation theorem guarantees, that the system energy is conserved if no resulting forces are acting on the system. For the homogenous equation in (2.6) this is the case for a configuration without damping and parametric excitation. Here the total energy is just a sum of the kinetic energy corresponding to the system inertias and the potential energy corresponding to the stiffness parameters. An introduction of damping parameters may destroy this energy conservation, if the damping elements are located outside of the system borders. In this case, the system starts to exchange energy with its environment and energy flows occur. Generally speaking, if the damping parameters are positive then the total system energy is reduced. The energy flow is negative if the system energy decreases and positive if the system energy increases.

With a parametric excitation the time history of a system parameter is prescribed. The realization of a parametric excitation requires a certain amount of energy flow. E.g. to vary the length of a beam or to move the pivot point of a pendulum, that depends on the actual position of the system. Hence, introducing a parametric excitation enables the system to exchange energy with its environment, too.

For a system, where damping coefficients and parametric excitation coexist, the question arises, how these two effects interact energetically, whether system energy is lost or gained. In the context of the presented and investigated phenomenon of vibration suppression, the total energy flow is negative, but it is non-trivial how the particular energy flows corresponding to damping elements and parametric excitation are distributed. Especially interesting is, whether the system looses its energy mainly by its positive damping or if it is possible to loose energy across the system border caused by parametric excitation. Knowing the distribution of the particular energy flows helps to develop an intuitive picture and deeper understanding how vibration suppression by parametric excitation is created.

9.1. Energy definition

In physics, the differential work is generally defined as the integral of the inner product of force and differential translation

$$W_{12} = \int_{\mathbf{s}_1}^{\mathbf{s}_2} \mathbf{f} \, d\mathbf{s}, \tag{9.1}$$

where $\mathbf{s}_1, \mathbf{s}_2$ are the position vectors at the starting and end point of the motion. The terms work and energy are used interchangeably in the following discussion. Power is the amount of work delivered per unit of time,

$$P = \frac{dW}{dt},$$

which is equivalent to the rate of energy transfer, the energy flow. For explicitly time-dependent forces, like for instance $\mathbf{f}(t) = k(t)\,\mathbf{s}(t)$, we can reformulate the path integral with implicit time dependent integral borders in (9.1) as a simple integral over time:

$$W_{12} = \int_0^t \mathbf{f}(\tau) \frac{d\mathbf{s}(\tau)}{d\tau} d\tau, \quad \text{with} \quad \mathbf{s}(0) = \mathbf{s}_1, \; \mathbf{s}(t) = \mathbf{s}_2. \tag{9.2}$$

The differential work performed become

$$dW = \mathbf{f}(t) \frac{d\mathbf{s}(t)}{dt} dt,$$

and the resulting power yield

$$P = \frac{dW}{dt} = \mathbf{f}(t) \frac{d\mathbf{s}(t)}{dt}.$$

Hence, the definition

$$P = \mathbf{f} \, \mathbf{v} \tag{9.3}$$

does not depend on the fact whether the force is time-dependent or not. This statement is shown for an exemplary system, for which the energy flow balance is developed directly from the equations of motion by multiplying with $\mathbf{v}$ and from the definition in (9.3), respectively.

We investigate a system with a kinematic constraint, a pendulum system in a gravity field with prescribed time-periodic vertical pivot motion

$$y_0 = A \cos t,$$

as shown in Fig. 9.1. Friction in the sliding and rotating joint is neglected.

1. Energy flow balance from equation of motion:
   The force due to gravity mg and the inertia force $m\ddot{y}_0$ leads to an effective gravity g_{eff}. The projection of the resulting force is

   $$m(g - \ddot{y}_0)\mathbf{e}_y = mg_{eff}\mathbf{e}_y,$$

   where $\mathbf{e}_y$ points in the direction of the gravity acceleration $\mathbf{g}$. The direction vector

   $$\mathbf{n} = \cos\varphi \, \mathbf{e}_x - \sin\varphi \, \mathbf{e}_y$$

9 Interaction of damping and parametric excitation 117

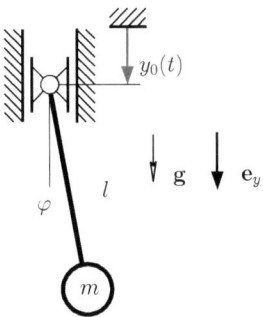

Figure 9.1.: Parametric excitation by time-periodic motion of the pivot point.

yields the effective force that causes the motion of the pendulum mass

$$mg_{eff}\mathbf{e}_y \mathbf{n} = -mg_{eff}\sin\varphi.$$

The equation of motion becomes

$$ml^2\ddot{\varphi} + mg_{eff}l\sin\varphi = 0.$$

If an additional torsional stiffness k is attached at the pivot point, then this equation alters and the energy flow balance obtained by multiplication with $\dot{\varphi}$ results in

$$ml^2\ddot{\varphi}\,\dot{\varphi} + k\varphi\dot{\varphi} + m(g - \ddot{y}_0)\,l\sin\varphi\,\dot{\varphi} = 0. \tag{9.4}$$

2. Alternative approach by evaluating (9.3):
   Applying (9.3) the energy flow with regard to moments resulting from inertia forces we obtain $(ml^2\ddot{\varphi})\,\dot{\varphi}$ and $(mg\,l\sin\varphi)\,\dot{\varphi}$, respectively. The energy flow resulting from the stiffness torque is $(k\varphi)\,\dot{\varphi}$. Finally, the energy flow due to the force acting on the sliding joint is

$$f(t)\,v(t) = ml^2\ddot{\varphi}\,\dot{\varphi} + k\varphi\dot{\varphi} + m(g - \ddot{y}_0)\,l\sin\varphi\,\dot{\varphi} = 0. \tag{9.5}$$

Thus, both approaches lead to the same result and (9.4), (9.5) are equivalent.

9.2. Interaction of damping and parametric excitation

Applying the general definition of the energy flow in (9.3) and the definition of work in (9.2) enables us to analyze the stored and emitted energy portions of an open system. The system under consideration exchanges energy with its environment due to dissipation, described by damping coefficients, and due to parametric excitation, a prescribed time-periodic evolution of one or more system parameters.

For the sake of simplicity we consider a simple system with two degrees of freedom as shown in Fig. 2.1, but with constant inertia, damping and stiffness parameters. Only a sine harmonic variation of the stiffness parameter k_{01} is introduced. Moreover, coupling by the damping parameter c_{12} is not considered. The equations of motion become

$$\begin{bmatrix} m_1 & 0 \\ 0 & m_2 \end{bmatrix} \ddot{\mathbf{x}}(t) + \begin{bmatrix} c_{01} & 0 \\ 0 & c_{02} \end{bmatrix} \dot{\mathbf{x}}(t) + \begin{bmatrix} k_{01}(1 + \varepsilon_{01} \sin(\omega t)) + k_{12} & -k_{12} \\ -k_{12} & k_{12} + k_{02} \end{bmatrix} \mathbf{x}(t) = \mathbf{0}, \quad (9.6)$$

with the position vector $\mathbf{x} = (x_1, x_2)^T$. Individual amounts of energy that are accumulated during a time interval $[0, t]$ are calculated by integration of the energy flow over the time t according to (9.2). The different locations where energy is concentrated in the considered system are the energy portions due to the inertia and damping forces,

$$W_i^m = \int_0^t m_i \ddot{x}_i \frac{dx_i}{d\tau} d\tau, \qquad W_{0i}^c = \int_0^t c_{0i} \dot{x}_i \frac{dx_i}{d\tau} d\tau, \quad (9.7)$$

the energy portions due to the forces resulting from the constant stiffness parameters,

$$W_{0i}^k = \int_0^t k_{0i} x_i \frac{dx_i}{d\tau} d\tau, \qquad W_{12}^k = \int_0^t k_{12}(x_2 - x_1) \frac{d}{d\tau}(x_2 - x_1) d\tau, \quad (9.8)$$

and finally the energy portion due to the force resulting from the time-periodic stiffness parameter $\varepsilon_{01} k_{01}$,

$$W_{01}^{k\sim} = \int_0^t \varepsilon_{01} k_{01} \sin(\omega \tau) x_i \frac{dx_i}{d\tau} d\tau. \quad (9.9)$$

The total system energy at time t is the sum of these energy portions and is constant due to the principle of energy conservation,

$$W_{tot} = \sum_i W_i^m + \sum_{i,j} \left(W_{ij}^k + W_{ij}^c \right) + W_{01}^{k\sim} = \text{const}. \quad (9.10)$$

The relation (9.10) represents the sum of energy portions that lies inside as well as outside the system borders. Therefore, the total amount energy remains constant, although energy flow across the system borders occur.

9.2.1. Negative damping

To make clear how vibration suppression is achieved in a dynamically unstable system at a parametric anti-resonance, the time series of physical parameters and energy portions are analyzed. We choose the following exemplary set of values for the system variables of system (9.6)

$$m_1 = 1, \ m_2 = 4, \ k_{01} = k_{12} = k_{02} = 1, \ c_{01} = 0.15, \ c_{02} = -0.07, \ \varepsilon_{01} = 0.3. \quad (9.11)$$

9 Interaction of damping and parametric excitation

The system is dynamically unstable due to the negative damping coefficient c_{02} and the positive damping coefficient c_{01} is not large enough to compensate the destabilizing effect of c_{02}.

Applying a transformation matrix $\mathbf{T}$ as outlined in (2.16), non-dimensional modal damping coefficients Θ_{11}, Θ_{22} can be obtained by the transformation performed in (2.15),

$$\begin{bmatrix} \Theta_{11} & \Theta_{12} \\ \Theta_{21} & \Theta_{22} \end{bmatrix} = \mathbf{T}^{-1} \begin{bmatrix} m_1 & 0 \\ 0 & m_2 \end{bmatrix}^{-1} \begin{bmatrix} c_{01} & 0 \\ 0 & c_{02} \end{bmatrix} \mathbf{T}.$$

The size of the damping coefficients c_{01}, c_{02} are chosen such, that one modal damping Θ_{ii} is negative and the other positive. Therefore, without parametric excitation, $\varepsilon_{01} = 0$, the system is unstable according to (3.14).

According to the analytical results in Chapter 3.1, a parametric anti-resonance frequency for the symmetric parametric excitation matrix in (9.6) is expected at

$$\omega = \sqrt{\frac{k_{12}}{m_1}} |\Omega_1 - \Omega_2|, \tag{9.12}$$

where Ω_1, Ω_2 are the natural frequencies of the conventional system without parametric excitation, $\varepsilon = 0$. The prefactor comes from the inverse transformation from non-dimensional to dimensional system parameters. The occurrence of a parametric anti-resonance frequency (9.12) demands the restriction that the first condition in (3.23) on page 31,

$$\Theta_{11} + \Theta_{22} > 0,$$

is satisfied. This is the case for the values chosen in (9.11). Furthermore, the occurrence of a parametric anti-resonance frequency (9.12) demands that the second condition in (3.23) is satisfied, too. This results in a critical value ε_{crit} for the amplification factor ε_{01} that has to be exceeded. To make the parametric anti-resonance exceptionally strong, we use a very high value for the amplification factor ε_{01}. For the parameter values chosen, the parametric anti-resonance frequency occurs at $\omega = 0.876$. With the initial conditions

$$x_1 = x_2 = 1, \quad \dot{x}_1 = \dot{x}_2 = 0, \tag{9.13}$$

the constant total energy is equal to the potential energy at the initial time and yields

$$W^{tot} = \frac{1}{2} \left(k_{01} x_1^2 + k_{12} (x_1 - x_2)^2 + k_{02} x_2^2 \right) \bigg|_{t=0} = 1.$$

The time series of the displacements, velocities and the resulting energies for the exemplary system in (9.6) with the initial conditions in (9.13) are plotted in Figs. 9.2-9.6 for two cases, for the conventional system, $\varepsilon_{01} = 0$, and for a system with activated parametric excitation, $\varepsilon_{01} > \varepsilon_{crit}$. Note, that in both cases the same system parameters are used.

Figure 9.2 shows time series of the system states. Starting from the initial conditions in (9.13) and $\varepsilon_{01} = 0$ the amplitudes of the displacements x_j become larger and larger. The system is unstable and the linear model predicts that the amplitudes of the displacements grow to infinity. On the other hand, the system is effectively damped for activated parametric excitation, $\varepsilon_{01} > \varepsilon_{crit}$, if the stability conditions in (3.23) are satisfied. Note, that the vibration amplitude of the light mass, x_1, first grows and then is reduced, while the vibration amplitude of the heavy mass, x_2, is reduced from the beginning. The same statements can be made for the velocities. Furthermore, the main frequencies of both vibration signals are increased, but this change is negligible for the slow heavy mass. Looking at a longer time scale for the case of a parametric anti-resonance the amplitudes of the displacements decrease until they completely disappear and both masses come to rest. Here the mechanism of damping by parametric excitation is effective.

For a real system large amplitudes are usually bounded by non-linearities being also present in the system, but the chosen linear model is quite sufficient for proving the effectiveness of damping by parametric excitation. If the system is positively damped and therefore stable, then the amplitudes of the displacements are low and non-linearities are negligible. In this case, the dynamics can be described sufficiently by the proposed linear model (9.6) in the vicinity of the equilibrium position. On the other hand, if parametric excitation is not capable of introducing additional damping into the system, the system is unstable. Now, long-term time series of the unstable system cannot be predicted appropriately by the chosen linear model, because now a linearization as performed in (2.1) is not allowed. While the vibration amplitudes of the fully non-linear system grow until a limit circle is reached, they increase exponentially and unlimited in the linear model. The linear model is meaningful only for the first time steps during the onset of instability.

The resulting time histories of the kinetic energy W_i^m are shown in Fig. 9.3. For $\varepsilon_{01} = 0$ these energy portions grow because the amplitudes of the displacement grow, too. For $\varepsilon_{01} > \varepsilon_{crit}$ the kinetic energy portions diminish because the amplitudes of the displacements diminish, too. For the heavy mass the kinetic energy decreases immediately, while for the light mass it first grows up to more than three times the value at $\varepsilon_{01} = 0$ and only after that starts to decrease. The vibration amplitudes of x_1 are increased in the initial time interval and have a higher frequency compared to the case of $\varepsilon_{01} = 0$. Both effects leads to a faster motion and lead to an increase of the energies of motion. After some time the system cease to vibrate and the energies of motion vanish.

Figure 9.4 presents the time series of the energies that are absorbed or emitted by the damping elements. For the positive damping coefficient c_{01} energy is transferred from the motion to the damping element. The system loses energy, because it is absorbed by the damping element and the sign of the energy is positive. On the other hand, for the negative damping coefficient c_{02} energy is transferred from the environment into the system. The system gains energy, but because from the viewpoint of the damping element energy is lost, the sign of the energy is negative. For $\varepsilon_{01} = 0$ the self-excitation continuously pumps energy into the system at a slightly

9 Interaction of damping and parametric excitation 121

faster rate than it is dissipated by the positive damping element. This mechanism leads to the increasing amplitudes in Fig. 9.2. On the other hand, for $\varepsilon_{01} > \varepsilon_{crit}$ both, the absorbed and emitted energy of the damping elements reaches a stationary value, because the motion of the system comes to rest, and any further energy transfer is stopped.

The time series of the potential energy corresponding to the constant stiffness elements and of the energy resulting from the parametric excitation are shown in Fig. 9.5. The potential energy is by definition always positive. Due to the chosen initial conditions in (9.13) the initial energy values corresponding to k_{01}, k_{12}, k_{02} are equal to $\frac{1}{2}, 0, \frac{1}{2}$. For $\varepsilon_{01} = 0$ the amplitudes of the potential energies grows because of the growing amplitudes in Fig. 9.2. For $\varepsilon_{01} > \varepsilon_{crit}$ the amplitudes of the displacement of the light mass decreases only after they initially grow, and the potential energy W_{01}^k initially grows, too. The amplitudes of the heavy mass always decreases, which leads to decreasing amplitudes of the potential energy W_{02}^k. Initially, the potential energy of the coupling stiffness element k_{12} increases, although the amplitudes of the heavy mass diminish immediately, because this potential depends on the difference of both amplitudes and the amplitudes of the light mass initially grow, too. All potential energies vanish, when the system comes to rest. Finally, the energy resulting from the parametric excitation is obviously zero for $\varepsilon_{01} = 0$. For activated parametric excitation, $\varepsilon_{01} \neq 0$, the energy is negative, which means that energy flows into the system. The system gains energy from parametric excitation, and the energy flow stops as soon as the system ceases to vibrate.

Figure 9.6 presents a comparison of the total amounts of the energy portions, which correspond to the different types of the system parameters. The sum of all energy portions resulting from the constant and time-periodic stiffness parameters is equal to W_{tot}^k and plotted in Fig. 9.5. The total amount of energy which is dissipated in or feeded by the constant damping parameters is equal to W_{tot}^c and shown in Fig. 9.4. Finally, W_{tot}^m is the sum of the kinetic energy portions from Fig. 9.3. It is worth to note, that the total energy W_{tot}, which is the sum of these energy portions, is constant at any time respecting the principle of energy conservation and that the value equals to its initial value of one.

In the conventional system, $\varepsilon_{01} = 0$, two devices cause an exchange of energy between the system and the environment, the both damping elements c_{01} and c_{02}. The positive damping parameter yields a loss in the kinetic energy and the negative damping element pumps energy into the system. For the exemplary values chosen the pumping mechanism is more powerful than the dissipation. Therefore, the system continuously gains energy that is conversed into kinetic energy and leads to growing amplitudes of the potential as well as of the kinetic energy. It is worth to note, that the total energy W_{tot}^c is positive at the beginning and becomes negative only after some time. This arises from the difference between the partial energy portions W_{01}^c and W_{02}^c in Fig. 9.4. In the first time steps, the energy dissipation works more effective than the pumping of energy, because the positive damping element is attached to the light mass m_1 that vibrates more easily and whose velocity varies more often in the beginning.

Besides both damping elements, activating a proper parametric excitation at a parametric anti-resonance and for $\varepsilon_{01} > \varepsilon_{crit}$ introduces a further exchange of energy between the system and its environment. As shown in Fig. 9.5 this energy is negative, which means that the parametric excitation is a mechanism that pumps energy into the system. Hence, introducing a parametric anti-resonance excitation in a vibrating systems seems to contribute to the dynamical destabilization of the system in the same way as the self-excitation modelled by the negative damping element. Indeed, parametric excitation pumps energy into the system, but contrary to the negative damping element this energy is mainly transformed to higher frequency vibrational motion near the positive damping element, than to a considerable increase of the amplitudes of the displacement. Illustratively speaking, this higher frequency helps to utilize the positive damping more effectively, while the operation of the negative damping element remains mainly unchanged. This interpretation is underpinned by the equal slopes of the time history of W_{02}^c in Fig. 9.4.

The accumulated energy that flows across the three open system borders is shown in the bottom figure in Fig. 9.6. A negative value means that energy is pumped into the system. For $\varepsilon_{01} = 0$ the energy takes positive values in the first time steps, and the vibration amplitudes are slightly reduced, as the time history of x_1 in Fig. 9.2 confirms. After that this energy becomes negative and starts to transfer energy into the motion. The amplitudes are increases. For $\varepsilon_{01} > \varepsilon_{crit}$ this behavior changes. Now, the accumulated energy is negative within the first time steps and the vibration amplitudes are amplified, as the time history of x_1 in Fig. 9.2 confirms. After a short time the dissipative mechanism resulting from the parametric excitation becomes effective. The system comes to rest when the initial potential energy of one is completely dissipated.

Summarizing, only the positive damping element is capable of dissipating energy, while two mechanisms feed the system continuously with energy, the parametric excitation and the self-excitation modelled by a negative damping element. If the frequency of the parametric excitation is tuned near a parametric anti-resonance frequency, then the kinetic energy in the system decreases until the system comes to rest. On the other hand, for deactivated parametric excitation the amount of energy that is pumped into the system by the negative damping element cannot be dissipated by the positive damping element, and the system is dynamically unstable. The different slopes of the total energy W_{tot}^c point out the efficiency of a properly introduced parametric excitation. For the same constant damping parameters, the slope becomes negative in case of an activated parametric excitation. Hence, the effective positive damping is amplified. Therefore, we can call this the mechanism of damping by parametric excitation.

9 Interaction of damping and parametric excitation

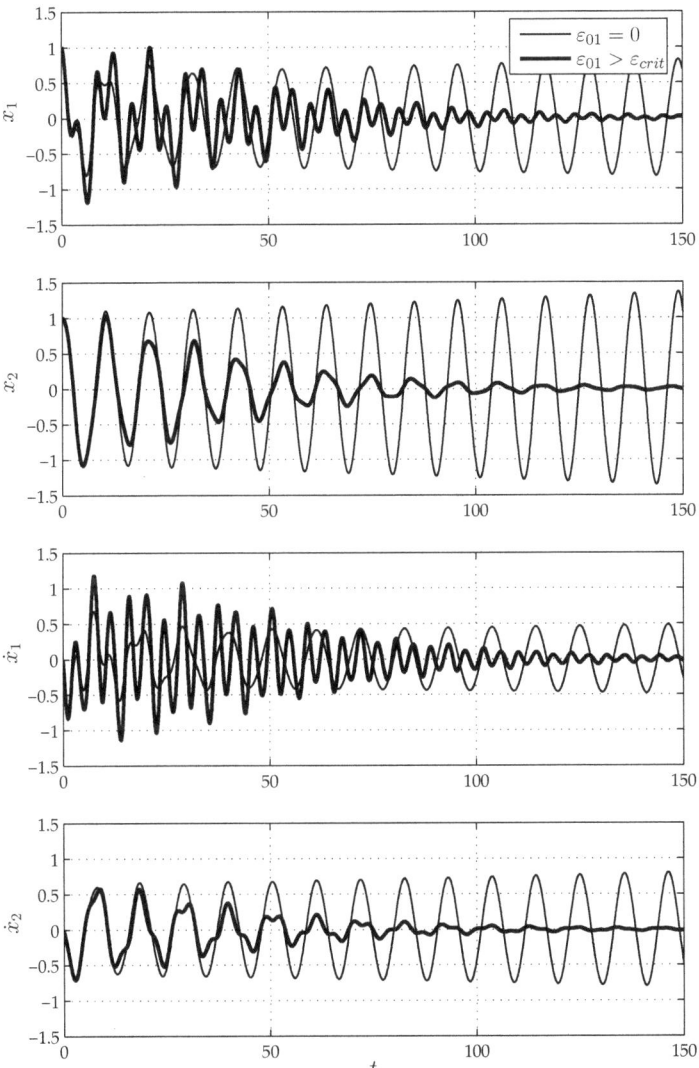

Figure 9.2.: Time series of system states for different amplitudes of parametric excitation ε_{01} and negative damping coefficient c_{02}.

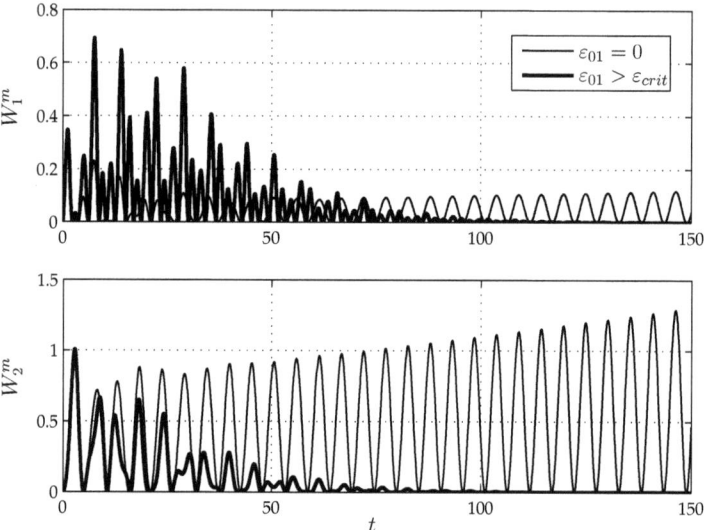

Figure 9.3.: Time series of energy of motion for different amplitudes of parametric excitation ε_{01} and negative damping coefficient c_{02}.

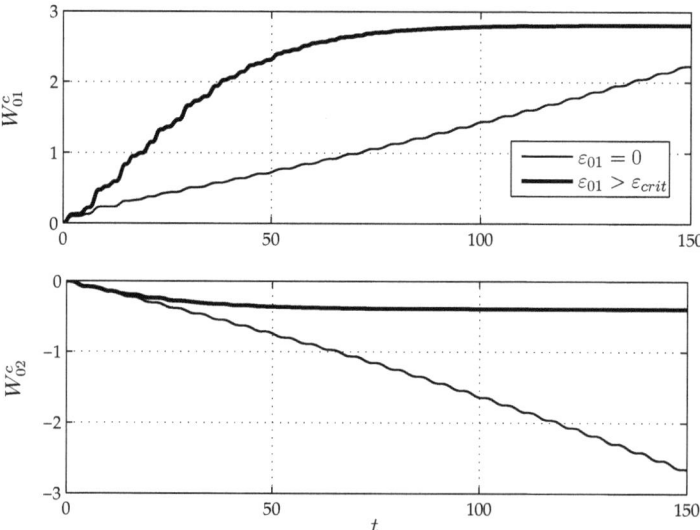

Figure 9.4.: Time series of damping energy for different amplitudes of parametric excitation ε_{01} and negative damping coefficient c_{02}.

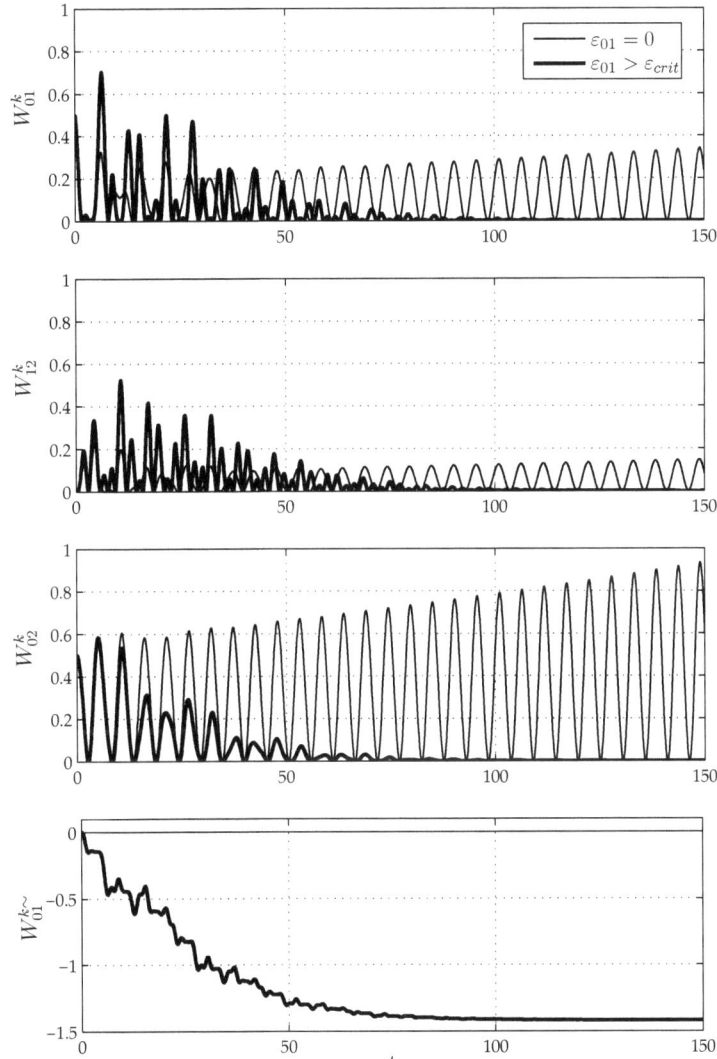

Figure 9.5.: Time series of potential energy for different amplitudes of parametric excitation ε_{01} and negative damping coefficient c_{02}.

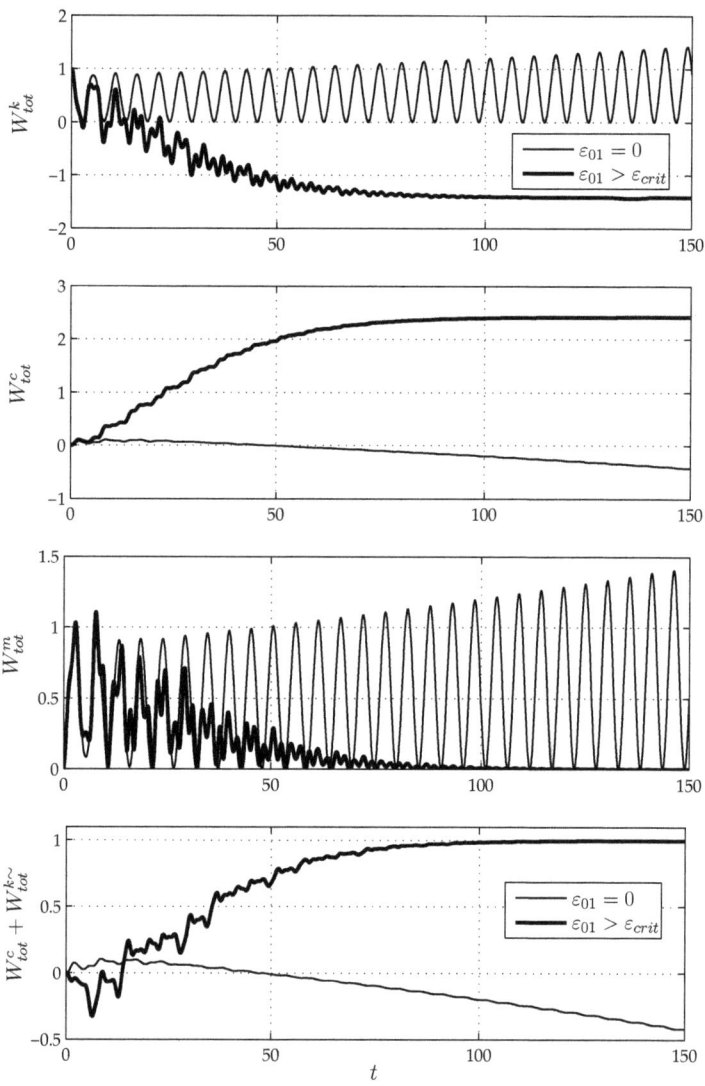

Figure 9.6.: Time series of total energy for different amplitudes of parametric excitation ε_{01} and negative damping coefficient c_{02}.

9.2.2. Positive damping

In the previous section we investigated the mechanism of damping by parametric excitation in the presence of a self-excitation, modelled by a negative damping element. The capability of the presented damping mechanism is not restricted to negative values of a damping parameter, although this is the working field where the effectiveness of the proposed method is very impressive. In this section we analyze the previously shown two-mass system with the same values for the system parameters as in (9.11), but now with a positive damping element $c_{02} = 0.07$.

The time series of the system states are shown in Fig. 9.7. For the system without parametric excitation, $\varepsilon_{01} = 0$, both damping coefficients are positive and the damping elements dissipates kinetic energy. For activated parametric excitation, $\varepsilon_{01} > \varepsilon_{crit}$, the amplitude of the displacement x_1 of the light mass are increased in the first time steps and then starts to decrease, while the amplitudes of the heavy are reduced immediately. The same situation as in the presence of a negative damping element in Fig. 9.2 occurs.

A comparison of the total partial energy portions is plotted in Fig. 9.8. The system possesses three open boundaries, where energy can flow across. The both damping elements always dissipate energy while the mechanism of parametric excitation always pumps energy into the system. The same time history is obtained as in Fig. 9.2 for a negative damping coefficient, except that now W_{tot}^c is always positive. Comparing the two initial slopes of W_{tot}^c for the cases with and without parametric excitation suggests that the introduced parametric excitation uses the damping elements 100% more efficiently than a passive system without parametric excitation. For the resulting energy that flows across the system boundary we have to add the energy that is pumped into the system by parametric excitation, as shown in the bottom figure in Fig. 9.8. It is important to mention that the parametric excitation has to persist for a certain time, because the parametric excitation amplifies the vibrational amplitudes in the first time steps, as indicated by an even negative total energy that flows across the system boundaries. Only after a certain time interval has expired, the parametric excitation starts to have a positive impact on the dynamical stabilization of the system.

By introducing a proper parametric excitation the system comes to rest in about half of the time compared to a passively damped system with $\varepsilon_{01} = 0$. Hence, we can interpret the mechanism of damping by parametric excitation as a selective energy insertion, that helps to improve, especially to amplify, the dissipation by the positive damping elements in the system.

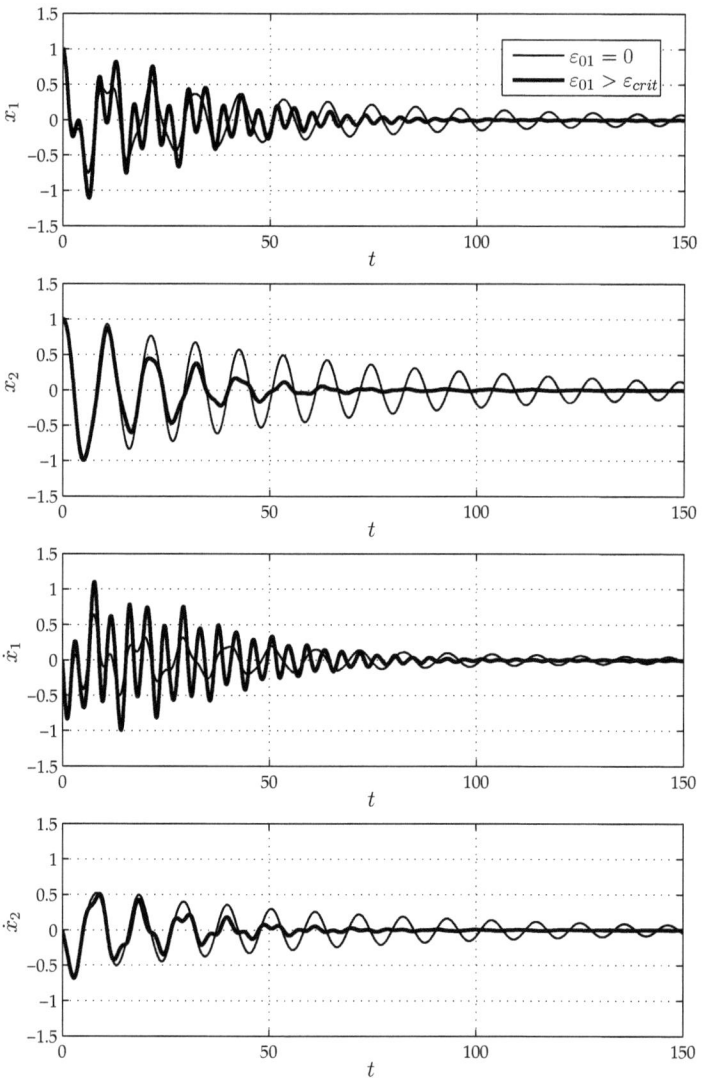

Figure 9.7.: Time series of system states for different amplitudes of parametric excitation ε_{01} and positive damping coefficient c_{02}.

9 Interaction of damping and parametric excitation

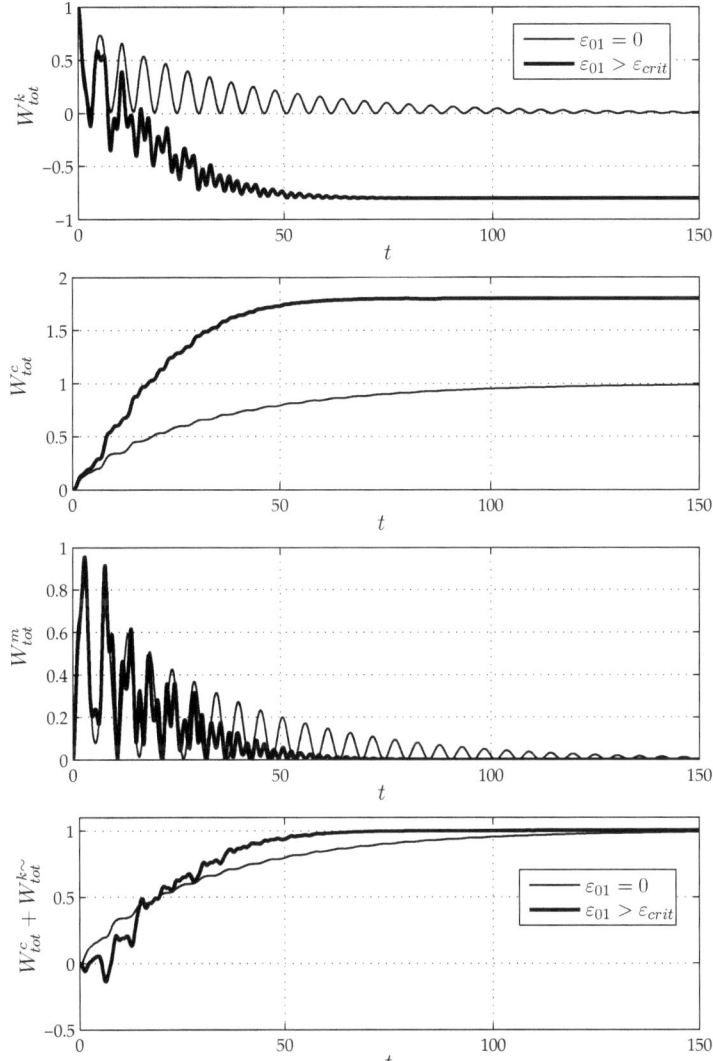

Figure 9.8.: Time series of total energy for different amplitudes of parametric excitation ε_{01} and positive damping coefficient c_{02}.

10. Optimization of periodic functions

The stability boundary curves determine values of the parametric excitation frequency η for which vibration suppression can be achieved. These curves define a dense frequency interval of η within which damping by parametric excitation may occur. The strongest damping is achieved if the value of η is chosen at the center η_0 of this interval. Applying the formulae derived in Part I induces that the parametric anti-resonance frequency η_0 is exactly equal to a parametric combination frequency of the system, as summarized in Tab. 3.1 on page 36. However, numerical calculations show that the real stability boundary curve is shifted slightly from the analytical predictions, see for example Fig. 12.3 on page 184. Nevertheless, the width of the numerically obtained frequency interval is equal to the analytical one (see Fig. 12.3). Hence, analyzing the analytical expressions in Part I is as meaningful as performing numerical analysis in order to optimize a desired system. Furthermore, due to the slight frequency shift the center of the stable frequency interval has to be calculated separately for each value set of system parameters in case of a numerical analysis, while the analytical expressions allow a direct optimization.

The optimization performed in this chapter is restricted to systems for which the stability boundary curves can be approximated adequately by a first order perturbation.

10.1. Optimal multi-location harmonic stiffness variation

In this section we analyze a mechanical system with two degrees of freedom with linear stiffness and damping coefficients as shown in Fig. 10.1. This investigation is an analytical generalization of the numerical studies performed in [25]. In addition to the system as in [25] a coupling of both masses through the damping coefficient c_{12} is added, to obtain the most general linear two degrees of freedom system.

Any of the stiffness coefficients is harmonically changed with the same frequency ω and the same periodic shape function, but with different relative phase angles:

$$k_{01}(t) = k_{01}(1 + \varepsilon_{01}\cos(\omega t + 0)), \qquad (10.1)$$
$$k_{12}(t) = k_{12}(1 + \varepsilon_{12}\cos(\omega t + \alpha_{12})) = k_{12} + \hat{k}_{12}\cos\alpha_{12}\cos(\omega t) - \hat{k}_{12}\sin\alpha_{12}\sin(\omega t),$$
$$k_{02}(t) = k_{02}(1 + \varepsilon_{02}\cos(\omega t + \alpha_{02})) = k_{02} + \hat{k}_{02}\cos\alpha_{02}\cos(\omega t) - \hat{k}_{02}\sin\alpha_{02}\sin(\omega t),$$

with the abbreviations $\hat{k}_{ij} = \varepsilon_{ij} k_{ij}$. The variation of the stiffness coefficient $k_{01}(t)$ is the reference variation. Using the same shape function for all time-periodic stiffness coefficients in (10.1) means, that the expressions ε_{kl} contain the corresponding prefactors from the leading Fourier coefficient of a chosen periodic shape function, according to the results obtained in the previous section and Chapter 3.3. Variables ε_{kl} are the amplitude amplification factors of the corresponding stiffness parameters $\hat{k}_{kl}$. The equations of motion can be written as

$$\mathbf{M}\ddot{\mathbf{x}}(t) + \mathbf{C}\dot{\mathbf{x}}(t) + \mathbf{K}(t)\mathbf{x}(t) = \mathbf{0}, \qquad (10.2)$$

where

$$\mathbf{M} = \begin{bmatrix} m_1 & 0 \\ 0 & m_2 \end{bmatrix}, \quad \mathbf{C} = \begin{bmatrix} c_{01} + c_{12} & -c_{12} \\ -c_{12} & c_{12} + c_{02} \end{bmatrix},$$

$$\mathbf{K}(t) = \begin{bmatrix} k_{01}(t) + k_{12}(t) & -k_{12}(t) \\ -k_{12}(t) & k_{12}(t) + k_{02}(t) \end{bmatrix}.$$

According to (10.1) the time-periodic stiffness matrix can be split into a constant part, the mean value, and two harmonic parts

$$\mathbf{K}(t) = \mathbf{K}_0 + \mathbf{K}_c \cos(\omega t) + \mathbf{K}_s \sin(\omega t), \qquad (10.3)$$

where

$$\mathbf{K}_0 = \begin{bmatrix} k_{01} + k_{12} & -k_{12} \\ -k_{12} & k_{12} + k_{02} \end{bmatrix},$$

$$\mathbf{K}_c = \begin{bmatrix} \hat{k}_{01} + \hat{k}_{12} \cos \alpha_{12} & -\hat{k}_{12} \cos \alpha_{12} \\ -\hat{k}_{12} \cos \alpha_{12} & \hat{k}_{12} \cos \alpha_{12} + \hat{k}_{02} \cos \alpha_{02} \end{bmatrix}, \qquad (10.4)$$

$$\mathbf{K}_s = \begin{bmatrix} -\hat{k}_{12} \sin \alpha_{12} & \hat{k}_{12} \sin \alpha_{12} \\ \hat{k}_{12} \sin \alpha_{12} & -\hat{k}_{12} \sin \alpha_{12} - \hat{k}_{02} \sin \alpha_{02} \end{bmatrix}.$$

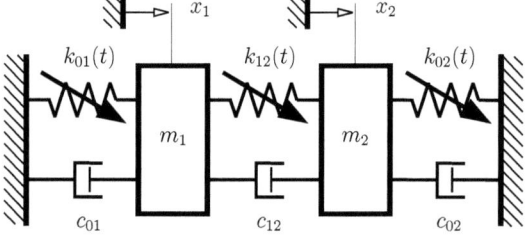

Figure 10.1.: Mechanical system with multi-location stiffness variation.

10 Optimal multi-location harmonic stiffness variation

Applying a transformation matrix $\mathbf{T}$ so that matrix $\mathbf{M}^{-1}\mathbf{K}_0$ becomes diagonal

$$\mathbf{T}^{-1}\mathbf{M}^{-1}\mathbf{K}_0\mathbf{T} = \begin{bmatrix} \Omega_1^2 & 0 \\ 0 & \Omega_2^2 \end{bmatrix} = \mathbf{\Omega}^2, \qquad \varepsilon\mathbf{\Theta} = \mathbf{T}^{-1}\mathbf{M}^{-1}\mathbf{C}\mathbf{T},$$

$$\varepsilon\mathbf{Q} = \mathbf{T}^{-1}\mathbf{M}^{-1}\mathbf{K}_c\mathbf{T}, \qquad \varepsilon\mathbf{P} = \mathbf{T}^{-1}\mathbf{M}^{-1}\mathbf{K}_s\mathbf{T}, \qquad (10.5)$$

transforms the equations of motion into its quasi-normal form

$$\ddot{\mathbf{x}}(t) + \varepsilon\mathbf{\Theta}\dot{\mathbf{x}}(t) + \mathbf{\Omega}^2\mathbf{x}(t) = -\varepsilon\left(\mathbf{Q}\cos(\omega t) + \mathbf{P}\sin(\omega t)\right)\mathbf{x}(t). \qquad (10.6)$$

Introducing phase relationships in a single harmonic stiffness variation corresponds to a general harmonic stiffness variation as investigated in Chapter 3.2 and which finally led to a stability width according to (3.40). Inserting the matrix entries from (10.4) into (10.5) reveals that for the diagonal inertia matrix $\mathbf{M}$ in (10.2) the relation

$$Q_{12}P_{21} - P_{12}Q_{21} = 0$$

is satisfied. In this case the qualitative behavior of the system is equivalent to the single harmonic stiffness variation in Chapter 3.1 and the formulae for the stability width simplify to

$$\sigma_1 = \frac{\Theta_{11} + \Theta_{22}}{2}\sqrt{-1 - \frac{Q_{12}Q_{21} + P_{12}P_{21}}{4\Omega_1\Omega_2\Theta_{11}\Theta_{22}}},$$

$$\omega_{crit} = |\Omega_1 - \Omega_2| \pm \varepsilon\sigma_1, \qquad (10.7)$$

that is similar to (3.24) but with the additional term $P_{12}P_{21}$.

In order to maximize the robustness with respect to changes of the frequency ω the stability interval $[|\Omega_1 - \Omega_2| - \varepsilon\sigma_1, |\Omega_1 - \Omega_2| + \varepsilon\sigma_1]$ has to be maximized. Allowing only changes of the phase angles (α_{12}, α_{02}) and amplitude amplification factors (ε_{01}, ε_{12}, ε_{02}) of the parametric excitations terms, the stability width $\Delta\omega$ has to be maximized:

$$\Delta\omega = 2\varepsilon\sigma_1 \to \max.$$

For the investigated system in Fig. 10.1 the system matrices in (10.4) are symmetric. In this case, as shown in Chapter 3.1, the parametric excitation term $Q_{12}Q_{21} + P_{12}P_{21}$ in (10.7) is always positive. If both modal damping parameters Θ_{ii} have the same sign, then σ_1, as defined above, becomes purely imaginary valued and the stability width is infinite. For such modal damping parameters a real-valued σ_1 can be achieved only by adopting the stability formulae as outlined in Section 8.3. Without any loss of generality, we restrict the further optimization to the case where one modal damping is negative. The modal damping parameters have different signs and σ_1 can become real-valued, too. To maximize a real-valued σ_1 under the restrictions from above, we have to maximize the parametric excitation term. We define a quality function

J proportional to the parametric excitation term as a function of the phase angles and the amplitude amplification factors

$$Q_{12}Q_{21} + P_{12}P_{21} \propto \frac{f(\alpha_{12}, \alpha_{02}, \varepsilon_{01}, \varepsilon_{12}, \varepsilon_{02})}{d_0 > 0} = J(\alpha_{12}, \alpha_{02}, \varepsilon_{01}, \varepsilon_{12}, \varepsilon_{02}) = J \to \max. \quad (10.8)$$

Inserting the matrix entries from (10.4) into (10.5) and (10.8) results, after a lengthy calculation, in

$$J = -(1-a)\varepsilon_{01}\varepsilon_{12}\cos\alpha_{12} - a\varepsilon_{01}\varepsilon_{02}\cos\alpha_{02} + a(1-a)\varepsilon_{12}\varepsilon_{02}\cos(\alpha_{12} - \alpha_{02})$$
$$+ \frac{1}{2}\left(\varepsilon_{01}^2 + (1-a)^2\varepsilon_{12}^2 + a^2\varepsilon_{02}^2\right) \to \max, \quad (10.9)$$

with the abbreviation

$$\boxed{a = Q^2 = \frac{m_1 k_{02}}{m_2 k_{01}} = \left(\frac{\omega_2}{\omega_1}\right)^2,} \quad (10.10)$$

where a is defined as a frequency ratio. Note that the quality function J does neither dependent on the coupling stiffness k_{12} nor on the damping parameters.

10.1.1. Optimal phase angles

First we search for optimal phase relations $(\alpha_{12}, \alpha_{02})$ for arbitrary amplitude amplification factors $(\varepsilon_{01}, \varepsilon_{12}, \varepsilon_{02})$. Due to the fact that the parameters $(\alpha_{12}, \alpha_{02})$ occur only linearly in the argument of cosine functions, the quality function J is periodic with respect to these phase angles with a period of 2π. Furthermore, the quality function is symmetric with respect to $0, \pi, 2\pi, ...$, as the cosine function is. Any possible values of the quality function lie in the domain $[0, \pi] \times [0, \pi]$. The first derivative of J in (10.9) yields the conditions for extremal values of the quality function. The horizontal tangential planes yield

$$\frac{\partial J}{\partial \alpha_{12}} = 0 : \quad (1-a)\varepsilon_{12}\left(\varepsilon_{01}\sin\hat{\alpha}_{12} - a\varepsilon_{02}\sin(\hat{\alpha}_{12} - \hat{\alpha}_{02})\right) = 0, \quad (10.11a)$$

$$\frac{\partial J}{\partial \alpha_{02}} = 0 : \quad a\varepsilon_{02}\left(\varepsilon_{01}\sin\hat{\alpha}_{02} + (1-a)\varepsilon_{12}\sin(\hat{\alpha}_{12} - \hat{\alpha}_{02})\right) = 0. \quad (10.11b)$$

Trivial solutions

For $a, \varepsilon_{01}, \varepsilon_{12}, \varepsilon_{02} > 0$, $a \neq 1$ we can rewrite the conditions above as

$$\sin(\hat{\alpha}_{12} - \hat{\alpha}_{02}) = \frac{1}{a}\frac{\varepsilon_{01}}{\varepsilon_{02}}\sin\hat{\alpha}_{12} = -\frac{1}{1-a}\frac{\varepsilon_{01}}{\varepsilon_{12}}\sin\hat{\alpha}_{02}, \quad (10.12)$$

which yields the trivial solutions

$$\begin{pmatrix}\hat{\alpha}_{12}\\ \hat{\alpha}_{02}\end{pmatrix} = \left\{\begin{pmatrix}0\\0\end{pmatrix}, \begin{pmatrix}0\\ \pi\end{pmatrix}, \begin{pmatrix}\pi\\0\end{pmatrix}, \begin{pmatrix}\pi\\ \pi\end{pmatrix}\right\}. \quad (10.13)$$

10 Optimal multi-location harmonic stiffness variation

This result is already clear from the symmetry properties of the quality function with respect to 0 and π. The symmetry leads to horizontal tangential planes at the domain borders. For $a = 1$ and also for $\varepsilon_{12} = 0$ the expressions in (10.9) simplify to

$$J = -a\varepsilon_{01}\varepsilon_{02}\cos\hat{\alpha}_{02} + \text{const},\qquad(10.14)$$

and the condition for extremal values becomes

$$\sin\alpha_{02} = 0:\quad \hat{\alpha}_{02} = n\pi,\quad \alpha_{12}\text{ any value},\qquad(10.15)$$

with $n \in \mathbb{Z}$. For $\varepsilon_{02} = 0$ (10.9) becomes

$$J = -(1-a)\varepsilon_{01}\varepsilon_{12}\cos\hat{\alpha}_{12} + \text{const},$$

and for $a = 0$ (10.9) becomes

$$J = -\varepsilon_{01}\varepsilon_{12}\cos\hat{\alpha}_{12} + \text{const}.$$

and the condition for extremal values are now

$$\sin\alpha_{12} = 0:\quad \hat{\alpha}_{12} = n\pi\quad \alpha_{02}\text{ any value}.\qquad(10.16)$$

Finally, for $\varepsilon_{01} = 0$ in (10.9) we obtain

$$J = a(1-a)\varepsilon_{12}\varepsilon_{02}\cos(\hat{\alpha}_{12} - \hat{\alpha}_{02}) + \text{const},$$

and the conditions for extremal values are

$$\sin(\hat{\alpha}_{12} - \hat{\alpha}_{02}) = 0:\quad \hat{\alpha}_{12} - \hat{\alpha}_{02} = n\pi.$$

In this case the stiffness coefficient k_{01} is kept constant. Introducing a new reference variation by defining $\alpha_{12} = 0$ gives again the solution $\hat{\alpha}_{02} = (-)n\pi$, as in the case $\varepsilon_{12} = 0$ in (10.15).

Non-trivial solution

For $a, \varepsilon_{01}, \varepsilon_{12}, \varepsilon_{02} > 0$, $a \neq 1$ the non-trivial solution derived from (10.12) is

$$\hat{\alpha}_{12} = \arcsin\left(-\frac{a}{1-a}\frac{\varepsilon_{02}}{\varepsilon_{12}}\sin\hat{\alpha}_{02}\right).$$

Inserting into (10.12) gives

$$\sin(\hat{\alpha}_{12} - \hat{\alpha}_{02}) = \sin\left(\arcsin\left(-\frac{a}{1-a}\frac{\varepsilon_{02}}{\varepsilon_{12}}\sin\hat{\alpha}_{02}\right) - \hat{\alpha}_{02}\right)$$

$$= -\sin\hat{\alpha}_{02}\left(\frac{a}{1-a}\frac{\varepsilon_{02}}{\varepsilon_{12}}\cos\hat{\alpha}_{02} + \sqrt{1 - \left(\frac{a}{1-a}\frac{\varepsilon_{02}}{\varepsilon_{12}}\right)^2\sin^2\hat{\alpha}_{02}}\right)$$

$$= -\frac{1}{(1-a)}\frac{\varepsilon_{01}}{\varepsilon_{12}}\sin\hat{\alpha}_{02},$$

by using again (10.12) for the last equation. Neglecting the trivial solutions $\alpha_{02} = n\pi$, that are already covered by the harmonic stiffness variation in Chapter 3.2, yields after a lengthy calculation finally

$$\cos \hat{\alpha}_{02} = \frac{\varepsilon_{01}^2 - (1-a)^2 \varepsilon_{12}^2 + a^2 \varepsilon_{02}^2}{2a\varepsilon_{01}\varepsilon_{02}}. \tag{10.17}$$

Determining the optimal phase angle $\hat{\alpha}_{12}$ from the condition above

$$\sin \hat{\alpha}_{12} = -\frac{a}{1-a} \frac{\varepsilon_{02}}{\varepsilon_{12}} \sqrt{1 - \cos^2 \hat{\alpha}_{02}}$$

results in very extensive terms, but applying the same procedure as for $\hat{\alpha}_{02}$ yields

$$\hat{\alpha}_{02} = \arcsin\left(-\frac{1-a}{a}\frac{\varepsilon_{12}}{\varepsilon_{02}} \sin \hat{\alpha}_{12}\right),$$

and respecting (10.12)

$$\sin(\hat{\alpha}_{12} - \hat{\alpha}_{02}) = \sin\left(\hat{\alpha}_{12} - \arcsin\left(-\frac{1-a}{a}\frac{\varepsilon_{12}}{\varepsilon_{02}} \sin \hat{\alpha}_{12}\right)\right)$$

$$= \sin \hat{\alpha}_{12} \left(\frac{1-a}{a}\frac{\varepsilon_{12}}{\varepsilon_{02}} \cos \hat{\alpha}_{12} + \sqrt{1 - \left(\frac{1-a}{a}\frac{\varepsilon_{12}}{\varepsilon_{02}}\right)^2 \sin^2 \hat{\alpha}_{12}}\right)$$

$$= \frac{1}{a}\frac{\varepsilon_{01}}{\varepsilon_{02}} \sin \hat{\alpha}_{12},$$

results in the much simpler analytical expression

$$\cos \hat{\alpha}_{12} = \frac{\varepsilon_{01}^2 + (1-a)^2 \varepsilon_{12}^2 - a^2 \varepsilon_{02}^2}{2(1-a)\varepsilon_{01}\varepsilon_{12}}. \tag{10.18}$$

Further useful relations are

$$\frac{\cos \hat{\alpha}_{02}}{\cos \hat{\alpha}_{12}} = \frac{(1-a)}{a}\frac{\varepsilon_{12}}{\varepsilon_{02}}\frac{\varepsilon_{01}^2 - (1-a)^2 \varepsilon_{12}^2 + a^2 \varepsilon_{02}^2}{\varepsilon_{01}^2 + (1-a)^2 \varepsilon_{12}^2 - a^2 \varepsilon_{02}^2} = \frac{(1-a)}{a}\frac{\varepsilon_{12}}{\varepsilon_{02}} A$$

$$= -\frac{\sin \hat{\alpha}_{02}}{\sin \hat{\alpha}_{12}} A = +\frac{1-a}{a}\frac{\varepsilon_{12}}{\varepsilon_{02}} A,$$

$$\tan \hat{\alpha}_{02} = -A \tan \hat{\alpha}_{12}.$$

With (10.18), (10.17) and

$$\sin \hat{\alpha}_{12} = \sqrt{1 - \cos^2 \hat{\alpha}_{12}} = -\frac{a}{1-a}\frac{\varepsilon_{02}}{\varepsilon_{12}}\sqrt{1 - \cos^2 \hat{\alpha}_{02}},$$

$$\cos(\hat{\alpha}_{12} - \hat{\alpha}_{02}) = \cos \hat{\alpha}_{12} \cos \hat{\alpha}_{02} \pm \sqrt{(1-\cos^2 \hat{\alpha}_{02})(1-\cos^2 \hat{\alpha}_{12})}$$

$$= \cos \hat{\alpha}_{12} \cos \hat{\alpha}_{02} \mathop{-}_{(+)} \left(-\frac{a}{1-a}\frac{\varepsilon_{02}}{\varepsilon_{12}}\right)(1-\cos^2 \hat{\alpha}_{02}),$$

the quality function at the non-trivial extremal phase angles becomes

$$J(\hat{\alpha}_{12}, \hat{\alpha}_{02}) = -\varepsilon_{01}^2 \pm \left(-\frac{a}{1-a}\frac{\varepsilon_{02}}{\varepsilon_{12}}\right)$$

$$+ \frac{1}{4\varepsilon_{01}^2}\left(\varepsilon_{01}^2 + (1-a)^2 \varepsilon_{12}^2 - a^2 \varepsilon_{02}^2\right)\left(\varepsilon_{01}^2 - (1-a)^2 \varepsilon_{12}^2 + a^2 \varepsilon_{02}^2\right)$$

$$\pm \frac{1}{4\varepsilon_{01}^2}\frac{1}{1-a}\frac{1}{a\varepsilon_{12}\varepsilon_{02}}\left(\varepsilon_{01}^2 - (1-a)^2 \varepsilon_{12}^2 + a^2 \varepsilon_{02}^2\right)^2. \tag{10.19}$$

10 Optimal multi-location harmonic stiffness variation

Note that performing this analysis the optimal quality function J is not longer explicitly dependent on a harmonic function and becomes a much simpler algebraic expression.

Maximum quality function

The quality function (10.9) is extremal at the above values of the phase angles. To decide whether the quality function is maximized the additional condition

$$\tilde{J}(\alpha_{12}, \alpha_{02}) = \begin{vmatrix} \dfrac{\partial^2 J}{\partial \alpha_{12} \partial \alpha_{12}} & \dfrac{\partial^2 J}{\partial \alpha_{12} \partial \alpha_{02}} \\ \dfrac{\partial^2 J}{\partial \alpha_{02} \partial \alpha_{12}} & \dfrac{\partial^2 J}{\partial \alpha_{02} \partial \alpha_{02}} \end{vmatrix}$$

$$= a(1-a)\varepsilon_{01}\varepsilon_{12}\varepsilon_{02}\left[-(1-a)\varepsilon_{12}\cos\alpha_{12}\cos(\alpha_{12}-\alpha_{02})\right.$$
$$\left. -a\varepsilon_{02}\cos\alpha_{02}\cos(\alpha_{12}-\alpha_{02}) + \varepsilon_{01}\cos\alpha_{12}\cos\alpha_{02}\right] > 0, \qquad (10.20)$$

has to be satisfied. With the abbreviation

$$c = a(1-a)\varepsilon_{01}\varepsilon_{12}\varepsilon_{02},$$

this expression has the following values at the different trivial phase angles from (10.13):

$$\tilde{J}(0,0) = c\left(+\varepsilon_{01} - \varepsilon_{12}(1-a) - a\varepsilon_{02}\right),$$
$$\tilde{J}(\pi,0) = c\left(-\varepsilon_{01} - \varepsilon_{12}(1-a) + a\varepsilon_{02}\right),$$
$$\tilde{J}(0,\pi) = c\left(-\varepsilon_{01} + \varepsilon_{12}(1-a) - a\varepsilon_{02}\right),$$
$$\tilde{J}(\pi,\pi) = c\left(+\varepsilon_{01} + \varepsilon_{12}(1-a) + a\varepsilon_{02}\right). \qquad (10.21)$$

For arbitrary but positive values of $a, \varepsilon_{01}, \varepsilon_{12}, \varepsilon_{02} \neq 0$, the following case differentiation can be made for the trivial solutions from (10.13)

$$a = 0 \quad \text{or} \quad \varepsilon_{02} = 0: \quad \alpha_{12} = \pi \quad \text{and} \quad \text{arbitrary } \alpha_{02},$$

$$0 < a < 1: \quad c > 1 \quad \text{and} \quad \begin{pmatrix} \alpha_{12} \\ \alpha_{02} \end{pmatrix}_{optimal} = \begin{pmatrix} \pi \\ \pi \end{pmatrix},$$

$$a = 1 \quad \text{or} \quad \varepsilon_{12} = 0: \quad \text{arbitrary } \alpha_{12} \quad \text{and} \quad \alpha_{02} = \pi,$$

$$a > 1: \quad c < 1 \quad \text{and} \quad \begin{pmatrix} \alpha_{12} \\ \alpha_{02} \end{pmatrix}_{optimal} = \begin{pmatrix} 0 \\ \pi \end{pmatrix}. \qquad (10.22)$$

This result can also easily be concluded from the quality function in (10.9) at the different extremal phase angles

$$J(0,0) = \frac{1}{2}\left(-\varepsilon_{01} + \varepsilon_{12}(1-a) + a\varepsilon_{02}\right)^2,$$

$$J(\pi,0) = \frac{1}{2}\left(-\varepsilon_{01} - \varepsilon_{12}(1-a) + a\varepsilon_{02}\right)^2,$$

$$J(0,\pi) = \frac{1}{2}\left(+\varepsilon_{01} - \varepsilon_{12}(1-a) + a\varepsilon_{02}\right)^2,$$

$$J(\pi,\pi) = \frac{1}{2}\left(+\varepsilon_{01} + \varepsilon_{12}(1-a) + a\varepsilon_{02}\right)^2. \tag{10.23}$$

The quality function is maximized if all terms in the bracket have equal sign. For $0 < a < 1$ equal signs are achieved for $J(\pi,\pi)$ and for $a > 1$ for $J(0,\pi)$, which coincides with the solution found in (10.22).

To obtain the optimal quality function for the non-trivial phase angles, we perform the following estimation, instead of expensively replacing the harmonic functions by their algebraic representations (10.17, 10.18): The possible maximum value of the quality function (10.9) is found by considering the boundedness of the cosine functions in the interval $[-1,+1]$ and assuming positive amplification factors $\varepsilon_{01}, \varepsilon_{12}, \varepsilon_{02}$ and a positive frequency ratio a

$$J^{\max} = |1-a|\,\varepsilon_{01}\varepsilon_{12} + a\varepsilon_{01}\varepsilon_{02} + a\,|1-a|\,\varepsilon_{12}\varepsilon_{02} + \frac{1}{2}\left(\varepsilon_{01}^2 + (1-a)^2\varepsilon_{12}^2 + a^2\varepsilon_{02}^2\right)$$

$$= \frac{1}{2}\left(\varepsilon_{01} + \varepsilon_{12}\,|1-a| + a\varepsilon_{02}\right)^2.$$

This expression was derived already for $J(0,\pi)$ and $J(\pi,\pi)$ respectively in (10.23), depending on the parameter value of a. Hence, the non-trivial extremal value $\tilde{J}(\hat{\alpha}_{12}, \hat{\alpha}_{02})$ from (10.19) can only be a side maximum.

Figures 10.2 and 10.3 show the quality function J as in (10.9) a function of the relative phase angles α_{12}, α_{02} for fixed amplitude amplification factors $\varepsilon_{01}, \varepsilon_{12}, \varepsilon_{02}$. Only positive amplification factors need to be considered, because a negative factor is already respected by a phase angle of π. The value ranges considered for the phase angles are normalized to the interval $[0,1]$. The variation of the amplitude amplification factors ε_{kl} in the particular sub-figure respects the constraint $\varepsilon_{01} + \varepsilon_{12} + \varepsilon_{02} = 0.6 = \text{const}$. The amplification factor ε_{01}, as the reference parametric excitation is kept constant. Hence, the constraint represents a linear dependency of the amplification factors ε_{12} and ε_{02} with a certain offset, $\varepsilon_{02} = 0.6 - \varepsilon_{12}$. Varying these amplification factors, while respecting the constraint condition above, deforms the resulting shape of the quality function.

The typical shapes of the quality function J for small values of a, $a < 1$, are plotted in Fig. 10.2. In this case the extremal values of the quality function occurs only at the trivial positions found in (10.13). Starting with Fig. 10.2a, a special case occurs where the amplification factor ε_{02} vanishes. Hence, the optimal phase angles are derived from (10.15) and simplify the dependance of the quality function on the phase angles, $J(\alpha_{12}, \alpha_{02}) = f(\alpha_{12})$. The independence of

10 Optimal multi-location harmonic stiffness variation

the optimum on the phase angle α_{02} results in a shape that is created by moving the remaining dependent part $f(\alpha_{12})$ parallel to the α_{02}-axis. For non-vanishing amplification factors ε_{kl} the optimal phase angles for the maximum value are found at the normalized position $(1,1)$, as derived analytically in (10.22), see Figs. 10.2bcd. Beside this optimal position the remaining three positions in (10.13) represent extremal positions, too. The relative location of the quality function at these suboptimal positions is examined in the following according to (10.23). The following relation between the function values at the positions $(1,0)$ and $(0,1)$, respectively, holds

$$J(\pi, 0) > J(0, \pi),$$
$$|-\varepsilon_{01} - \varepsilon_{12}(1-a) + a\varepsilon_{02}| > |+\varepsilon_{01} - \varepsilon_{12}(1-a) + a\varepsilon_{02}|$$

if

$$\frac{\varepsilon_{02}}{\varepsilon_{12}} < \frac{1-a}{a} (>0) \qquad (10.24)$$

is satisfied. The condition in (10.24) is fulfilled for Fig. 10.2b, $0.33 < 3$, and for Fig. 10.2c, $1 < 3$. In these cases the function value at $(1,0)$ is always higher than at $(0,1)$. In Fig. 10.2d the critical case occurs where both extrema have the same value, because the amplitude ratio $\varepsilon_{02}/\varepsilon_{12}$ is of equal size as the term $(1-a)/a$ containing the frequency ratio, $3 = 3$. On the other hand, the relation between the function values at the positions $(0,0)$ and $(0,1)$

$$J(0,0) < J(0, \pi),$$
$$|-\varepsilon_{01} + \varepsilon_{12}(1-a) + a\varepsilon_{02}| < |+\varepsilon_{01} - \varepsilon_{12}(1-a) + a\varepsilon_{02}|$$

is fulfilled if

$$\frac{\varepsilon_{01}}{\varepsilon_{12}} > 1 - a (> 0) \qquad (10.25)$$

is valid. Evaluating the condition (10.25) gives $0.66 > 0.75$ for Fig. 10.2b, $1 > 0.75$ for Fig. 10.2c, and $2 > 0.75$ for Fig. 10.2d. Hence, only in Fig. 10.2b J is slightly higher in position $(0,0)$ than in position $(0,1)$, as a closer look on the function shape verifies. For the special case where $\varepsilon_{12} = 0$ is satisfied, see Fig. 10.2e, the optimal phase angles are derived from (10.16). Now, in comparison with Fig. 10.2a the dependency of the quality function simplifies to $J(\alpha_{12}, \alpha_{02}) = g(\alpha_{12})$ and the complex shape degenerates to a shape that is created by shifting the curve $g(\alpha_{02})$ parallel to the α_{12}-axis.

If at least one amplification factor, ε_{12} or ε_{02}, vanishes, see Fig. 10.2a or e, then the function value J at the position $(1,1)$ remains optimal, but the dependance on the phase angles is simplified and the resulting shape degenerates. Performing the linear adaptation of the amplification factors $\varepsilon_{01}, \varepsilon_{12}, \varepsilon_{02}$ respecting the mentioned constraint reveals that the shape of the quality function J is transformed from one degenerated shape $f(\alpha_{12})$, through the general shapes in Figs. 10.2bcd, into the degenerated shape $g(\alpha_{02})$.

The typical shapes of the quality function J for big values of a, $a = 4 > 1$, are plotted in Fig. 10.3. In this case the extremal values of the quality function occurs at the trivial positions found in (10.13) as well as at the previously found non-trivial positions. Starting with the degenerate case in Fig. 10.3a, where $\varepsilon_{02} = 0$ is valid, the dependency of the quality function simplifies to $J(\alpha_{12}, \alpha_{02}) = f(\alpha_{12})$, according to (10.22). The function shapes for non-vanishing amplification factors ε_{kl} are shown in Figs. 10.3bcd. The optimum is found at the normalized position $(0, 1)$ according to the analytical result in (10.22). The relation between the function value at the position $(1, 0)$ and the optimal value at $(1, 1)$ is

$$J(\pi, 0) > J(\pi, \pi) \quad \text{for} \quad \frac{\varepsilon_{01}}{\varepsilon_{12}} < a - 1 \, (> 0) . \tag{10.26}$$

Evaluating the condition in (10.26) gives $0.66 < 3$ for Fig. 10.3b, $1 < 3$, for Fig. 10.3c, and $2 < 3$ for Fig. 10.3. Hence, for the sketched cases the function value at $(1, 1)$ is increasing from Fig 10.3a to Fig. 10.3d, but the value is always smaller than the value at $(1, 0)$, because the condition (10.26) is satisfied. On the other hand, the relation between the function values at the positions $(0, 0)$ and $(1, 1)$ is

$$J(0, 0) < J(\pi, \pi) \quad \text{for} \quad \frac{\varepsilon_{02}}{\varepsilon_{12}} > \frac{a-1}{a} \, (> 0) . \tag{10.27}$$

Evaluating the condition at the right hand side gives $0.33 < 0.75$ for Fig. 10.3b, but already in case of equal amplification this condition is violated, $1 \not< 0.75$ in Fig. 10.3c, and the declination angle of the function shape changes its sign. This comes visible clear in Fig. 10.3d where (10.27) gives $3 \not< 0.75$. Further adapting of the amplification factors leads to the degenerated case in Fig. 10.3e, where the dependency of the quality function simplifies to $J(\alpha_{12}, \alpha_{02}) = g(\alpha_{02})$.

Recapitulating the case of a large frequency ratio, $a > 1$, shows that the quality function of $J(\alpha_{12}, \alpha_{02})$ is optimal at the phase angles $(0, \pi)$, for general nonzero amplification factors $\varepsilon_{12}, \varepsilon_{02}$ as well as for the special cases of vanishing amplification factors. Recapitulating the case of a small frequency ratio, $a < 1$, reveals the same picture, but now the phase angles for the optimal function value of J are placed at (π, π). The question is, how does the shape of the function shape look like for the critical case $a = 1$? The special value $a = 1$ represents a symmetric arrangement with respect to the masses, $m_1 = m_2$, and the stiffness coefficients, $k_{01} = k_{02}$, that are attached to the inertial reference frame. Due to the previous results, the optimal value of the quality function is found at least at the phase angles $(0, \pi)$ and (π, π). Evaluating the quality function defined in (10.9) at this critical value of a we obtain

$$J|_{a=1} = \frac{1}{2} \left(\varepsilon_{01}^2 + \varepsilon_{02}^2 \right) - \varepsilon_{01} \varepsilon_{02} \cos \alpha_{02} . \tag{10.28}$$

Thus, the quality function loses its dependency on ε_{12} as well as on α_{12}, and is symmetric in $\varepsilon_{01}, \varepsilon_{02}$. For fixed amplification factors ε_{kl} the resulting function shape of J simplifies to the degenerated shapes similar to 10.2e or Figs. 10.3e, because the special cases $a = 1$ and $\varepsilon_{12} = 0$ coincide, as it was shown in (10.22). Plotting the quality function for different combinations of amplification factors as in Figs. 10.2 and 10.3 would always lead to a similar function shape.

10 Optimal multi-location harmonic stiffness variation

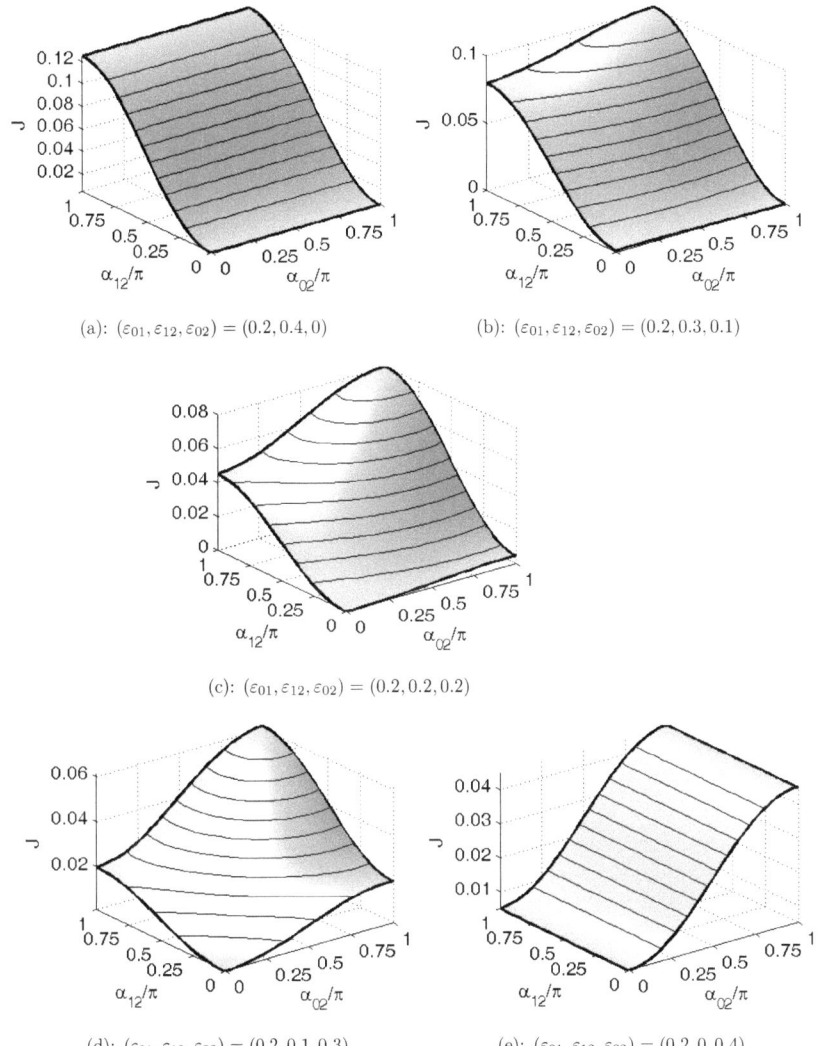

(a): $(\varepsilon_{01}, \varepsilon_{12}, \varepsilon_{02}) = (0.2, 0.4, 0)$

(b): $(\varepsilon_{01}, \varepsilon_{12}, \varepsilon_{02}) = (0.2, 0.3, 0.1)$

(c): $(\varepsilon_{01}, \varepsilon_{12}, \varepsilon_{02}) = (0.2, 0.2, 0.2)$

(d): $(\varepsilon_{01}, \varepsilon_{12}, \varepsilon_{02}) = (0.2, 0.1, 0.3)$

(e): $(\varepsilon_{01}, \varepsilon_{12}, \varepsilon_{02}) = (0.2, 0, 0.4)$

Figure 10.2.: Quality function for $a = 0.25(< 1)$ in case of different amplification factors ε_{01}, ε_{12}, ε_{02}.

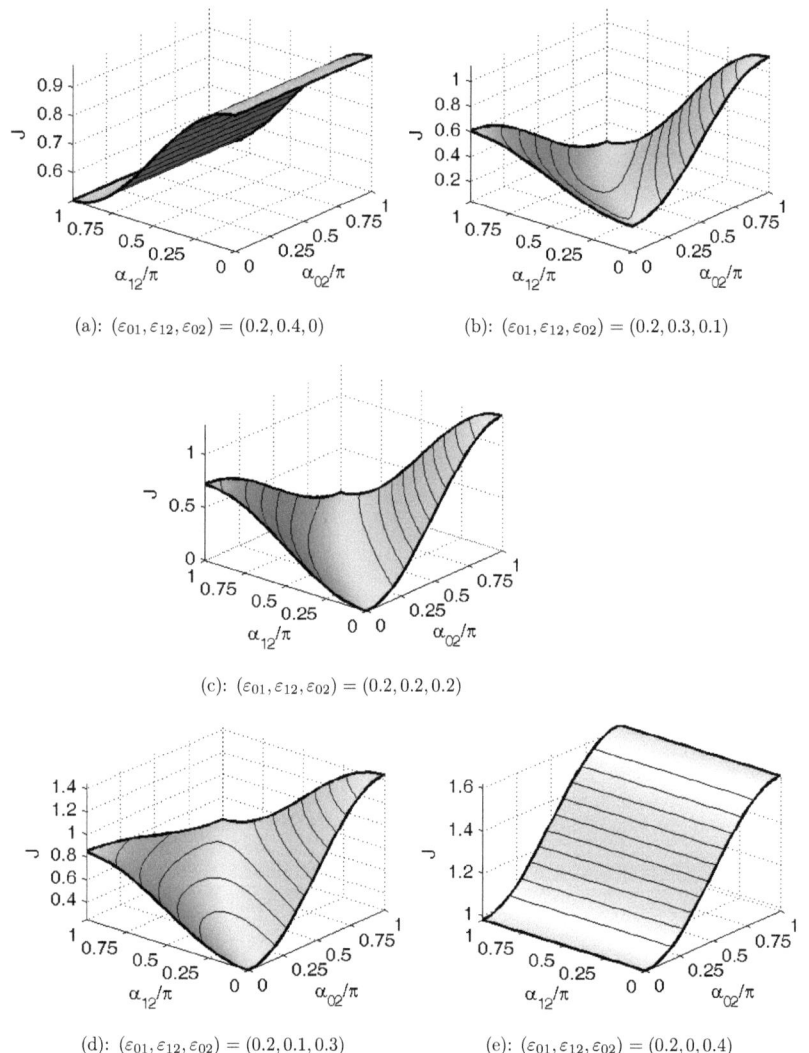

Figure 10.3.: Quality function for $a = 4(>1)$ in case of different amplification factors ε_{01}, ε_{12}, ε_{02}.

10 Optimal multi-location harmonic stiffness variation

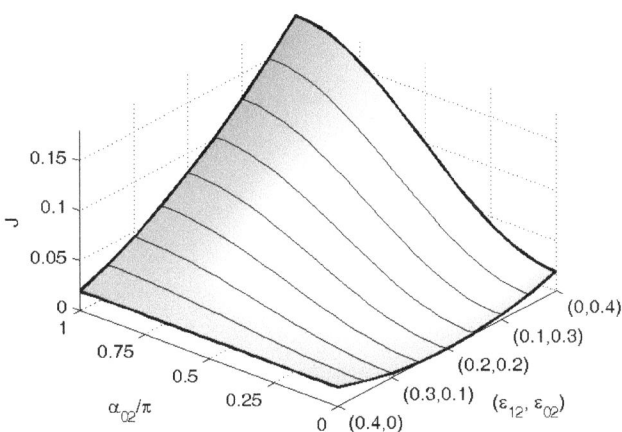

Figure 10.4.: Quality function for $a = 1$ and fixed amplification factor $\varepsilon_{01} = 0.2$.

Figure 10.4 shows the quality function J in dependency of the phase angle α_{02} and the amplification factor ε_{02} for fixed reference amplification factor ε_{01}. Respecting the constraint $\varepsilon_{01} + \varepsilon_{12} + \varepsilon_{02} = 0.6 =$ const and keeping ε_{01} fixed means that for increasing ε_{02} a decrease of ε_{12} would occur. This circumstance is pointed out by introducing a bracket notation $(\varepsilon_{12}, \varepsilon_{02})$. For $\varepsilon_{02} = 0$ the quality function in (10.28) even loses its dependency on the second phase angle α_{02}, which results in a straight line at the position $(0.4, 0)$. In this case the quality function is optimal for arbitrary phase angles α_{12}, α_{02} and the function shape further degenerates from the shape in Fig. 10.3e to a simple plane in the α_{12}, α_{02}-domain. Increasing ε_{02} the cosine term in (10.28) demands a phase angle of $\alpha_{02} = \pi$ to compensate its negative sign. For non-optimal phase angle $\alpha_{02} = 0$ the function value first decreases for increasing ε_{02} reaching the minimum of the function J and increases afterwards. In this case the function values are equal at the considered starting position $\varepsilon_{02} = 0$ and the final position $\varepsilon_{02} = 0.4$, respectively, because the following relation holds:

$$J|_{\varepsilon_{02}, \alpha_{02}=0} = \frac{1}{2}\varepsilon_{01}^2, \qquad J|_{\varepsilon_{02}, \alpha_{02}=0} = \frac{1}{2}(\varepsilon_{01} - \varepsilon_{02})^2.$$

Hence, for our specific starting and final position both expressions coincide, and the function shape becomes symmetric in between these points.

Concluding this special case of the critical frequency ration $a = 1$ shows, that except for the special case where $\varepsilon_{02} = 0$ the optimal function value of J is found at the phase angle $\alpha_{02} = \pi$. The phase angle α_{12} is arbitrary, as shown in (10.22). Due to the symmetry of J in the variables $\varepsilon_{01}, \varepsilon_{02}$, a variation of ε_{01} is equivalent to a variation of ε_{02}.

Special case of equal amplification factors

For the special case where all amplification factors are equal, $\varepsilon_{01} = \varepsilon_{12} = \varepsilon_{02} = \varepsilon$, the expression for extremal quality function in (10.20) simplifies to

$$\tilde{J}(\alpha_{12}, \alpha_{02})|_{\varepsilon_{01}=\varepsilon_{12}=\varepsilon_{02}=\varepsilon} = a(1-a)\varepsilon[\cos\alpha_{12}\cos\alpha_{02}$$
$$-(1-a)\cos\alpha_{12}\cos(\alpha_{12}-\alpha_{02}) - a\cos\alpha_{02}\cos(\alpha_{12}-\alpha_{02})], \qquad (10.29)$$

and the solutions derived in (10.23) leads to

$$J(0,0) = 0, \qquad J(\pi,0) = 2(1-a)^2\varepsilon^2,$$
$$J(0,\pi) = 2a^2\varepsilon^2, \qquad J(\pi,\pi) = 2\varepsilon^2. \qquad (10.30)$$

Inserting equal amplitudes into (10.18) and (10.17) yields

$$\cos\hat{\alpha}_{12} = \cos\hat{\alpha}_{02} = 1.$$

In this case no local extremal values occur and the maximum values are found on the interval borders, and we directly conclude, that only the trivial solutions found in (10.13) can be optimal.

Figure 10.5 shows exemplarily the dependency of the quality function J from (10.29) on the phase angles for different values of the frequency ratio a defined in (10.10). The characteristic topology of the quality function J for small values of a, $a = 0.25 < 1$, is shown in Fig. 10.5a. As predicted by (10.13) the global maximum value for small a is found at the position (π, π). Additionally two local maxima occur at $(\pi, 0)$ and $(0, \pi)$ in the domain $[0, 2\pi) \times [0, 2\pi)$. Increasing a, as in Fig. 10.5b, the global maximum remains at the same position, but the relative values of the local maxima change. Now, the local maximum at $(\pi, 0)$ is lower than the local maximum at $(0, \pi)$. According to (10.30) both local maxima are equal if

$$J(\pi,0) = J(0,\pi): \qquad (1-a)^2 = a^2 \quad \Rightarrow \quad a = \frac{1}{2}.$$

The resulting function shape is similar to Fig. 10.2d. Further increasing of a until its critical value $a = 1$, as in Fig. 10.5c, shows that at this specific value the global maximum is independent on the phase angle α_{12}, as it was revealed earlier for the trivial solutions. This critical value represents the transition of the characteristic topology of the quality function between low and high values of a. For high values of a, $a > 1$, as in Fig. 10.5d, the global maximum changes from the position (π, π) to $(0, \pi)$ according to (10.13). Now, two local maxima occur at $(\pi, 0)$ and (π, π) in the domain $[0, 2\pi) \times [0, 2\pi)$. Increasing a, as in Fig. 10.5f, the location of the global maximum remains unchanged, but the relative values of the local maxima change. Again, according to (10.30) both local maxima are equal if

$$J(\pi,0) = J(\pi,\pi): \qquad (-1+a)^2 = 1 \quad \Rightarrow \quad a = 2,$$

as shown in Fig. 10.5e.

10 Optimal multi-location harmonic stiffness variation

Note that, while the function value at the global maximum takes on different numerical values for high frequency ratios, $a > 1$, the global maximum remains unchanged for low frequency ratios, $a \leq 1$. According to (10.30) the optimal value $J(\pi,\pi) = 2\varepsilon^2$ is independent on the parameter a for low values of a, while for high values the optimal value $J(0,\pi) = 2a^2\varepsilon^2$ is a function of a^2.

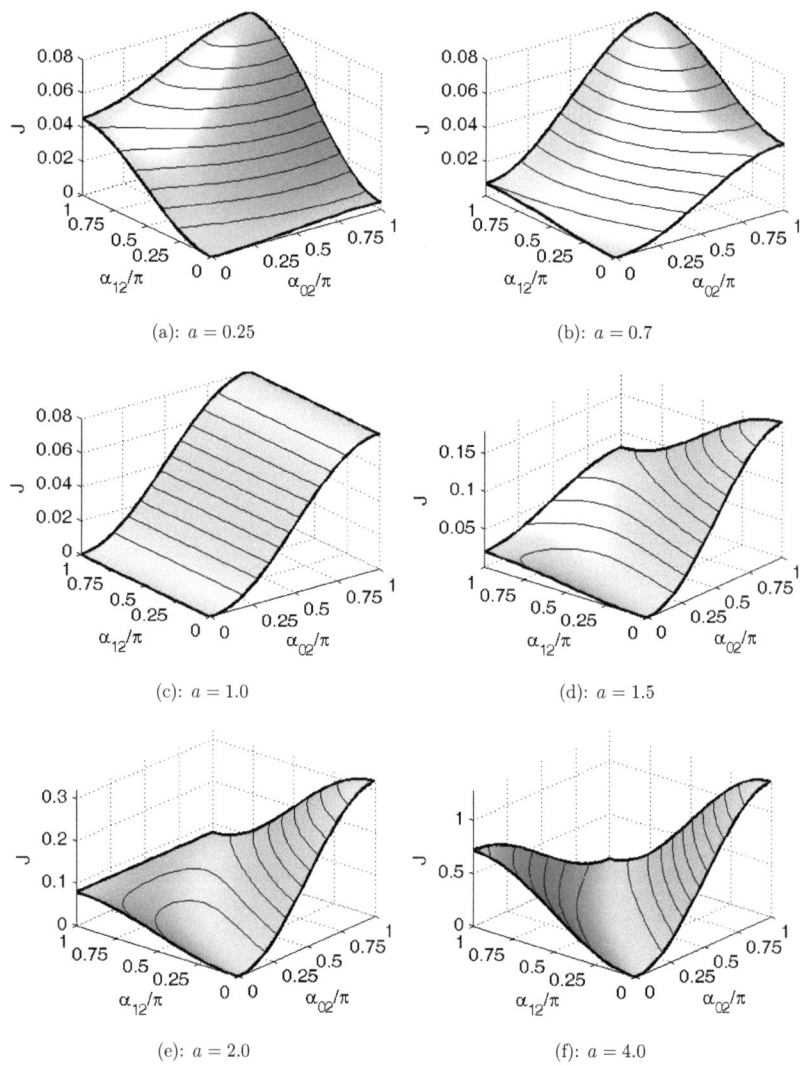

Figure 10.5.: Quality function in dependency of the phase angles α_{12}, α_{02} for different parameter values a in case of equal amplification factors
$\varepsilon_{01} = \varepsilon_{12} = \varepsilon_{02} = 0.2$.

10 Optimal multi-location harmonic stiffness variation 147

10.1.2. Optimal amplification factors

In the previous section the amplitude amplification factors $\varepsilon_{01}, \varepsilon_{12}, \varepsilon_{02}$ respect the constraint $\varepsilon_{01} = \varepsilon_{12} = \varepsilon_{02} = \text{const}$. Comparing the exemplary function shapes of the quality function J for small frequency ratios, $a < 1$, in Fig. 10.2 reveals, that maximizing J for fixed amplification factor ε_{01} demands to maximize ε_{12} and to minimize ε_{02}, while respecting the constraint. Hence, the optimal amplification factors for fixed ε_{01} are $\varepsilon_{12} \mapsto \varepsilon_{12\,\text{max}}$ and $\varepsilon_{02} = 0$. On the other hand, comparing the exemplary function shapes for big frequency ratios, $a > 1$, in Fig. 10.3 shows that maximizing J for fixed ε_{01} asks for $\varepsilon_{12} = 0$ and $\varepsilon_{02} \mapsto \varepsilon_{02\,\text{max}}$. These results are quite obvious and only valid for fixed amplification factor ε_{01} and numerical values as used in Figs. 10.2 and 10.3.

In general, realizing a device for parametric excitation the power supply is limited by a mechanical or electrical load, as the maximum force or the maximum current. In this section we are searching for an optimal distribution of the amplification factors $\varepsilon_{01}, \varepsilon_{12}, \varepsilon_{02}$ under the new constraint condition, that the maximum value of the potential energy is restricted

$$h(\varepsilon_{01}, \varepsilon_{12}, \varepsilon_{02}) = \varepsilon_{01} k_{01} + \varepsilon_{12} k_{12} + \varepsilon_{02} k_{02} = C = \text{const}, \tag{10.31}$$

which is an expansion of the constraint used in the previous section. The scalar C represents a boundary due to a limited power supply. Applying for a multi-variable optimization the method of Lagrange multipliers to $J(\varepsilon_{01}, \varepsilon_{12}, \varepsilon_{02}, \alpha_{12}, \alpha_{02})$ in (10.9) with the constraint condition $h(\varepsilon_{01}, \varepsilon_{12}, \varepsilon_{02})$ from (10.31) yields

$$\hat{\varepsilon}_{01} - (1-a)\hat{\varepsilon}_{12}\cos\hat{\alpha}_{12} - a\hat{\varepsilon}_{02}\cos\hat{\alpha}_{02} + \lambda k_{01} = 0,$$
$$-(1-a)\hat{\varepsilon}_{01}\cos\hat{\alpha}_{12} + a(1-a)\hat{\varepsilon}_{02}\cos(\hat{\alpha}_{02} - \hat{\alpha}_{12}) + (1-a)^2\hat{\varepsilon}_{12} + \lambda k_{12} = 0,$$
$$-a\hat{\varepsilon}_{01}\cos\hat{\alpha}_{02} + a(1-a)\hat{\varepsilon}_{12}\cos(\hat{\alpha}_{02} - \hat{\alpha}_{12}) + a^2\hat{\varepsilon}_{02} + \lambda k_{02} = 0,$$
$$(1-a)\hat{\varepsilon}_{12}(\hat{\varepsilon}_{01}\sin\hat{\alpha}_{12} - a\hat{\varepsilon}_{02}\sin(\hat{\alpha}_{12} - \hat{\alpha}_{02})) = 0, \tag{10.32a}$$
$$a\hat{\varepsilon}_{02}(\hat{\varepsilon}_{01}\sin\hat{\alpha}_{02} + (1-a)\hat{\varepsilon}_{12}\sin(\hat{\alpha}_{12} - \hat{\alpha}_{02})) = 0, \tag{10.32b}$$
$$\hat{\varepsilon}_{01} k_{01} + \hat{\varepsilon}_{12} k_{12} + \hat{\varepsilon}_{02} k_{02} = C,$$

for optimal amplification factors $\hat{\varepsilon}_{01}, \hat{\varepsilon}_{12}, \hat{\varepsilon}_{02}$ and phase angles $\hat{\alpha}_{12}, \hat{\alpha}_{02}$, where the quality function $J(\varepsilon_{01}, \varepsilon_{12}, \varepsilon_{02}, \alpha_{12}, \alpha_{02})$ is extremal under the constraint (10.31). The optimal phase angles $\hat{\alpha}_{12}, \hat{\alpha}_{02}$ are derived from (10.32a) and (10.32b) in analogy to (10.11a) and (10.11b) in the previous section, because the constraint condition in (10.31) is independent of the phase angles α_{12}, α_{02}. Respecting these optimal phase angles from (10.23) the remaining linear equations are

$$\begin{bmatrix} 1 & \mp(1-a) & a & k_{01} \\ \mp(1-a) & (1-a)^2 & \mp(1-a)a & k_{12} \\ a & \mp(1-a)a & a^2 & k_{02} \\ k_{01} & k_{12} & k_{02} & 0 \end{bmatrix} \begin{bmatrix} \hat{\varepsilon}_{01} \\ \hat{\varepsilon}_{12} \\ \hat{\varepsilon}_{02} \\ \lambda \end{bmatrix} = \begin{bmatrix} 0 \\ 0 \\ 0 \\ C \end{bmatrix}, \tag{10.33}$$

where the upper sign corresponds to $(0, \pi)$ with $a > 1$ and the lower sign to (π, π) with $a < 1$. Expressing (10.33) as a matrix equation $\mathbf{Ax} = \mathbf{b}$, the coefficient matrix $\mathbf{A}$ is symmetric and has a rank of three, and also in case of equal stiffness parameters, $k_{01} = k_{12} = k_{02} = k$. This matrix is singular, $\det\{\mathbf{A}\} = 0$. Consequently, the method of Lagrange multipliers leads to the same optimal phase angles as in (10.31), but is not applicable to our quality function to achieve optimal amplification factors ε_{kl}.

The geometrical interpretation of the occurrence of a singular coefficient matrix $\mathbf{A}$ is explained in the following paragraph. The quality function $J(\varepsilon_{01}, \varepsilon_{12}, \varepsilon_{02}, \hat{\alpha}_{12}, \hat{\alpha}_{02})$ is strictly monotonically increasing with increasing amplification factors ε_{kl},

$$\frac{\partial J}{\partial \varepsilon_{01}} = \varepsilon_{01} + |1 - a|\, \varepsilon_{12} + a\varepsilon_{02} > 0,$$
$$\frac{\partial J}{\partial \varepsilon_{12}} = |1 - a|\, \frac{\partial J}{\partial \varepsilon_{01}} > 0,$$
$$\frac{\partial J}{\partial \varepsilon_{02}} = a\frac{\partial J}{\partial \varepsilon_{01}} > 0,$$

for positive amplification factors. The constraint condition $h(\varepsilon_{01}, \varepsilon_{12}, \varepsilon_{02}) = C$ represents a plane area in the parameter space $\varepsilon_{01}, \varepsilon_{12}, \varepsilon_{02}$. The function h is plane is also strictly monotonically increasing in terms of the amplification factors, because the stiffness parameters k_{kl} possess in physical meaningful systems only positive values:

$$\frac{\partial h}{\partial \varepsilon_{01}} = k_{01} > 0, \qquad \frac{\partial h}{\partial \varepsilon_{12}} = k_{12} > 0, \qquad \frac{\partial h}{\partial \varepsilon_{02}} = k_{02} > 0.$$

Hence, both the quality function and the constraint condition are monotonically increasing in dependency of the amplification factors. A geometrical meaning of the method of Lagrange multipliers is, that the gradients of the quality function J and the constraint condition $h - C$ are parallel vectors at the maximum of J. Introducing an unknown scalar λ, the gradient of $K = J + \lambda(h - C)$ vanishes for some value of λ and K must be stationary, where the multiplier λ is a new variable, at a local extremum. The method of Lagrange multipliers can be used to optimize rather general functions, but the main assumption is, that the optimum must lie inside the parameter domain. Therefore, if both the quality function as well as the constraint condition are strictly monotonically increasing, then the extremal values are placed without exception at the boundary of the parameter domain, which causes a singular coefficient matrix $\mathbf{A}$. Consequently, using this method does not help in achieving optimal amplification factors, but it reveals, that the extremal values of J are located on the boundary lines resulting from the constraint h.

Boundary maxima

The amplification factors $\varepsilon_{01}, \varepsilon_{12}, \varepsilon_{02}$ and the frequency ratio a are defined in the interval $[0, \infty)$. The quality function $J(\varepsilon_{01}, \varepsilon_{12}, \varepsilon_{02}, \alpha_{12}, \alpha_{02})$ at the optimal positions found in (10.13), $(0, \pi)$

10 Optimal multi-location harmonic stiffness variation

or (π, π), simplifies to

$$J(\varepsilon_{01}, \varepsilon_{12}, \varepsilon_{02}) = \frac{1}{2}\left(\varepsilon_{01} + |1-a|\,\varepsilon_{12} + a\varepsilon_{02}\right)^2,$$

according to (10.23).

The constraint from (10.31) represents a plane in the parameter domain $\varepsilon_{01}, \varepsilon_{12}, \varepsilon_{02}$, which is positioned by using the constant potential C, see Fig. 10.6. Increasing C starting from zero shifts the plane area away from the origin. Reaching the maximal value of C, $C = C_{\max}$, the perpendicular distance between the plane and the origin is maximal, too. Searching for the optimal quality function J under the constraint $h = C$, for arbitrary but limited potential C, $C \leq C_{\max}$, means comparing the values of J for any parameter combination $\varepsilon_{01}, \varepsilon_{12}, \varepsilon_{02}$, that lie inside the thick-lined body sketched in Fig. 10.6 on the left hand side.

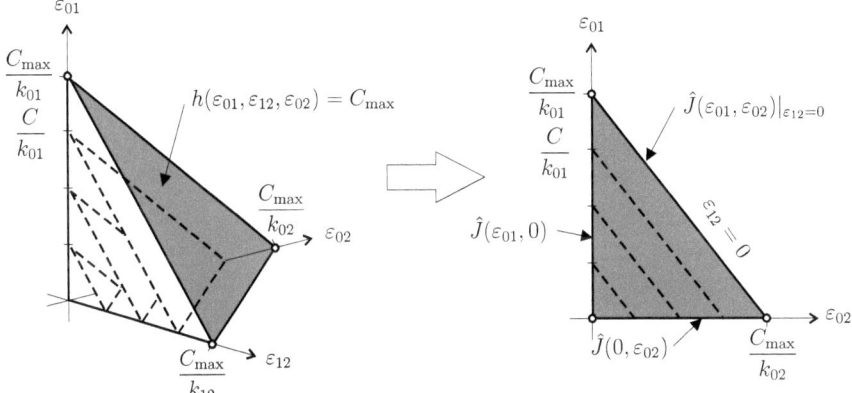

Figure 10.6.: Geometrical interpretation of constraint condition (10.31) – boundary maxima.

From the previous section we know that the optimum is found somewhere on the boundary of this parameter body. Since J is strictly monotonically increasing for increasing amplification factors, the optimum lies on the triangle $h = C_{\max}$. Substituting one parameter by using the constraint condition in (10.31) for positive stiffness parameters k_{01}, k_{12}, k_{02}, and consequently positive system potential C,

$$\varepsilon_{12} = \frac{C - \varepsilon_{01} k_{01} - \varepsilon_{02} k_{02}}{k_{12}},$$

results in the mapping

$$J(\varepsilon_{01}, \varepsilon_{12}, \varepsilon_{02}) \mapsto \hat{J}(\varepsilon_{01}, \varepsilon_{02}) = \frac{1}{2}\left(\varepsilon_{01} + \frac{C - \varepsilon_{01} k_{01} - \varepsilon_{02} k_{02}}{k_{12}}|1-a| + a\varepsilon_{02}\right)^2. \quad (10.34)$$

The new quality function $\hat{J}$ means evaluating the quality function J on the plane $h = C$. From a geometric point of view, the three dimensional parameter domain is mapped to a two dimensional parameter plane, by projecting the shaded triangles $h = C$ onto the shaded ε_{01}-ε_{02}-coordinate plane looking in opposite direction to the ε_{12}-axis, as shown in Fig. 10.6 on the right hand side. Now, searching the optimum requires to compare the values of $\hat{J}$ for any parameter combination $\varepsilon_{01}, \varepsilon_{02}$. Respecting the constraint condition limits the intervals of ε_{kl} for fixed positive value of C $[0, C/k_{kl}]$.

Analyzing the quality function $\hat{J}$ for arbitrary amplification factors, but fixed frequency ratio and stiffness parameters shows, that the sign of the declination of the tangential plane is independent on the amplification factors:

$$\frac{\partial \hat{J}}{\partial \varepsilon_{01}} = \sqrt{2\hat{J}} \left(1 - |1-a| \frac{k_{01}}{k_{12}} \right),$$

$$\frac{\partial \hat{J}}{\partial \varepsilon_{02}} = \sqrt{2\hat{J}} \left(-|1-a| \frac{k_{02}}{k_{12}} + a \right). \tag{10.35}$$

The tangential plane is horizontal if the conditions $m_1 = m_2$ and $|k_{02} - k_{01}| = k_{12}$ are satisfied. For this special case $\hat{J}$ is simply constant, because this result is valid for arbitrary amplification factors, and the entire quality function degenerates to a horizontal plane. Excluding this special case the relations in (10.35) reveals that $\hat{J}$ is either always strictly monotonically increasing or strictly monotonically decreasing in dependency of ε_{01} and ε_{02}. Hence, for $m_1 \neq m_2$ or $|k_{02} - k_{01}| \neq k_{12}$, the global extremal value of the quality function J is found at the boundary of the parameter domain $\varepsilon_{01}, \varepsilon_{12}, \varepsilon_{02}$, the boundary plane $h = C$, or more precisely, at the boundary of this boundary plane, the boundary lines:

$$\hat{J}(\varepsilon_{01}, 0) = \frac{1}{2} \left(\varepsilon_{01} + \frac{C - \varepsilon_{01} k_{01}}{k_{12}} |1-a| \right)^2,$$

$$\hat{J}(0, \varepsilon_{02}) = \frac{1}{2} \left(\frac{C - \varepsilon_{02} k_{02}}{k_{12}} |1-a| + a\varepsilon_{02} \right)^2, \tag{10.36}$$

$$\hat{J}(\varepsilon_{01}, \varepsilon_{02})\Big|_{\varepsilon_{12}=0} = \frac{1}{2} \left(\varepsilon_{01} + a\varepsilon_{02} \right)^2.$$

The upper three relations for the boundaries can be reformulated by using the original quality function J in (10.9) as $J(\varepsilon_{01} = 0, \varepsilon_{12}, \varepsilon_{02})$, $J(\varepsilon_{01}, \varepsilon_{12} = 0, \varepsilon_{02})$ and $J(\varepsilon_{01}, \varepsilon_{12}, \varepsilon_{02} = 0)$. Both representations are drawn in Fig. 10.6. Analyzing the slope of these boundary lines we obtain the same equation as in (10.35) and the resulting conditions in (10.37), except for the boundary line $\varepsilon_{12} = 0$. Using the directional derivative of $\hat{J}$ along the vector $\mathbf{h} = (-k_{02}, k_{01})^T$ leads to

$$\frac{\partial \hat{J}(\varepsilon_{01}, \varepsilon_{02})}{\partial \mathbf{h}}\bigg|_{\varepsilon_{12}=0} = \frac{\sqrt{2\hat{J}\big|_{\varepsilon_{12}=0}}}{\sqrt{k_{01}^2 + k_{02}^2}} [-k_{02}, \ k_{01}] \begin{bmatrix} 1 \\ a \end{bmatrix}$$

$$= \sqrt{\frac{2\hat{J}\big|_{\varepsilon_{12}=0}}{k_{01}^2 + k_{02}^2}} k_{02} \left(\frac{m_1}{m_2} - 1 \right),$$

10 Optimal multi-location harmonic stiffness variation

where

$$\left.\frac{\partial \hat{J}}{\partial \mathbf{h}}\right|_{\varepsilon_{12}=0} \lessgtr 0 \quad \text{for} \quad \frac{m_1}{m_2} \lessgtr 1,$$

or reformulated in absolute values of $\hat{J}$

$$\hat{J}(\varepsilon_{01}, 0) \gtrless \hat{J}(0, \varepsilon_{02}) \quad \text{for} \quad \frac{m_1}{m_2} \lessgtr 1. \tag{10.37}$$

Read these equations by respecting only the upper or only the lower inequality signs. Note that ε_{12} is zero if the parameters ε_{01} or ε_{02} are maximal, $\varepsilon_{01} = \varepsilon_{01\,\text{max}}$ or $\varepsilon_{02} = \varepsilon_{02\,\text{max}}$. Excluding the special case of equals sign in (10.37), these boundary lines, as their carrier plane $h = C$, are strictly monotonically for fixed system parameters. A direct conclusion from the gradient of the carrier plane to the slope of the boundary lines is not possible, because the carrier plane is not restricted to have the same monotonically behavior in both parameters ε_{01} and ε_{02}, respectively, as can be seen from (10.35).

Concluding, to find the global maximum of the quality function J simply the values of the mapped function $\hat{J}$ at the following three parameter combinations, indicated by circles in Fig. 10.6, need to be compared:

$$\begin{aligned}
\hat{J}(\varepsilon_{01\,\text{max}}, 0) &= \frac{1}{2}(\varepsilon_{01\,\text{max}})^2, \\
\hat{J}(0, \varepsilon_{02\,\text{max}}) &= \frac{1}{2}(a\varepsilon_{02\,\text{max}})^2, \\
\hat{J}(0, 0) &= \frac{1}{2}(|1-a|\,\varepsilon_{12})^2,
\end{aligned} \tag{10.38}$$

while the maximal possible values for the amplification factors $\varepsilon_{kl\,\text{max}}$ are equal to C/k_{kl}, according to the constraint condition in (10.31). Note that the expression $\hat{J}(0,0)$ is equivalent to $J(0, \varepsilon_{12\,\text{max}}, 0)$ but is different from $J(0, 0, 0)$, see Fig. 10.6.

Comparing two boundary points from (10.38) that are capable of being optimal yields

$$\hat{J}(\varepsilon_{01\,\text{max}}, 0) \gtrless \hat{J}(0, \varepsilon_{02\,\text{max}}) \quad \text{for} \quad \varepsilon_{01\,\text{max}} \gtrless a\varepsilon_{02\,\text{max}} > 0, \tag{10.39}$$

keeping in mind that the expression in the brackets in (10.38) are always positive. Substituting $\varepsilon_{kl\,\text{max}}$ by C/k_{kl} and a by its definition in (10.10) shows, that both the condition in (10.39) and the critical case in (10.37) coincide. Comparing the other two boundary points from (10.38) that are capable of being optimal, we obtain

$$\hat{J}(0, \varepsilon_{02\,\text{max}}) \gtrless \hat{J}(0, 0) \quad \text{for} \quad a\varepsilon_{02\,\text{max}} \gtrless \mp(1-a)\varepsilon_{12\,\text{max}} > 0. \tag{10.40}$$

Finally, comparing the last two extremal positions in (10.38) results in

$$\hat{J}(\varepsilon_{01\,\text{max}}, 0) \gtrless \hat{J}(0, 0) \quad \text{for} \quad \varepsilon_{01\,\text{max}} \gtrless \mp(1-a)\varepsilon_{12\,\text{max}} > 0. \tag{10.41}$$

10 Optimal multi-location harmonic stiffness variation

Reformulating the conditions in (10.39-10.41) by using non-vanishing physical parameters simplifies to

$$\varepsilon_{01\,\text{max}} \gtreqless a\varepsilon_{02\,\text{max}} : \quad \frac{m_1}{m_2} \lesseqgtr 1,$$

$$a\varepsilon_{02\,\text{max}} \gtreqless |1-a|\,\varepsilon_{12\,\text{max}} : \quad ak_{12} \gtreqless |1-a|\,k_{02}, \tag{10.42}$$

$$\varepsilon_{01\,\text{max}} \gtreqless |1-a|\,\varepsilon_{12\,\text{max}} : \quad k_{12} \gtreqless |1-a|\,k_{01},$$

with the frequency ratio a as defined in (10.10).

10.1.3. Summary for optimal phase and amplification factors

The optimal distribution of the amplification factors and the phase angles in order to maximize the quality function $J(\varepsilon_{01}, \varepsilon_{12}, \varepsilon_{02}; \alpha_{12}, \alpha_{02})$ as defined in (10.8) under the constraint $h(\varepsilon_{01}, \varepsilon_{12}, \varepsilon_{02}) = C$ from (10.31), are found in (10.22) and (10.42) respectively. The optimization of the phase angles is decoupled from the optimizing process of the amplification factors, because the constraint is not influenced by the phase angles. The optimal phase angles for arbitrary amplification factors are

$$(\alpha_1, \alpha_2) = \left\{ \begin{pmatrix} 0 \\ \pi \end{pmatrix}, \begin{pmatrix} \pi \\ \pi \end{pmatrix} \right\}, \tag{10.43}$$

with the vector of phase angles $\alpha_i = (\alpha_{12}, \alpha_{02})^T$. With the frequency ratio a as defined in (10.10) α_1 is optimal for $a > 1$ and α_2 for $a < 1$. The optimal distribution of the amplitude amplification factors at these phase angles are one of the following amplification vectors

$$(\mathbf{e}_1, \mathbf{e}_2, \mathbf{e}_3) = \left\{ \begin{pmatrix} \varepsilon_{01\,\text{max}} \\ 0 \\ 0 \end{pmatrix}, \begin{pmatrix} 0 \\ \varepsilon_{12\,\text{max}} \\ 0 \end{pmatrix}, \begin{pmatrix} 0 \\ 0 \\ \varepsilon_{02\,\text{max}} \end{pmatrix} \right\}, \tag{10.44}$$

with $\mathbf{e}_i = (\varepsilon_{01}, \varepsilon_{12}, \varepsilon_{02})^T$. The conditions in (10.42) decide which of these parameter combinations is optimal. Rewriting the last two conditions we obtain

$$\frac{k_{12}}{k_{01}} \gtreqless |1-a| : \quad \frac{\frac{k_{12}}{k_{02}}}{\left|\frac{k_{01}}{k_{02}} - \frac{m_1}{m_2}\right|} \gtreqless 1,$$

$$a\frac{k_{12}}{k_{02}} \gtreqless |1-a| : \quad \frac{\frac{k_{12}}{k_{01}}}{\left|\frac{k_{02}}{k_{01}} - \frac{m_2}{m_1}\right|} \gtreqless 1.$$

10 Optimal multi-location harmonic stiffness variation

This leads quite naturally to a definition of characteristic parameters. Introducing the following set of non-dimensional parameters

$$R_{12} = \frac{k_{12}}{k_{01}}, \quad R_{02} = \frac{k_{02}}{k_{01}}, \quad M = \frac{m_1}{m_2}, \tag{10.45}$$

where $MR_{02} = a$, the crucial conditions can be rewritten more compactly as

$$\boxed{\begin{aligned} J(\mathbf{e}_1) &\gtreqless J(\mathbf{e}_3): & & M \lesseqgtr 1, \\ J(\mathbf{e}_3) &\gtreqless J(\mathbf{e}_2): & \frac{MR_{12}}{|1-a|} &= \frac{MR_{12}}{|1-MR_{02}|} \gtreqless 1, \\ J(\mathbf{e}_1) &\gtreqless J(\mathbf{e}_2): & \frac{R_{12}}{|1-a|} &= \frac{R_{12}}{|1-MR_{02}|} \gtreqless 1. \end{aligned}} \tag{10.46}$$

The choice of the three non-dimensional parameters M, R_{12} and a would also be reasonable. It is quite remarkable, that the position of the global maximum at the boundary is determined by only three parameters. The achievable maximum value of the quality function is determined by

$$\boxed{J_{opt} = \tfrac{1}{2} \left(\max \left\{ \varepsilon_{01\,\max},\ |1-a|\,\varepsilon_{12\,\max},\ a\varepsilon_{02\,\max} \right\} \right)^2.} \tag{10.47}$$

Sometimes only the simplified system with $k_{02} = 0$ is considered, see Fig. 10.1. For vanishing stiffness parameter k_{02} the amplification factor ε_{02} does not exist, $\varepsilon_{02} = 0$, and in (10.46) only the last condition is relevant

$$R_{12} = \frac{k_{12}}{k_{01}} \gtreqless 1, \quad \text{for} \quad J(\mathbf{e}_1) \gtreqless J(\mathbf{e}_2). \tag{10.48}$$

It is important to mention, that the optimal parameter distribution is independent of any damping parameter, as it is the quality function in (10.8) itself. Consequently, whether a negative damping is place parallel to the stiffness that is periodically varied or not, does not influence the gained optimal parameter distribution at all. Furthermore, note that, according to (10.46), for $\varepsilon_{12} = 0$ the optimal parametric excitation is situated at the smaller mass, independent of the stiffness ratios R_{12} and R_{02}.

Equal stiffness parameters

Applying $k_{01} = k_{12} = k_{02} = k$, an assumption also made for the numerical studies in [25], the non-dimensional parameters simplify to $R_{12} = R_{02} = 1$, and the conditions in (10.46) becomes

$$\begin{aligned} J(\mathbf{e}_1) &\gtreqless J(\mathbf{e}_3): & & M \lesseqgtr 1, \\ J(\mathbf{e}_3) &\gtreqless J(\mathbf{e}_2): & \frac{1}{\left|1-\tfrac{1}{M}\right|} &\gtreqless 1, & M &\gtreqless \frac{1}{2}, \\ J(\mathbf{e}_1) &\gtreqless J(\mathbf{e}_2): & \frac{1}{|1-M|} &\gtreqless 1, & M &\lesseqgtr 2. \end{aligned} \tag{10.49}$$

10 Optimal multi-location harmonic stiffness variation

The detailed correlation between the extremal value $J(\mathbf{e}_i)$ along the number line of the mass ratio is shown in Fig. 10.7. For a mass ratio lower than one, the global optimum is placed at $\mathbf{e}_1$. While the distribution $\mathbf{e}_2$ is suboptimal for ratios lower than $\frac{1}{2}$, the distribution $\mathbf{e}_1$ is suboptimal for ratios between $\frac{1}{2}$ up to 1. Increasing M the global optimum switches to $\mathbf{e}_3$. While the parameter set $\mathbf{e}_1$ is suboptimal up to a mass ratio of 2, $\mathbf{e}_2$ is suboptimal for ratios greater than 2. Consequently, the distribution $\mathbf{e}_2$ can only be suboptimal. Summarizing this special case of equal stiffness parameters the globally optimal parameter distribution is

$$\mathbf{e}_1, \alpha_2, J_{opt} = \frac{1}{2}\varepsilon_{01\,\max}^2 \quad \text{for} \quad M < 1, \quad \text{and} \quad \mathbf{e}_3, \alpha_1, J_{opt} = \frac{1}{2}\varepsilon_{02\,\max}^2 \quad \text{for} \quad M > 1, \quad (10.50)$$

found in (10.43), (10.44) and (10.47). Note that for this special case the relation $a = M$ holds.

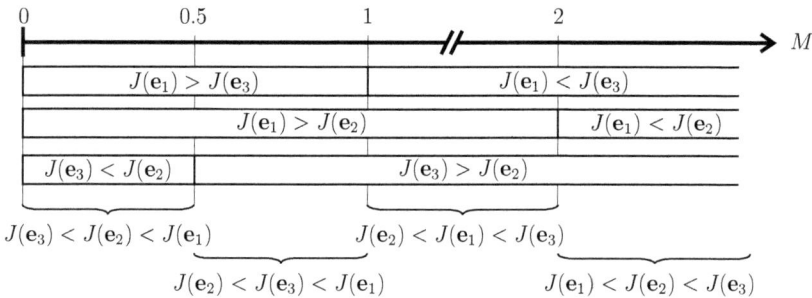

Figure 10.7.: Relations between extremal values in dependency of the mass ratio in case of equal stiffness parameters.

Examples of how the quality function J looks like for different frequency ratios a and a vanishing amplification factor $\varepsilon_{01} = 0$ ($< \varepsilon_{01\,\max}$) are presented in Fig. 10.8. For the chosen system potential $C_{\max} = 0.3$ the maximal possible value of the remaining amplification factors $\varepsilon_{12}, \varepsilon_{02}$ is 0.3. The constraint $h = C$ represents, for fixed ε_{01}, a straight line that is parameterized by the value of C. Evaluating J on this line leads to the non-trivial thick-lined boundary curves that lie on the resulting function shapes. The remaining two trivial boundaries are $J(0.1, \varepsilon_{12}, 0)$ and $J(0.1, 0, \varepsilon_{02})$. The optimal phase angles for $a < 1$ are α_2, that is indicated by $J(\pi, \pi)$ in Figs. 10.8ab. The gradient of the non-trivial boundary line is negative for $a < \frac{1}{2}$ and positive for $a > \frac{1}{2}$, according to (10.49). One must not mix this case with condition (10.37) which is only valid for maximal amplification factors, while here we assume $\varepsilon_{01} = 0$ ($< \varepsilon_{01\,\max}$). Applying the relations of (10.49) the optimal function value for $a = 0.25 < \frac{1}{2}$ in Fig. 10.8a is $J(0, 0.3, 0) = 0.003$ and for $a = 0.75 > \frac{1}{2}$ in Fig. 10.8b is $J(0, 0, 0.3) = 0.025$. Note that, although the maximal value of J at the extremal position $\varepsilon_{12} = \varepsilon_{02} = 0.3$ remains the same for $a \leq 1$, this point does not satisfy the constraint in (10.31), and is therefore out of interest.

10 Optimal multi-location harmonic stiffness variation

For the critical case $a = 1$ the function values J are equal at α_1 and α_2 respectively. As demonstrated in (10.22) the quality function becomes independent on ε_{12}, see Fig. 10.8c and the optimum is $J(\mathbf{e}_3)$. Increasing a, $a > 1$, the shape of J does not change and only the absolute values are influenced, see Fig. 10.8de. Again $J(\mathbf{e}_3) > J(\mathbf{e}_2)$ is valid, according to (10.49). Now, the optimal phase angles are α_1 indicated by $J(0,\pi)$.

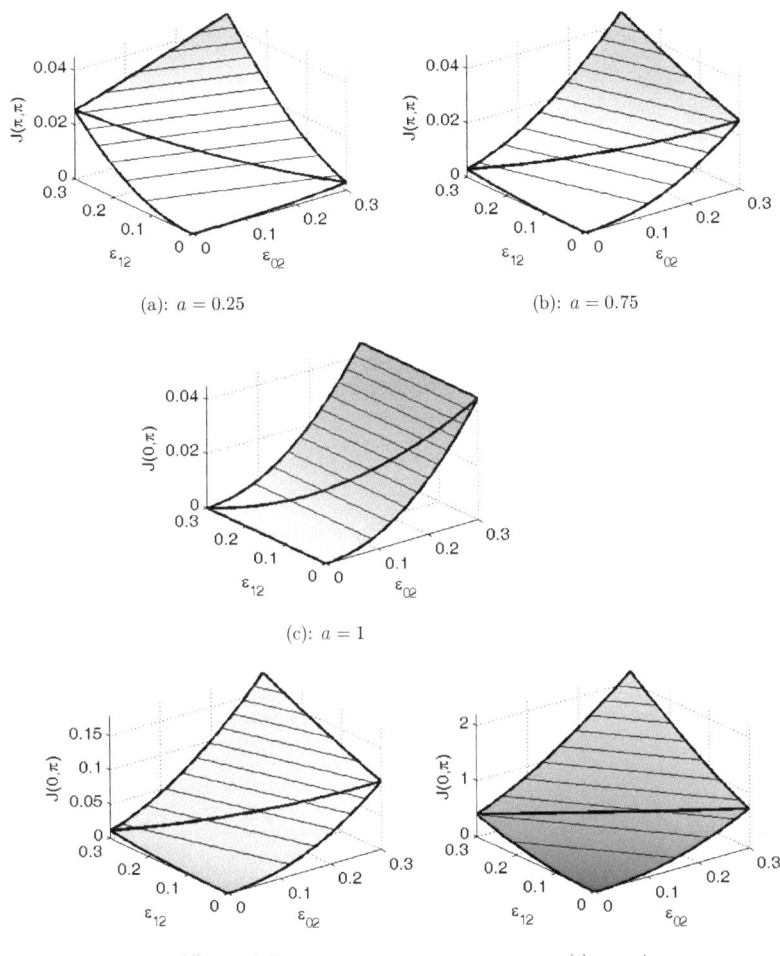

(a): $a = 0.25$

(b): $a = 0.75$

(c): $a = 1$

(d): $a = 1.5$

(e): $a = 4$

Figure 10.8.: Boundary maxima of quality function for different frequency ratios a but fixed $\varepsilon_{01} = 0$, $C_{\max} = 0.3$, and equal stiffness parameters $k_{01} = k_{12} = k_{02} = 1$.

Examples of how the quality function J looks like for a fixed value of $a = 0.75 > \frac{1}{2}$, but different values of ε_{01} are presented in Fig. 10.9. Due to $a < 1$ the optimal phase angles are α_2, that is indicated by $J(\pi, \pi)$. Figure 10.9a shows the domain and boundary planes of the parameter values $\varepsilon_{01}, \varepsilon_{12}, \varepsilon_{02}$. The edges of the boundary planes are thick-lined. The possible maximum values of the amplification factors ε_{kl} for positive system potential $C_{\max}$, $C_{\max} > 0$, are $C_{\max}/k_{kl}$. For $\varepsilon_{01} = 0$ the supposed values of the remaining amplification factors $\varepsilon_{12}, \varepsilon_{02}$ lie in the shaded plane. Increasing the fixed value of ε_{01}, while respecting the constraint $h = C_{\max}$, reduces the possible values for $\varepsilon_{12}, \varepsilon_{02}$ as indicated by the dashed lined planes, until the maximal value of $\varepsilon_{01} = \varepsilon_{01\max}$ is reached, and $\varepsilon_{12}, \varepsilon_{02}$ vanish. Figures 10.9bcd demonstrate the shapes of J for the different values of ε_{01} that are indicated in Fig. 10.9a with the corresponding boundary curves resulting from evaluating the constraint. Increasing ε_{01} means decreasing the maximal values of $\varepsilon_{12}, \varepsilon_{02}$ and the boundary curve is shifted on the corresponding shape of J towards the origin. The adjective corresponding is important, because the scales of the J-axis are different.

Comparing the maximal values of J for different fixed values of ε_{01} in Figs. 10.9bcd points out, that the value range of the remaining amplification factors $\varepsilon_{12}, \varepsilon_{02}$ is reduced for higher values of ε_{01} due to the constraint in (10.31), but the numerical value of J is increased. The maximal value for $\varepsilon_{01} = 0$ is $J(0, 0, 0.3) = 0.25$, as shown in Fig. 10.8b, and becomes larger for higher values of ε_{01}: $J(0.1, 0, 0.2) = 0.031$ in Figs. 10.9b, $J(0.2, 0, 0.1) = 0.038$ in Figs. 10.9c and $J(0.3, 0, 0) = 0.045$ in Figs. 10.9d. These values are maximal for fixed ε_{01}, and therefore represent local maximal values of J. Which of these local maxima is globally optimal is decided by the condition in (10.50). The optimal values for the numerical example with $a = 0.75$ is found at $\mathbf{e}_1$, which confirms the result in Fig. 10.9.

10 Optimal multi-location harmonic stiffness variation

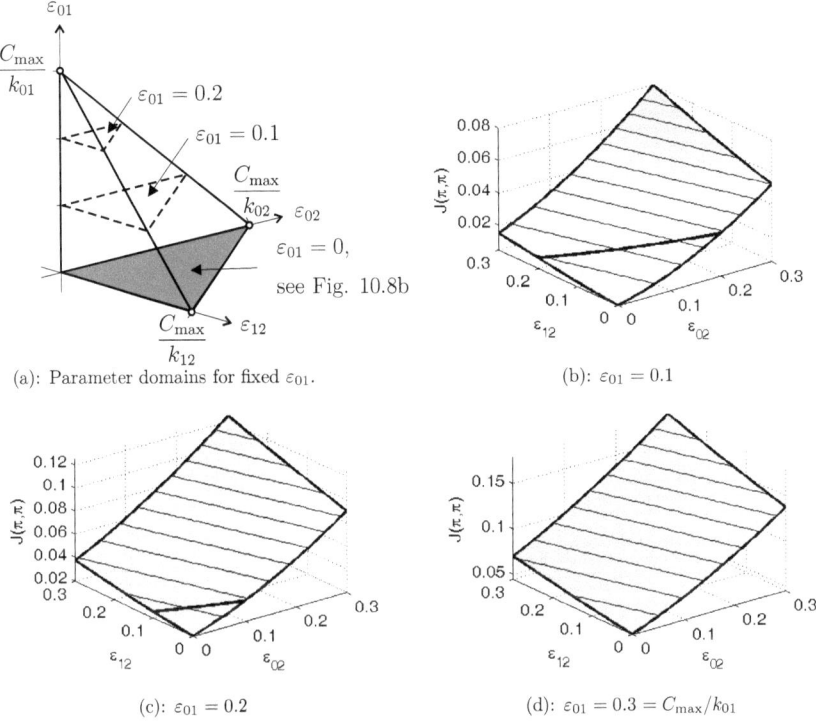

(a): Parameter domains for fixed ε_{01}.

(b): $\varepsilon_{01} = 0.1$

(c): $\varepsilon_{01} = 0.2$

(d): $\varepsilon_{01} = 0.3 = C_{\max}/k_{01}$

Figure 10.9.: Boundary maxima of quality function for different amplification factors ε_{01}, but fixed frequency ratio $a = 0.75$ and $C_{\max} = 0.3$, and equal stiffness parameters $k_{01} = k_{12} = k_{02} = 1$.

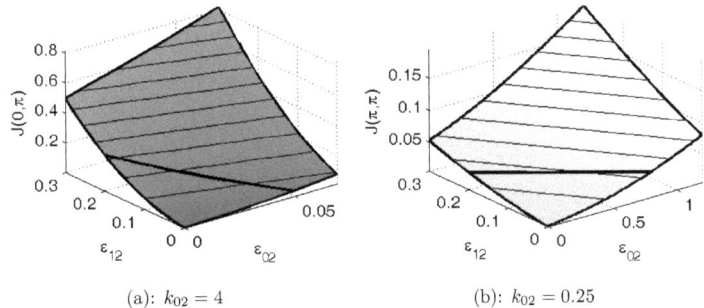

(a): $k_{02} = 4$

(b): $k_{02} = 0.25$

Figure 10.10.: Boundary maxima of quality function for $M = 1$, $\varepsilon_{01} = 0.1$, $C_{\max} = 0.3$, $k_{01} = k_{12} = 1$, but different values of k_{02}.

Equal mass ratio

Applying the relation $m_1 = m_2$, $M = 1$, the conditions in (10.46) shows, that $J(\mathbf{e}_1) = J(\mathbf{e}_3)$ is valid and only one condition from (10.49) remains:

$$J(\mathbf{e}_2) \gtreqless J(\mathbf{e}_1), J(\mathbf{e}_3) : \quad \frac{R_{12}}{|1 - R_{02}|} = \frac{k_{12}}{|k_{02} - k_{01}|} \gtreqless 1. \tag{10.51}$$

If $|k_{02} - k_{01}| > k_{12}$ is satisfied then $J(\mathbf{e}_2)$ is optimal. For $|k_{02} - k_{01}| < k_{12}$ the parameter distributions $\mathbf{e}_1$ and $\mathbf{e}_3$ are optimal simultaneously. The optimal phase angles are

$$\alpha_1 \quad \text{for} \quad \frac{k_{02}}{k_{01}} > 1 \quad \text{and} \quad \alpha_2 \quad \text{for} \quad \frac{k_{02}}{k_{01}} < 1, \tag{10.52}$$

according to (10.43), respecting that $a = k_{02}/k_{01}$. Exemplary shapes of J in case of $|k_{02} - k_{01}| > k_{12}$ are shown in Fig. 10.10a and in case of $|k_{02} - k_{01}| < k_{12}$ in Fig. 10.10b. Note that, due to the occurrence of different numerical values for the stiffness parameters, $k_{01} = k_{12} = 1$ and $k_{02} = 4$ or 0.25, respecting the constraint in (10.31) results in different maximal values of the amplification factors. For a chosen value of the system potential C, $C_{\max} = 0.3$, and fixed amplification factor $\varepsilon_{01} = 0.1$ in Fig. 10.10 these values are $\varepsilon_{12\max} = (C_{\max} - \varepsilon_{01}k_{01})/k_{12} = 0.2$ and $\varepsilon_{02\max} = (C_{\max} - \varepsilon_{01}k_{01})/k_{02} = 0.05$ in Fig. 10.10a and $\varepsilon_{02\max} = 0.8$ in Fig. 10.10b. As predicted in (10.51) the sign of the directional derivative of the non-trivial boundary curve changes. Therefore the maximal value of J at $\varepsilon_{01} = 0.1$, is found at $\mathbf{e}_2$, $J(0.1, 0.2, 0) = 0.245$, in Fig. 10.10a and at $\mathbf{e}_3$, $J(0.1, 0, 0.8) = 0.045$, in Fig. 10.10b.

10.1.4. Generalization of results

By the constraint (10.31) a certain level of potential energy is allocated to the system. In this section we consider additional constraint on the amplification factors ε_{kl}. If due to constructional limitations the optimal design found in (10.50) with

$$\varepsilon_{01\max} = \frac{C_{\max}}{k_{01}} \quad \text{or} \quad \varepsilon_{02\max} = \frac{C_{\max}}{k_{02}}$$

cannot be achieved, then the results from the previous section have to be adapted and generalized statements are obtained. In addition to the constraint in (10.31), the amplification factors ε_{kl} are assumed to be limited by known values of $\varepsilon_{kl,d}$:

$$\varepsilon_{kl,d} \leq \varepsilon_{kl\max} = \frac{C_{\max}}{k_{kl}}. \tag{10.53}$$

If an amplification factor $\varepsilon_{kl,d}$ has no design limitation, then its size is equal to $\varepsilon_{kl\max}$. These restrictions reduce the parameter space of the amplification factors $\varepsilon_{01}, \varepsilon_{12}, \varepsilon_{02}$ and the presented parameter space in Fig. 10.9a is adapted to Fig. 10.11ab. Each additional limitation from (10.53) introduces a boundary plane at the critical value $\varepsilon_{kl,d}$ that is parallel to two coordinate axes.

10 Optimal multi-location harmonic stiffness variation

In case of small limitations the parameter body is sketched in Fig. 10.11a. In this context the term small means that each border line of the constraint $h = C$ forms, at least partly, the boundary of the parameter body, that is indicated by thick lines. This results in rather complex boundary shapes. The introduction of the design limitations (10.53) does not influence the monotonic behavior of the quality function J. The sign of the gradients are conserved, and the optimal function value of J is found again on the boundary lines, or more precisely on the corners of these boundary lines that are indicated by circles. Hence, considering rather small limitations for all three amplification factors, we obtain now six candidates for an optimal distribution of the parameters ε_{kl}, instead of the three distributions found in (10.44), where the additional limitations in (10.53) have not been an issue. These six corner points lie on the boundary plane due to the constraint $h = C$.

If $\varepsilon_{kl,d}$ is significantly smaller than the optimal values in (10.44) the picture changes. Significantly smaller means that not all boundary lines of the constraint are part of the border of the parameter space, see Fig. 10.11b. Now, the boundary degenerates partly to a simple cuboid in the vicinity of such a restrictive limitation, while in the area where the additional limitation is small or non-existing, the boundary is determined by the constraint in (10.31) and remains non-trivial. If at least two amplification factors are limited by such a small value of $\varepsilon_{kl,d}$, then the number of candidates for optimal parameter distribution reduces to five or less, as shown in the figure. For the considered example the example considered the remaining five corner points on the boundary plane are defined by the constraint $h = C$ as shown in Fig. 10.11b.

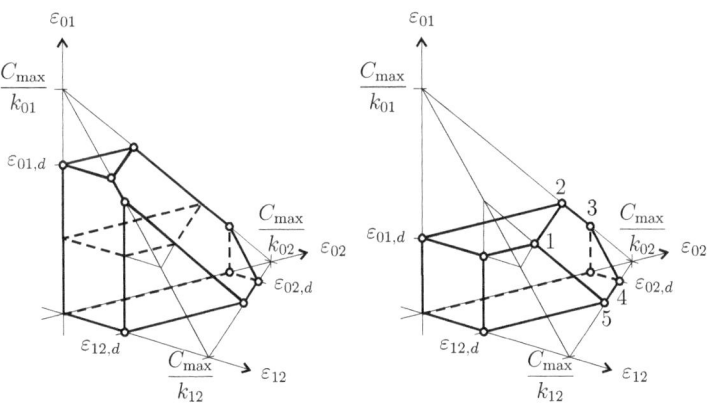

(a): Six candidates for optimum.   (b): Five candidates for optimum.

Figure 10.11.: Borders of parameter domain in case of design limitations.

For very restrictive design limitations, where

$$\varepsilon_{01,d}k_{01} + \varepsilon_{12,d}k_{12} + \varepsilon_{02,d}k_{02} < C$$

is valid, the border is a perfect cuboid. The optimal value is found on the corner opposite to the origin of the coordinate system,

$$J(\varepsilon_{01,d}, \varepsilon_{12,d}, \varepsilon_{02,d}), \qquad (10.54)$$

because the quality function J is an additive combination of weighted, but positive expressions. If one design limitation is relaxed, then $\varepsilon_{kl,d}$ is getting closer to the optimal value of $\varepsilon_{kl\,\mathrm{max}}$. Now, the boundary planes due to the design limitation (10.53) intersect with the constraint in (10.31) and two additional corners on the plane $h = C$ are generated. Any existing intersection point, that fulfills the constraint on the plane $h = C$, is better than corners of the remaining border lines that are not part of the constraint. The optimal value of J is found at one of the following positions:

$$\left\{ \begin{pmatrix} \varepsilon_{01,d} \\ C - \dfrac{\varepsilon_{01,d}}{k_{01}} \\ 0 \end{pmatrix}, \begin{pmatrix} \varepsilon_{01,d} \\ 0 \\ C - \dfrac{\varepsilon_{01,d}}{k_{01}} \end{pmatrix}, \begin{pmatrix} 0 \\ C - \dfrac{\varepsilon_{02,d}}{k_{02}} \\ \varepsilon_{02,d} \end{pmatrix} \right\}.$$

Note that the case of only two candidates does not occur. Further relaxing of the design limitation $\varepsilon_{kl,d}$ towards the (sub)optimal value $\varepsilon_{kl\,\mathrm{max}}$ raises the number of positions that have to be considered for optimization up to the number of six, as sketched in Fig. 10.11a. For the graph presented in Fig. 10.11b we obtain the following set of positions:

$$\left\{ \begin{pmatrix} \varepsilon_{01,d} \\ \varepsilon_{12,d} \\ C - \dfrac{\varepsilon_{01,d}}{k_{01}} - \dfrac{\varepsilon_{12,d}}{k_{12}} \end{pmatrix}, \begin{pmatrix} \varepsilon_{01,d} \\ 0 \\ C - \dfrac{\varepsilon_{01,d}}{k_{01}} \end{pmatrix}, \begin{pmatrix} C - \dfrac{\varepsilon_{02,d}}{k_{01}} \\ 0 \\ \varepsilon_{02,d} \end{pmatrix}, \begin{pmatrix} 0 \\ C - \dfrac{\varepsilon_{02,d}}{k_{02}} \\ \varepsilon_{02,d} \end{pmatrix}, \begin{pmatrix} 0 \\ \varepsilon_{12,d} \\ C - \dfrac{\varepsilon_{12,d}}{k_{12}} \end{pmatrix} \right\}.$$

If only one amplification is limited by design, for instance $\varepsilon_{01,d}$, then four candidates for optimal parameter distribution exist:

$$\left\{ \begin{pmatrix} \varepsilon_{01,d} \\ C - \dfrac{\varepsilon_{01,d}}{k_{01}} \\ 0 \end{pmatrix}, \begin{pmatrix} \varepsilon_{01,d} \\ 0 \\ C - \dfrac{\varepsilon_{01,d}}{k_{01}} \end{pmatrix}, \begin{pmatrix} 0 \\ \varepsilon_{12,\mathrm{max}} \\ 0 \end{pmatrix}, \begin{pmatrix} 0 \\ 0 \\ \varepsilon_{02,\mathrm{max}} \end{pmatrix}, \right\}.$$

In (10.46) we found, that, depending on the system parameters, either a maximal amplification ε_{01} or ε_{02} is optimal. The question arises, whether a constellation exists where maximizing ε_{12} is demanded. If the condition $\varepsilon_{01,d}k_{01} + \varepsilon_{12,d}k_{12} + \varepsilon_{02,d}k_{02} < C$ holds then the maximal value of the quality function is trivial to find, see (10.54). For $\varepsilon_{01,d}k_{01} + \varepsilon_{12,d}k_{12} + \varepsilon_{02,d}k_{02} \geq C$ the constraint $h = C$ intersects the cuboid $[0, \varepsilon_{01,d}] \times [0, \varepsilon_{12,d}] \times [0, \varepsilon_{02,d}]$, which results in a boundary being a polygon line as shown in Fig. 10.11b. If additionally $\varepsilon_{01,d}k_{01} + \varepsilon_{12,d}k_{12} < C$

10 Optimal multi-location harmonic damping variation

is fulfilled, then adapting (10.39-10.41) by using the design limitations $\varepsilon_{kl,d}$ instead of $\varepsilon_{kl\,\text{max}}$ reveals, that the case of maximizing ε_{12} occurs if the conditions

$$g > |1-a|\,\varepsilon_{12,d} + a\frac{C-\varepsilon_{12,d}k_{12}}{k_{02}},$$

$$g > \varepsilon_{01,d} + a\frac{C-\varepsilon_{01,d}k_{01}}{k_{02}},$$

$$g > \frac{C-\varepsilon_{02,d}k_{02}}{k_{01}} + a\varepsilon_{02,d},$$

$$g > |1-a|\frac{C-\varepsilon_{02,d}k_{02}}{k_{12}} + a\varepsilon_{02,d},$$

are satisfied, where

$$g = \varepsilon_{01,d} + |1-a|\,\varepsilon_{12,d} + a\frac{C-\varepsilon_{01,d}k_{01}-\varepsilon_{12,d}k_{12}}{k_{02}},$$

as sketched in Fig. 10.11b. In this case the maximal value is

$$J\left(\varepsilon_{01,d},\varepsilon_{12,d},\frac{C-\varepsilon_{01,d}k_{01}-\varepsilon_{12,d}k_{12}}{k_{02}}\right).$$

Otherwise, if $\varepsilon_{01,d}k_{01} + \varepsilon_{12,d}k_{12} > C$ is fulfilled, then the corresponding expression g on the left hand side of the inequality above changes to

$$g = \varepsilon_{01,d} + |1-a|\frac{C-\varepsilon_{01,d}k_{01}}{k_{12}} \quad \text{or} \quad g = \frac{C-\varepsilon_{12,d}k_{12}}{k_{01}} + |1-a|\,\varepsilon_{12,d}.$$

With the help of the criteria obtained in this section, we are capable of optimizing a system with general time-harmonic variation of one or more stiffness coefficients under the constraint condition (10.31).

10.2. Optimal multi-location harmonic damping variation

Section 10.1 investigates the optimal distribution of phase angles and amplifications factors in case of a multi-located time-harmonic stiffness variation. In this section we examine a multi-located time-harmonic variation of the damping parameters. The explicit expressions of the damping parameters are obtained by mapping $k_{kl} \mapsto c_{kl}$ in (10.1). In analogy to Section 10.1 again the interaction term vanishes, $R_{12}S_{21} - R_{21}S_{12} = 0$ is satisfied, and the corresponding stability conditions are obtained from the conditions found in (4.14) and (4.17), whereas both conditions coincide

$$\sigma_{1,2} = \pm\frac{(\Theta_{11}+\Theta_{22})}{2}\sqrt{-\frac{\Theta_{11}\Theta_{22}-\frac{1}{4}(R_{12}R_{21}+S_{12}S_{21})}{\Theta_{11}\Theta_{22}}},$$

using the parametric excitation term $R_{12}R_{21} + S_{12}S_{21}$. Searching for the optimal distribution of phase angles and amplification factors for otherwise fixed system parameters results in maximizing the parametric excitation term. Introducing a quality function J proportional to the parametric excitation term

$$R_{12}R_{21} + S_{12}S_{21} = \frac{J}{2m_2^2 k_{02}^2},$$

gives, after a lengthy calculation, the result

$$J = -(1-a) L_{01}L_{12}\varepsilon_{01}\varepsilon_{12} \cos\alpha_{12} - aL_{01}L_{02}\varepsilon_{01}\varepsilon_{02} \cos\alpha_{02}$$
$$+ a(1-a) L_{12}L_{02}\varepsilon_{12}\varepsilon_{02} \cos(\alpha_{12} - \alpha_{02})$$
$$+ \frac{1}{2}\left(L_{01}^2\varepsilon_{01}^2 + (1-a)^2 L_{12}^2\varepsilon_{12}^2 + a^2 L_{02}^2\varepsilon_{02}^2\right) \to \max, \quad (10.55)$$

with the frequency ratio

$$a = \frac{m_1}{m_2}\frac{k_{02}}{k_{01}}, \quad (10.56)$$

that is identical to (10.10), and the non-dimensional damping stiffness ratios

$$L_{kl} = \frac{c_{kl}}{k_{kl}}. \quad (10.57)$$

Comparing the quality function as defined in (10.8) and (10.55) shows, that the quality function for harmonic damping variation differs from the quality function for harmonic stiffness variation defined in (10.8) and therefore cannot be obtained by simple performing the mapping relation $k_{kl} \mapsto c_{kl}$. For $c_{kl} = k_{kl}$ both function are equivalent and the same relations for optimal parameter distribution as in the previous section are obtained. For general damping and stiffness parameters a closer look at (10.55) reveals, that by introducing the transformations

$$\boxed{L_{kl}\varepsilon_{kl} \mapsto \varepsilon_{kl},} \quad (10.58)$$

identical quality functions are obtained by (10.8) and (10.55). Hence, performing the mapping relations in (10.58), we end up with the same results and statements as obtained in the previous chapter.

The optimal distributions of the phase angles and the amplification factors are found according to (10.46) and (10.47), with the definitions in (10.43) and (10.44). If, in addition to the constraint condition, design limitations have to be respected, the statements in Section 10.1.4 have to be considered.

10.3. Optimal shape of functions for stiffness variation

The method of suppressing self-excited and damped vibrations has been studied extensively for the case of time-harmonic variations of the system parameters. Following [9], in this chapter we examine the efficiency of different time-periodic, but not necessarily harmonic variations,

10 Optimal shape of functions for stiffness variation

and give approximated solutions even for non-smooth parameter variations. The following study is restricted to stiffness variations only, but more than one stiffness parameter may be varied. The corresponding analytical conditions for vibration suppression were obtained in Chapter 3.3 by a first order perturbation method. The case of different phase angles is not considered, $\alpha_{12} = \alpha_{02} = 0$, but the results obtained so far can be easily adapted to the case of non-vanishing phase angles as in Chapter 3.2 or to the case of the most general system as presented in Chapter 7.

Previous investigations showed, that it is advantageous to increase the interval of vibration suppression, σ, for several reasons. First of all, an increased stability interval also increases the robustness of the method, since fluctuations of system parameters would have less or no consequences on the performance. Additionally, a wider interval positively affects the transient behavior of vibration suppression. Therefore, we aim for the maximization of the suppression interval and consider a stiffness variation as *optimal* if it maximizes σ for a fixed n:

$$\sigma \to \max. \tag{10.59}$$

The case of one modal damping coefficient being negative,

$$\Theta_{11}\Theta_{22} < 0, \tag{10.60}$$

is of practical importance, since self-excitation occurs in this case.

Any stiffness coefficient in the system shall be considered for a periodical change with the same frequency and phase angle and the same periodic shape function:

$$k_{kl}(t) = k_{kl}\left(1 + \varepsilon_{kl}k_{kl}^{var}(t)\right), \tag{10.61}$$

where k_{kl} is the constant part and k_{kl}^{var} the time-periodic part of the of the stiffness parameter, and ε_{kl} the corresponding amplitude amplification factor, as already introduced in the previous sections. The parametric excitation frequency η is normalized to one, $\eta t \mapsto t$. We search for an optimal shape function for stiffness variations $k_{kl}(t)$, which can be a regular function or a function with a finite number of finite discontinuities in the interval $[0, 2\pi]$, to which the following constraint conditions shall apply, see Fig. 10.12:

1. periodic variation: $k_{kl}(0) = k_{kl}(2\pi)$,

2. zero mean value variation k_{kl}^{var} and

3. variation in between a stiffness bandwidth defined as

$$k_{kl,\min} = k_{kl}\left(1 - \varepsilon_{kl}\right) \quad \text{and} \quad k_{kl,\max} = k_{kl}\left(1 + \varepsilon_{kl}\right).$$

The first condition is proposed since the quenching effect is only known for periodic variations. The second constraint condition guarantees that the mean value of the stiffness parameter k_{kl}

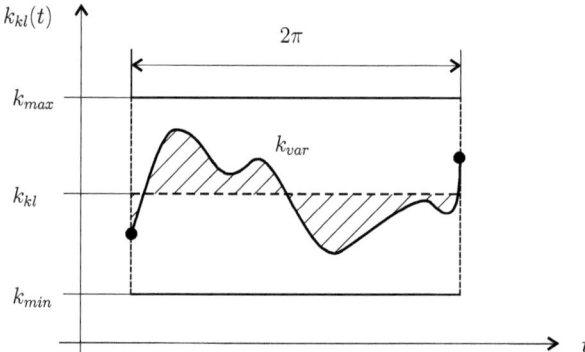

Figure 10.12.: General time-periodic stiffness variation.

remains unchanged by introducing a time-periodic function $k_{kl}^{var}(t)$ in the equations of motion. A non-zero mean value of the time dependent part would cause a shift of the natural frequencies Ω_1, Ω_2. This shift can be treated in analogy to the following analysis, but would not change the main result, and is therefore omitted at this point. The third constraint condition is a physical requirement to avoid solutions which result in non-feasible infinite stiffness amplitudes. All physically reasonable signals meet these conditions. Note that the tangent at 0 and the tangent at 2π can possess different gradients, hence the function can be non-smooth on the interval boundaries, as indicated in Fig. 10.12. Furthermore, the time evolution can be even non-smooth in the entire interval.

The Dirichlet Fourier series conditions suggest that all practical signals have a Fourier series associated to its function representation. Without loosing generality we restrict the analysis to *even* functions. The further analysis is carried out only for Fourier cosine series

$$k_{var}(t) = \sum_{n=1}^{\infty} a_n \cos nt, \qquad (10.62)$$

restricted to the constraint conditions 1-3 as listed above. Vibration suppression of a dynamically unstable system is achieved only in case of parametric anti-resonance, as summarized in Tab. 3.1 on page 36. The parametric anti-resonance can only occur if the normalized frequency nt of the parametric excitation is near to a parametric combination frequency $|\Omega_1 \mp \Omega_2|/n$ of the order n. That results in the relations for the stability interval

$$\begin{gathered}\Theta_{11} + \Theta_{22} > 0, \\ \sigma^{(n)} = \pm \frac{\Theta_{11} + \Theta_{22}}{2n} \sqrt{-\frac{\Delta^{(n)}}{\Theta_{11}\Theta_{22}}} \qquad n = 1, 2, \ldots,\end{gathered} \qquad (10.63)$$

10 Optimal shape of functions for stiffness variation

according to (3.58), and for the case of a stable point $\sigma^{(n)} = 0$ in

$$\Delta^{(n)} = \Theta_{11}\Theta_{22} + \frac{a_n^2}{4\Omega_1\Omega_2}Q_{12}Q_{21} > 0 \qquad n = 1, 2, \ldots \tag{10.64}$$

$$\Theta_{11} + \Theta_{22} > 0,$$

Note that, any of the n harmonic components of the Fourier series in (10.62) possesses its won stability width $\sigma^{(n)}$. Hence, the optimization in (10.59) has to be adapted to

$$\sigma^{(n)} \to \max \qquad \text{for fixed } n = 1, 2, \ldots \tag{10.65}$$

Equation (10.63) determines the boundary between stable and unstable system parameter values. To decide which side of the boundary is stable and which is unstable we need the additional condition in (10.64). If the second condition in (10.64) is fulfilled for a certain value of n, then we have a parametric anti-resonance near $\eta_0^{(n)}$ with a width of $\sigma^{(n)}$ obtained from (10.63). Otherwise, there is no suppression possible and the vibration amplitudes grow without restriction. Note that for the case of $n = 1$ and $a_1 = 1$ the result of (10.63) is equivalent to the conditions derived in (3.24, 3.29) and (10.64) to the conditions in (3.23, 3.28).

Searching for a maximal suppression interval we first have to fulfill the stability conditions. Inserting the second stability condition for a stable point from (10.64) into (10.63) implicates, that the radical of the square root is positive and consequently the values of $\sigma^{(n)}$ are all real-valued. Inserting (10.62) into (10.61) gives

$$k_{kl}(t) = k_{kl}\left(1 + \sum_{n=1}^{\infty}(\varepsilon_{kl}a_n)\cos(nt)\right),$$

and we recognize, that the existing amplification factors ε_{kl} are themselves amplified by the Fourier coefficients a_n. The amplification factors ε_{kl} are diminished in case of $a_n < 1$, and enlarged in case of $a_n > 1$. For the critical case $a_n = 1$ ε_{kl} remain unchanged. The search for an optimal $\sigma^{(n)}$ with a fixed n is equivalent to maximizing $\Delta^{(n)}$ in (10.64) and this simplifies the task finally to

$$a_n = \frac{1}{\pi}\int_0^{2\pi} k_{var}(t)\cos nt\, dt \to \max, \qquad n = 1, 2, \ldots \tag{10.66}$$

We expand a given function $k_{var}(t)$ into Fourier series and determine its Fourier coefficients. The largest interval of vibration suppression is located near a parametric anti-resonance with a specific value n for which the Fourier coefficient is maximal. In most cases only the *leading* Fourier coefficient a_1 is decisive, because the stability interval $\sigma^{(n)}$ in (10.63) is inverse proportional to n. Fourier series and their leading coefficient for exemplary shapes of the $k_{var}(t)$ are listed in Tab. 10.1. The reference shape is the cosine function as investigated in the literature

166   10 Optimal shape of functions for stiffness variation

and the present work, with an amplification ratio of one, $a_1 = 1$. Applying a triangular excitation would mean an amplification loss of 19% and for the case of a parabola shape we gain as less as 3%. Using an impulse or trapezoid shape leads to a gain of up to 27% depending on parameter φ. Hence, the greatest enhancement compared with a stiffness variation of a classical harmonic shape is achieved for a simple rectangular shape for $n = 1$ and $\varphi = 0$

$$a_1 = \frac{4}{\pi} = 1.27. \tag{10.67}$$

This means that a benefit of 27% can be achieved just by adjusting the shape of stiffness variation function, but without increasing the peak values $k_{\min}$, $k_{\max}$. An additional advantage of a Fourier coefficient $a_1 > 1$ ca be seen from (10.64). Since the excitation amplitude amplifies the positive term in (10.64) quadratically, see (3.60), and compensates the negative damping term more effectively. The optimal result obtained in (10.67) is an analytical verification and generalization of the numerical studies performed in [38], but therein only one stiffness parameter was varied periodically in time.

For a system with two degrees of freedom with time-periodic stiffness parameters $k_{kl}(t)$ the optimal function is a simple rectangular shape, or as known in control theory, an open-loop bang-bang control. Even within this subgroup of functions the most simple ones with zero-crossing at the half period π, or for non-normalized frequency η at the time π/η, is the optimal one. Impulse-like shapes, where the fast dynamics of the signal are concentrated in a very small time interval, are not optimal.

For the optimal symmetric time-periodic rectangular function

$$k_{kl}^{var}(t) = \text{rect}(t) = \begin{cases} +1 & 0 < t \leq \frac{T}{2} \\ -1 & \frac{T}{2} < t \leq T \end{cases}, \quad T = \frac{2\pi}{\eta}, \tag{10.68}$$

we obtain the Fourier series expansion of the stiffness parameters

$$k_{kl}(t) = k_{kl}(1 + \varepsilon \, \text{rect}(\eta \tau)) \approx k \left(1 + \varepsilon \frac{4}{\pi} \sum_{n=1}^{\infty} \frac{(-1)^{n+1}}{2n-1} \cos((2n-1)\eta \tau)\right), \tag{10.69}$$

according to Tab. 10.1 or (3.51), with arbitrary frequency η and the Fourier coefficients

$$a_n = \frac{4}{\pi} \frac{(-1)^{n+1}}{2n-1}. \tag{10.70}$$

The usage of the $\approx$ sign in (10.69) means applying the Dirichlet Fourier conditions, where near any isolated discontinuity the function is approximated by a continuous shape, whereas the function value lies in between the both limiting values before and after the discontinuity. Evaluating the formulae obtained in previous chapters for the optimal function in (10.68) gives the following relations. By setting

$$\eta_0^{(n)} = \frac{|\Omega_2 \mp \Omega_1|}{2n-1} > 0 \quad \text{or} \quad \varpi_j \mp \varpi_i = (2n-1) = m,$$

10 Optimal shape of functions for stiffness variation

where $n = 1, 2, 3, \ldots$ and $\eta_0^{(1)} = \eta_0 = |\Omega_2 \mp \Omega_1|$ the first order approximation of the stability boundary in the frequency domain is

$$\eta^{(n)} = \eta_0^{(n)} + \varepsilon \sigma^{(n)},$$

in analogy to (3.4) and the stability width yields

$$\sigma_{1,2}^{(n)} = \pm \frac{\Theta_{11} + \Theta_{22}}{2(2n-1)} \sqrt{-\frac{\Theta_{11}\Theta_{22} - \frac{1}{4\Omega_1\Omega_2}\widehat{Q}_{12}^{(n)}\widehat{Q}_{21}^{(n)}}{\Theta_{11}\Theta_{22}}},$$

according to (3.58), with the abbreviations

$$\widehat{Q}_{12}^{(n)} = \frac{(-1)^{n+1}}{(2n-1)} \frac{4}{\pi} Q_{12} \quad \text{and} \quad \widehat{Q}_{21}^{(n)} = \frac{(-1)^{n+1}}{(2n-1)} \frac{4}{\pi} Q_{21},$$

using the Fourier coefficients a_n from (10.70). Examining the parametric excitation term resulting from (10.69)

$$\widehat{Q}_{12}^{(n)} \widehat{Q}_{21}^{(n)} = a_n^2 Q_{12} Q_{21} = \frac{16}{\pi^2 (2n-1)^2} Q_{12} Q_{21},$$

reveals that the sign depends only on the $Q_{12}Q_{21}$-term and not on the Fourier coefficients a_n of (10.70), which confirms the result obtained in (3.60). For $n = 1$ the corresponding parametric combination resonances are $\eta_0 = |\Omega_1 \mp \Omega_2|$, and the cross-coupling coefficients of the parametric excitation matrix become

$$\widehat{Q}_{12} = \frac{4}{\pi} Q_{12} \quad \text{and} \quad \widehat{Q}_{21} = \frac{4}{\pi} Q_{21}.$$

Compared with the results obtained in Chapter 3.1 for a stiffness variation by a cosine function only, both coefficients are amplified by $4/\pi$. A result that verifies the conclusions obtained by extensive numerical studies for a bang-bang controller in [38].

We relax the additional constraint conditions in search for an optimal function of $k_{kl}(t)$ by considering also shapes with non-zero mean value over one period. In this case the original stiffness parameters k_{kl} are adapted by the particular mean values k_{kl}^m to the new stiffness parameters $\tilde{k}_{kl}$,

$$\tilde{k}_{kl} = k_{kl} + \varepsilon k_{kl}^m, \tag{10.71}$$

where k_{kl}^m is equivalent to the Fourier coefficient a_0,

$$a_0 = \frac{1}{\pi} \int_0^{2\pi} k_{var}(t)\, dt,$$

the area enclosed by the function and the time axis. For non-zero mean values the stiffness matrix $\mathbf{K}$

$$\mathbf{K} = \begin{bmatrix} k_{01} + k_{12} & -k_{12} \\ -k_{12} & k_{12} + k_{02} \end{bmatrix}$$

Table 10.1.: Fourier coefficients for different periodic function shapes.

shape	Fourier coefficients	leading coefficient
triangle k_{var}	$a_m = \dfrac{1}{m^2}\dfrac{8}{\pi^2}$ $m = 2n-1,\ n = 1, 2, \ldots$	$a_1 = 0.81$
cosine k_{var}	$a_n = \begin{cases} 1, & n = 1 \\ 0, & n > 1 \end{cases}$ $n = 1, 2, \ldots$	$a_1 = 1$
parabola k_{var}	$a_m = \dfrac{1}{m^3}\dfrac{32}{\pi^3}$ $m = 2n-1,\ n = 1, 2, \ldots$	$a_1 = 1.032$
impulse k_{var}	$a_m = \dfrac{\cos m\varphi}{m}\dfrac{4}{\pi}$ $m = 2n-1,\ : n = 1, 2, \ldots$	$a_1 = 1.27\cos\varphi$
trapezoid k_{var}	$a_m = \dfrac{\sin m\varphi}{m^2\varphi}\dfrac{4}{\pi}$ $m = 2n-1,\ n = 1, 2, \ldots$	$a_1 = 1.27\dfrac{\sin\varphi}{\varphi}$

changes to the new stiffness matrix

$$\tilde{\mathbf{K}} = \begin{bmatrix} k_{01} + k_{01}^m + k_{12} + k_{12}^m & -k_{12} - k_{12}^m \\ -k_{12} - k_{12}^m & k_{12} + k_{12}^m + k_{02} \end{bmatrix}.$$

Hence, the original natural frequencies Ω_1, Ω_2 of $\mathbf{M}^{-1}\mathbf{K}$ differ from the natural frequencies $\tilde{\Omega}_1, \tilde{\Omega}_2$ of $\mathbf{M}^{-1}\tilde{\mathbf{K}}$.

Part III.

Comparison with numerical results

11. Numerical methods of investigation

In this chapter we will briefly review two methods for the numerical stability analysis of systems as defined by (2.6). These methods avoid a time consuming numerical integration of the system equations over many periods of the parametric excitation. The stability results are also useful as a reference for approximate analytical stability analysis.

11.1. Floquet theory

The stability of the system dynamics can be investigated by means of Floquet theory. Applying Floquet theory transforms a linear time-periodic system into a linear time-invariant system by using a Lyapunov transformation. Hence, the stability of the former system can be inferred from that of the latter system. Below is a brief review of Floquet theory and related technical terms based on [60] and [29].

The considered general n-dimensional equations of motion (2.6) with the position vector $\mathbf{x}(t)$ define a system of linear differential equations with periodic coefficients, which can be transformed to a $2n$-dimensional linear time-periodic system of first order differential equations

$$\dot{\mathbf{y}}(t) = \mathbf{A}(t)\mathbf{y}(t), \quad \mathbf{A}(t+T) = \mathbf{A}(t) \quad t \geq t_0,$$
$$\mathbf{y}(t_0) = \mathbf{y}_0,$$
(11.1)

with the state vector $\mathbf{y}(t) \in \mathbb{R}^{2n}$ and the piecewise continuous and periodic matrix $\mathbf{A}(t) \in \mathbb{R}^{2n \times 2n}$ with period T. In most cases, the conditions of the Picard-Lindelöf theorem for existence and uniqueness of initial value problems are trivially satisfied by (11.1), see [29]. Hence, there exist unique solutions of (11.1) for arbitrarily given initial conditions $\mathbf{y}_0 \in \mathbb{R}^{2n}$. The set of the solutions of (11.1) form an $2n$-dimensional linear space. Let $\mathbf{y}_1(t), \mathbf{y}_2(t), \ldots, \mathbf{y}_{2n}(t)$ be $2n$ linearly independent solutions, then

$$\mathbf{Y}(t) = [\mathbf{y}_1(t), \mathbf{y}_2(t), \ldots, \mathbf{y}_{2n}(t)] \tag{11.2}$$

is called the fundamental matrix. If the fundamental matrix is equal to the unity matrix at the initial time $t = t_0$, $\mathbf{Y}(t_0) = \mathbf{I}_{2n}$, then $\mathbf{Y}(t)$ is called the principal fundamental matrix or the state transition matrix for (11.1). The state transition matrix is denoted by $\mathbf{\Phi}(t, t_0)$, where the second argument indicates dependency on initial conditions. Any solution of (11.1) can be expressed as $\mathbf{\Phi}(t,t_0)\mathbf{c}$, where $\mathbf{c} \neq \mathbf{0}$, $\mathbf{c} \in \mathbb{R}^{2n}$ is a constant vector. In particular, for $\mathbf{y}_0$ the solution of (11.1) is given by

$$\mathbf{y}(t) = \mathbf{\Phi}(t,t_0)\mathbf{y}_0. \qquad (11.3)$$

The state transition matrix evaluated at $t = T$, $\mathbf{\Phi}(T,0)$, is called the monodromy matrix.

Floquet's theorem postulates that each fundamental matrix of (11.1), and consequently the state transition matrix $\mathbf{\Phi}(t,t_0)$, can be written as the product of two $2n \times 2n$-matrices

$$\mathbf{\Phi}(t,t_0) = \mathbf{L}(t,t_0) e^{(t-t_0)\mathbf{F}}, \qquad \mathbf{L}(t,t_0) = \mathbf{L}(t+T,t_0), \qquad (11.4)$$

where $\mathbf{L}$ is a $2n \times 2n$-matrix-valued function, which is periodic with period T, and $\mathbf{F}$ is a constant $2n \times 2n$-matrix. The eigenvalues of $e^{T\mathbf{F}}$ are called the characteristic multipliers. The Lyapunov transformation

$$\mathbf{z}(t) = \mathbf{L}^{-1}(t,t_0)\mathbf{y}(t)$$

transforms the time-periodic system (11.1) into a linear time-invariant system with constant coefficients

$$\begin{aligned}\dot{\mathbf{z}}(t) &= \mathbf{F}\mathbf{z}(t), \quad t \geq t_0, \\ \mathbf{z}(t_0) &= \mathbf{y}_0.\end{aligned} \qquad (11.5)$$

By defining

$$t = t_0: \quad \mathbf{\Phi}(t_0,t_0) = \mathbf{L}(t_0,t_0)$$

we prepare $2n$ sets of initial condition vectors $\mathbf{k}_i$, so that an identity matrix is formed

$$\mathbf{\Phi}(t_0,t_0) = \mathbf{L}(t_0,t_0) = [\mathbf{k}_1(t_0),\mathbf{k}_2(t_0),\ldots,\mathbf{k}_{2n}(t_0)] = \mathbf{I}_{2n}, \qquad (11.6)$$

which can be substituted in (11.4) and leads to

$$\mathbf{F} = \frac{1}{T}\ln\left(\mathbf{\Phi}(t_0+T,t_0)\right). \qquad (11.7)$$

As a result of (11.7), the stability of the time-periodic system (11.1) can be determined either from the eigenvalues of the Floquet exponent matrix $\mathbf{F}$ or from the monodromy matrix $\mathbf{\Phi}(T,0)$. Starting from independent sets of initial conditions defined in (11.6), the monodromy matrix is calculated numerically by repeated integration of the system equations (11.1) over one period T. By solving these $2n$ initial value problems over one period T the monodromy matrix $\mathbf{\Phi}(T,0)$ is obtained, where the eigenvalues of the monodromy matrix

$$\Lambda = \text{eig}\{\mathbf{\Phi}(T,0)\}, \qquad (11.8)$$

determine the stability of the system dynamics. The system is asymptotically stable if and only if all of the eigenvalues are less than one in magnitude

$$\max\{|\lambda_1|,|\lambda_2|,\ldots,|\lambda_{2n}|\} = \begin{cases} < 1 \text{ asymptotically stable system,} \\ > 1 \text{ unstable system.} \end{cases} \qquad (11.9)$$

This approach gives reliable results and is computationally very efficient, which is a necessity for extensive parameters studies. The procedure presented here was implemented by using the software package ACSL ([45]). The results obtained are treated graphically by applying the software package MATLAB ([44]).

11.2. Path following

The procedure introduced in the previous section is capable of determining the stability of a system with fixed parameter values. If we are interested in the stability border as a function of two or more system parameters, then we have to repeat this procedure for each parameter combination. Depending on the desired resolution of the stability border a large number of integrations may be necessary. In this case it might be more convenient to use a path following method in combination with an iterative mapping as described in the following.

The equations under consideration are a system of coupled differential equations of first order as presented in (11.1) resulting from (2.6). The period T of the system matrix results from the frequency ω of the parametric excitation defined in (2.6). Introducing the time transformation

$$t \mapsto \frac{\tau}{\omega} \qquad (11.10)$$

will transform the system (11.1) with a frequency dependent period T and the state variable $\mathbf{y}$ into a system with a constant period of 2π. This enables a mapping of the new state variables $\bar{\mathbf{y}}$ as

$$\bar{\mathbf{y}}(0) \mapsto \bar{\mathbf{y}}(2\pi) \qquad (11.11)$$

with a constant period of 2π. Determining the state vector at the end of a period can be performed by using an integration algorithm for ordinary differential equations like RADAU5. Applying the mapping (11.11) enables the implementation of an iterated map in a bifurcation continuation software like CONTENT or MATCONT, see [33]. The mapping (11.11) transforms the non-autonomous system of equations in (11.1) into an autonomous system of equations, that are able to be handled by CONTENT.

A Neimarck-Sacker point is a bifurcation in the parameter space, at which the conjugate complex eigenvalues of the system lie on the unit circle. Hence, at this special point, the mean values of the vibrational amplitudes of the system neither decrease nor increase, they remain unchanged. Examples of numerical simulations at three different parametric excitation frequencies in the neighborhood of a Neimark-Sacker point are presented in Fig. 11.1. Figure 11.1a shows the time history of a vibration amplitude at a parameter set in the unstable parameter domain, where the vibration amplitudes grow exponentially. On the other hand, Figure 11.1c illustrates the time evolution at a parameter set in the stable parameter domain, where the vibrational amplitudes decrease exponentially. Finally, for a parameter combination, where a

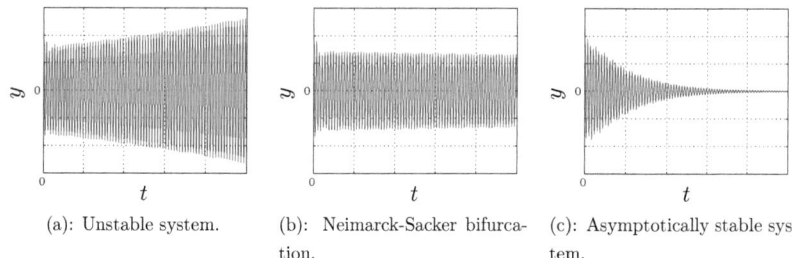

(a): Unstable system.   (b): Neimarck-Sacker bifurcation.   (c): Asymptotically stable system.

Figure 11.1.: Time series of a vibration amplitude in the neighborhood of a Neimarck-Sacker bifurcation.

Neimarck-Sacker bifurcation occurs, the system is at its stability border and the mean values of the vibrational amplitudes are constant, as shown in Fig. 11.1b.

Starting with arbitrary initial conditions we are searching for fixed points of the map in dependency of the parametric excitation frequency ω, until we find a Neimarck-Sacker point. Such a Neimarck-Sacker represents a point on the stability border, that can be used as a starting point for path following of the neighbored Neimarck-Sacker points. The set of all Neimarck-Sacker points represents the stability border of the system in (11.1) in the entire parameter space. Performing this procedure may drastically decrease the number of initial value problems necessary to be solved in order to find the stability border in the parameter space. The main restriction of the path following method is the requirement that the stability boundary is simply connected. Local inclusions of the stability boundary in the parameter space, as shown in Fig. 12.5a on page 188, are therefore hard to find by this method.

12. Minimum systems

In the previous chapters analytical stability investigations were performed of the equations of motion as defined in (2.6). By applying a perturbation technique we derived rather general conditions for stability, that lead to a deeper understanding of vibration suppression by parametric excitation. With the help of these conditions, the difference between the parametric resonance and the parametric anti-resonance was explained, see for example Tab. 3.1 on page 36. We identified the criteria for the occurrence of a parametric anti-resonance that permitted to optimize the design of a parametric excitation in order to obtain the maximally achievable equivalent damping.

In the following sections the minimum systems as introduced in Sections 2.4.1 to 2.4.4 are analyzed numerically and analytically.

12.1. Harmonic stiffness variation

In this section we analyze two minimum systems with two degrees of freedom as introduced in Sections 2.4.1 and 2.4.2 that are dynamically unstable due to a negative damping coefficient. It is shown that these systems can be stabilized and vibrations can be suppressed by introducing a parametric excitation. The parametric excitation is realized by a prescribed time-periodic variation of a stiffness coefficient, which can be interpreted as an open-loop vibration control. Both, analytical and numerical methods are applied to calculate ranges for control and system parameters where vibration suppression is effective. First we apply the numerical tool outlined in Section 11. Stability charts and stability bodies, as introduced in [19], are presented to visualize the rather complex domains of stability obtained in extensive parameter studies. Then we compare these results with the analytical results obtained from analytical formulae as derived in the Chapters 3.1 and 3.2.

The normal form of the equations of motion of the considered systems has the following structure

$$\ddot{\mathbf{z}} + \mathbf{\Theta}\dot{\mathbf{z}} + \left(\mathbf{\Omega}^2 + \varepsilon \mathbf{Q} \cos \eta \tau\right) \mathbf{z} = \mathbf{0}, \qquad (12.1)$$

with the position vector $\mathbf{z} = (z_1, z_2)^T$ and a small amplification factor ε. Note that the modal damping matrix $\mathbf{\Theta}$ in (12.1) is rescaled in comparison to the equations introduced in Chapter 2.3,

$$\varepsilon \mathbf{\Theta} \quad \text{in Chapter 2.3} \quad \mapsto \quad \mathbf{\Theta} \quad \text{in (12.1)}. \qquad (12.2)$$

The presented results are obtained by expanding equations (12.1) to a first order system as in (11.1) and carrying out the procedure as outlined in Section 11.1. By following this method the computational effort to determine the stability of the two mass system with a time-varying stiffness parameter consists of just four numerical integration runs with the equivalent first order system over one period T of the parametric excitation $\cos(\eta\tau)$ as defined in (12.1), and the calculation of all eigenvalues of the (4×4) monodromy matrix (11.8), that are used for the stability criterion in (11.9). This numerical method is used in order to verify the results of the analytical predictions presented in the previous chapters.

Choosing system parameters which have to be time-periodic results in the parametric excitation matrix $\mathbf{Q}$ in (12.2). In opposite to Chapter 10.1, just one system parameter is varied periodically for the systems introduced in Sections 2.4.1 and 2.4.2. This variation corresponds to the first relation in (10.1). Hence, just two parameters remain for the design of a proper open-loop control: the amplitude amplification factor ε and the frequency of parametric excitation η. The frequency is determined by the symmetry property of the parametric excitation matrix $\mathbf{Q}$. The original system matrices (2.21) or (2.32) are symmetric before transformation into the normal form (12.1). Hence, as explained in Chapter 2.3, the parametric excitation term is positive, $Q_{12}Q_{21} > 0$. According to Tab. 3.2 on page 38 such systems may possess a parametric anti-resonance at the frequency $\eta \approx |\Omega_1 - \Omega_2|$ and a parametric resonance at the frequency $\eta \approx \Omega_1 + \Omega_2$, depending mainly on the damping coefficients. Furthermore, if the parametric excitation frequency η is close to one of the frequencies $2\Omega_1$ and $2\Omega_2$ a parametric resonance may occur, depending on the conditions presented in (3.17). Therefore, only one design parameter for vibration control remains, the amplitude amplification factor ε.

The analytical predictions in Chapters 3.1 and 3.2 were derived by performing a first order approximation. However, we can conclude that the frequencies $|\Omega_1 - \Omega_2|/n$ with $n \in \mathbb{N}$ correspond to a parametric anti-resonance and the frequencies $(\Omega_1 + \Omega_2)/n$, $2\Omega_1/n$ and $2\Omega_2/n$ to a parametric resonance. This would arise from higher order approximations of order n, similar to the determination of instability tongues of higher order for the damped Mathieu equation, see [59].

For a certain parameter set the system is stable without open-loop control if the system parameters satisfy the stability conditions $\Theta_{11} > 0$, $\Theta_{22} > 0$, according to (3.14). Introducing a parametric excitation in such systems can amplify the existent damping to a higher value, as described thoroughly in Chapter 8. The mean value of these damping coefficients is proportional to the maximal equivalent damping achievable, as it was shown in the Figures 8.1 and 8.2. Especially when the values of the modal damping coefficients Θ_{11} and Θ_{22} are very different, the equivalent damping can be much higher than the original damping.

The mechanism of vibration suppression by parametric excitation is impressive especially in the case where one modal damping coefficient Θ_{ii} becomes negative and the conditions in (3.14) are not satisfied anymore. In the case of positive damping coefficients κ_1, κ_2 as introduced in

(2.23), both modal damping coefficients Θ_{11} and Θ_{22} are positive, too. On the other hand, if one of the damping coefficients κ_i becomes negative while the other remains positive, the modal damping coefficients Θ_{ii} can remain positive, depending on the mass ratio M and the frequency ratio Q^2 according to the expressions in (2.29) or (2.39). In this case the system is dynamically stable despite of a negative damping κ_i, because the positive damping coefficient is capable of compensating the destabilizing effect due to the negative damping coefficient. In this case the system operates at its damping reserve. If the negative damping coefficient exceeds a certain range that depends on M and Q^2, then one of the modal damping coefficients Θ_{ii} becomes negative and the system without open-loop control becomes dynamically unstable. The critical values of the characteristic system parameters for which a system with fixed parameters κ_1, κ_2 starts to be dynamically unstable follows from

$$\varepsilon = 0: \quad \Theta_{11} = 0 \quad \text{or} \quad \Theta_{22} = 0 \quad \rightarrow \quad \begin{cases} M_{crit} & \text{for fixed } Q \\ Q_{crit} & \text{for fixed } M. \end{cases} \quad (12.3)$$

If a characteristic system parameter M or Q crosses its critical value found in (12.3), then the system without parametric excitation, $\varepsilon = 0$, becomes dynamically unstable. By activating a parametric excitation, $\varepsilon \neq 0$, it is possible to suppress the vibrations of systems with a negative modal damping coefficient Θ_{ii}, too. Nevertheless, note that according to the first Routh-Hurwitz condition in (3.23), the condition

$$\Theta_{11} + \Theta_{22} > 0 \quad (12.4)$$

have to be satisfied, which implies that it is not possible to stabilize system configurations for which both modal damping coefficients are negative, $\Theta_{11} < 0$ and $\Theta_{22} < 0$ simultaneously. At least one modal damping coefficient has to be positive. Furthermore, according to (12.4) the size of the positive modal damping coefficient has to compensate the negative modal damping coefficient. Note that the condition (12.4) does not imply the same restriction for the damping coefficients κ_i. In addition to the condition (12.4), the second Routh-Hurwitz condition (3.23) has to be satisfied, too. Rescaling the modal damping coefficients according to (12.2) gives the critical value for the most important control parameter, the amplitude amplification factor ε,

$$\varepsilon\sigma = 0 \quad \text{and} \quad \eta = |\Omega_1 - \Omega_2|: \quad \varepsilon^2 \Theta_{11}\Theta_{22} + \frac{Q_{12}Q_{21}}{4\Omega_1\Omega_2} = 0 \quad \rightarrow \quad \varepsilon_{crit}. \quad (12.5)$$

The control parameter ε has to exceed this critical value in order to obtain vibration suppression for a control frequency η at or near the parametric anti-resonance frequency $|\Omega_1 - \Omega_2|$.

12.1.1. System 1

In this section we analyze the method of vibration suppression for System 1 as introduced in Section 2.4.1. The equations of motion in normal form are given by (12.1), with system matrices and corresponding coefficients defined in (2.28), (2.29) and (2.30).

The following non-dimensional characteristic system parameters, as defined in (2.23), are used as default values:

$$M = 0.5, \quad Q = 2, \quad \kappa_1 = -0.01, \quad \kappa_2 = 0.14. \tag{12.6}$$

The damping coefficients κ_i satisfy the main stability condition (12.4),

$$\Theta_{11} + \Theta_{22} = \kappa_1 + \kappa_2 > 0, \tag{12.7}$$

according to (2.29). Note that the damping coefficient κ_1 is chosen to be negative and causes the conventional system without open-loop control ($\varepsilon = 0$) to become unstable if $M < M_{crit} = 2.58$ or $Q > Q_{crit} = 1.73$. The critical system parameters M_{crit} or Q_{crit} are determined by inserting (12.6) into (12.3) and represent the stability border for the system without open-loop control. With parametric excitation activated ($\varepsilon \neq 0$) it is possible to stabilize the system beyond the critical system parameters M_{crit} or Q_{crit} within a certain range of the system and control parameters. The original stability boundary is deformed, which creates an additional stability region.

Applying the numerical method from Section 11.1, three-dimensional stability bodies are obtained for fixed values of system parameters. Using stability bodies is a very compact way to describe graphically the regions of stability in dependency of three system and/or control parameters. This tool was introduced for the first time in [19]. Inside the stability body all eigenvalues of the monodromy matrix lie inside the unit circle and the system is stable, see (11.9). The boundary of this stability body represents value sets of parameters for which at least one eigenvalue is equal to one, while the remaining eigenvalues are smaller than one. Stability bodies obtained for fixed values of system parameters in (12.6) are presented in Figs. 12.1. These stability bodies show parameter values in dependency of the control parameters ε and η and system parameters M and Q, respectively. The stability boundary is indicated by the shaded surface. Combinations of ε, η, M or Q, which are enclosed by the shaded surface lead to a stable system. The thick-lined curves on this surface represent levels of equal values of the amplification factor ε. As one can see from these diagrams, highly complex results are obtained. For example the quite smooth body of Fig. 12.1a is cut in two pieces by a gap of instability starting at $\eta \approx 1.8$ and propagating into the region of stability.

As outlined in the preface, the parametric anti-resonance frequency of the system investigated is located at $|\Omega_1 - \Omega_2|$. With the parameter values in (12.6) we obtain

$$\Omega_1 = 0.93, \quad \Omega_2 = 2.15, \tag{12.8}$$

according to (2.26) and (2.27), and the parametric anti-resonance frequency becomes

$$\eta_0 = |\Omega_1 - \Omega_2| = 1.23. \tag{12.9}$$

The unstable conventional system ($\varepsilon = 0$) with the parameter set chosen in (12.6) lies outside the stability bodies in Figs. 12.1, which indicates a dynamically unstable system. The stability

12 Harmonic stiffness variation

body of the control parameters ε, η and the mass ratio M, as presented in Fig. 12.1a, does not reach up to the level $\varepsilon = 0$. Hence, any system is dynamically unstable for mass ratios lower than the critical mass ratio $M_{crit} = 2.58$ and other parameters chosen as in (12.6). The critical mass ratio M_{crit} is outside of the displayed range.

The same statements can be made for the stability body of the control parameters ε, η and the frequency ratio Q presented in Fig. 12.1b. The parameter set chosen in (12.6) for the unstable conventional system ($\varepsilon = 0$) corresponds to a point that lies above the critical frequency ratio $Q_{crit} = 1.73$ and therefore outside of the stability body. Systems with a frequency ratio Q below the critical value Q_{crit} are represented by the black area and are stable although $\kappa_1 < 0$. The system with the parameters as chosen in (12.6) has a frequency ratio $Q = 2$ which corresponds to a point outside of the stability body.

Activating the parametric excitation ($\varepsilon \neq 0$) deforms the straight line of the stability border at $\varepsilon = 0$ and $Q_{crit} = 1.73$ in Fig. 12.1b to a curved stability boundary as indicated by the thick-lined curves. Each curve corresponds to a fixed level of the amplitude amplification factor ε. The deformation of the stability boundary can lead to stability gain or loss. Parameter ranges of the parametric excitation frequency η for which the stability border is located beneath the critical parameter value Q_{crit}, e.g. near $\eta \approx 2.7$, are ranges of stability loss. On the other hand, ranges of η for which the stability border is raised to higher values than Q_{crit}, e.g. near $\eta \approx 1$, are ranges of stability gain. Based on the analytical predictions in Chapter 3.1 the major stability gain is expected to occur near the parametric anti-resonance frequency $\eta \approx 1.23$, according to (12.9). A further, but small stability gain is achieved near the combination frequency of the second order, $|\Omega_1 - \Omega_2|/2$, which corresponds to a parametric anti-resonance frequency, too. Note that the stability may not only occur exactly at the frequency $|\Omega_1 - \Omega_2|$, in addition a narrow but dense frequency interval of stability is established. Furthermore, note that selecting a parametric excitation frequency near the anti-resonance frequency is not sufficient to stabilize a system. In addition to the proper selection of the frequency, the amplitude amplification factor ε also has to exceed a critical value ε_{crit}. The amplification factor ε has to be large enough in order to shift the original system without parametric excitation ($\varepsilon = 0$) from the unstable region into the stable region inside the stability bodies, see Figs. 12.1. For the system parameters chosen in (12.6) the critical value of the amplification factor can be estimated from the analytical analysis, see the final formula in (12.5):

$$\varepsilon_{crit} = 0.093. \qquad (12.10)$$

For this example, this result matches perfectly the numerical value. A parametric excitation with $\varepsilon < \varepsilon_{crit}$ is not capable of stabilizing the system. By crossing the stability boundary at ε_{crit}, the system is stabilized when the control parameters satisfy the conditions $\varepsilon > \varepsilon_{crit}$ and $\eta \approx |\Omega_1 - \Omega_2|$.

If the control parameters ε and η of the parametric excitation are properly designed such that the system with the parameters in (12.6) is stabilized, then the vibrations are suppressed successfully. The stabilization of an originally unstable system is equivalent to vibration suppression, and so-called *damping by parametric excitation* occurs. As shown in Figs. 12.1 the unstable system can be stabilized for almost any values of the system parameters M and Q. Summarizing the method of damping by parametric excitation, the following restrictions for

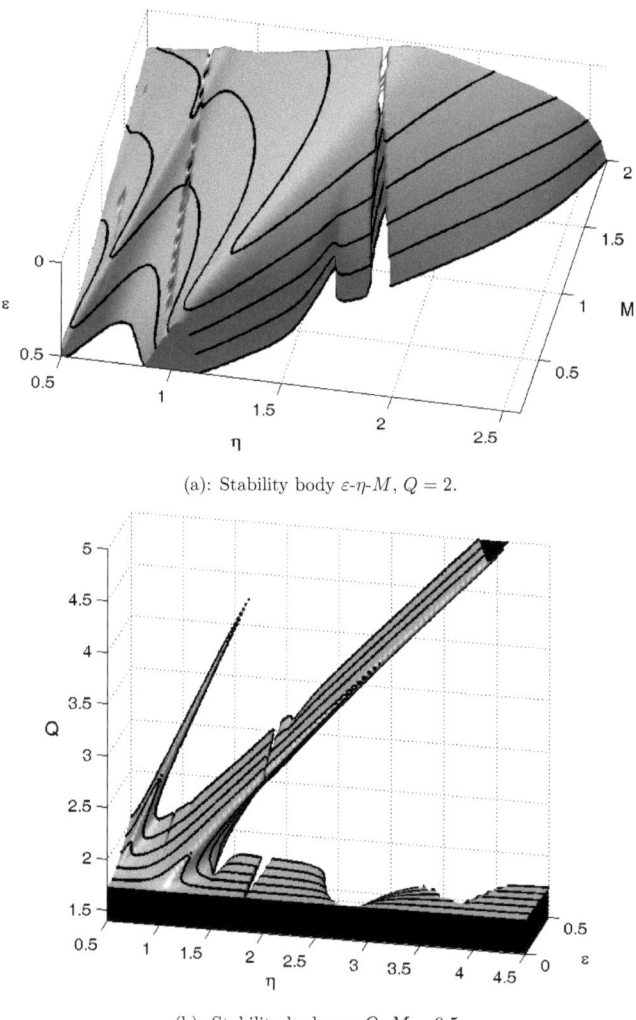

(a): Stability body ε-η-M, $Q = 2$.

(b): Stability body ε-η-Q, $M = 0.5$.

Figure 12.1.: Stability bodies of System 1 for the control parameters ε and η.

12 Harmonic stiffness variation

the control parameters ε and η can be concluded:

$$\varepsilon > \varepsilon_{crit} \quad \text{and} \tag{12.11}$$

$$\eta_{min} < \eta < \eta_{max} \quad \text{with} \quad \eta \approx \frac{|\Omega_1 - \Omega_2|}{n} \quad \text{and } n \in \mathbb{N}. \tag{12.12}$$

The first restriction demands that the amplitude amplification factor ε is large enough, while the second requires that the parametric excitation frequency η is located within a dense frequency interval around the parametric anti-resonance frequency $|\Omega_1 - \Omega_2|/n$.

For a detailed investigation of the structure of these stability bodies, we analyze just a slice of each body at a constant level of ε. The stability chart in Fig. 12.2a corresponds to the result in Fig. 12.1a for $\varepsilon = 0.3$. The shaded area indicates parameter values of the control parameter ε and the system parameter M for which the system is stable. This result, obtained by the numerical method, can be compared with the analytical results derived in Chapter 3.1 as shown in Fig. 12.2b. The further discussion refers to Fig. 12.2c, which is a combination of the figures above.

A key to the interpretation of the results are the parametric resonance frequencies of the first kind $2\Omega_1/n$ and $2\Omega_2/n$ and those of the second kind $|\Omega_1 \pm \Omega_2|/n$, with $n \in \mathbb{N}$. Resonance frequencies are plotted as functions of the mass ratio M, using dashed lines. According to the detailed analytical stability analysis in Chapter 3.1, each frequency line represent a skeleton line η_0 of stability intervals. The stability intervals of the frequencies of first order, $n = 1$, are calculated by a perturbation method of first order. This local stability analysis predicts intervals of the parametric excitation frequency η of the form

$$\eta_0 - \varepsilon\sigma_1 \leq \eta \leq \eta_0 + \varepsilon\sigma_1, \tag{12.13}$$

with the detuning factor σ_1 introduced in (3.4), similar to condition (12.12). Explicit expressions for the detuning factors, are summarized in Section 3.1.2 and can be found for $\eta_0 = 2\Omega_i$ in (3.19), for $\eta_0 = |\Omega_1 - \Omega_2|$ in (3.24) and for $\eta_0 = \Omega_1 + \Omega_2$ in (3.29). The detuning factor describes a stability boundary in the parameter space next to the frequency line η_0. For the frequencies of the first kind these boundaries enclose always unstable parameter values as it is shown in Fig. 3.1 on page 29, while for the frequencies of the second kind these boundaries may surround also stable parameter values, as it is summarized in Tab. 3.1 on page 36. For the system parameters considered in (12.6), the frequencies $2\Omega_1$ and $\Omega_1 + \Omega_2$ lead to an unstable system, while the frequency $|\Omega_1 - \Omega_2|$ represent the skeleton line of the main area of vibration suppression. To obtain explicit expressions at frequencies of the second order, $n = 2$, a perturbation analysis of second order is required. Due to the fact that the stability interval, which corresponds to the detuning factor σ_1, of the anti-resonance frequencies of higher order, $n \geq 2$, is always smaller than the interval of the main anti-resonance, a higher order perturbation is not performed.

As predicted by the analytical analysis, the main area of stability occurs near the parametric anti-resonance frequency $|\Omega_1 - \Omega_2|$. Hence, the frequencies that lie inside the interval defined in

(12.13) correspond to a dynamically stable system. The boundaries of this interval are plotted as a function of the mass ratio M and result in the thin solid lines that lie symmetrical to the skeleton line $|\Omega_1 - \Omega_2|$. The stability boundary gets wider for increasing M, until the critical value $M_{crit} = 2.58$ is reached. At $M = M_{crit}$ a modal damping parameter Θ_{ii} vanishes according to (12.3) and the detuning factor σ_1 in (12.13) becomes infinite, according to the formulae in (3.24). It is remarkable, that although the analytical results represent approximations for small ε, even for the rather large value $\varepsilon = 0.3$ the analytical and numerical results agree amazingly

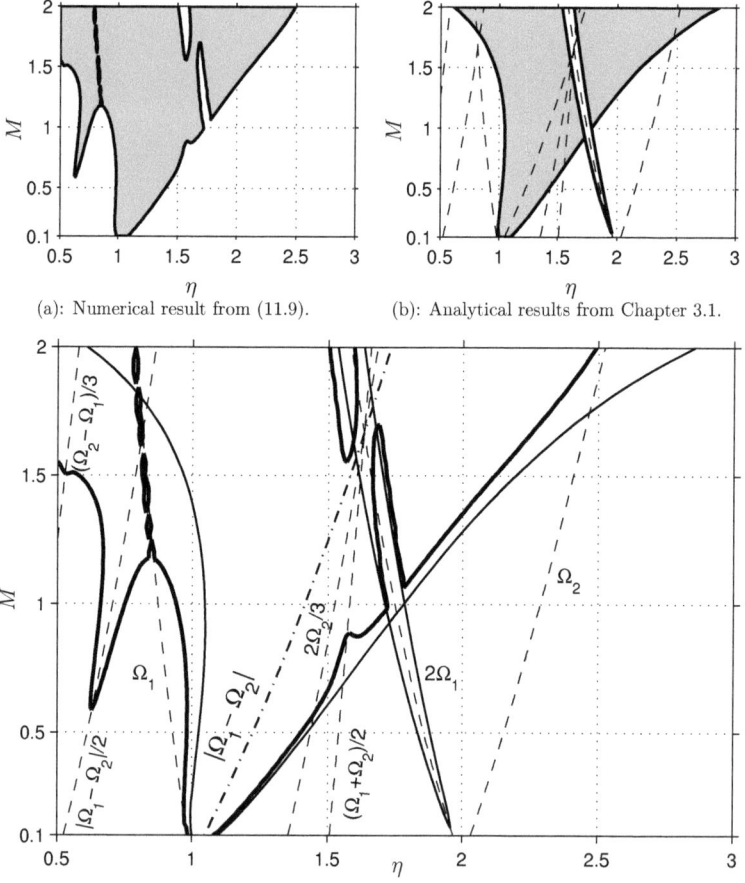

(a): Numerical result from (11.9).

(b): Analytical results from Chapter 3.1.

(c): Comparison of analytical (thin solid lines) and numerical results (thick solid lines).

Figure 12.2.: Two dimensional slice of the stability body from Fig. 12.1a for the control parameter value $\varepsilon = 0.3$ and the parameter set as defined in (12.6).

12 Harmonic stiffness variation

well if the isolated contribution of the parametric anti-resonance frequency η_0 is considered. An additional narrow stability area is caused by the second order anti-resonance frequency $|\Omega_1 - \Omega_1|/2$. Even a parametric anti-resonance frequency of third order is recognizable, but the deformation of the stability boundary is negligible. Analytical results for the intervals of stability and instability were derived from a local analysis near a certain parameter frequency. The transition between these frequencies cannot be calculated by this method. However, there is always a smooth shape that connects two stable frequency intervals that, in the worst case, can be interrupted by an intersection with an instability interval. Hence, the transition of stability boundary of the anti-resonance frequency $|\Omega_1 - \Omega_2|$ to the stability boundary of $|\Omega_1 - \Omega_2|/2$ cannot be reproduced correctly. The size of the stability area in Fig. 12.2c near $\eta \approx 1$ is even underestimated.

The previously mentioned gap in Fig. 12.1a turns out to be an area of instability caused by the parametric resonance frequency $2\Omega_1$. According to the analytical analysis, in this case the boundaries (12.13) enclose a region of instability. Its boundaries are plotted as thin solid lines following the frequency line $2\Omega_1$ in Fig. 12.2c. The analytical solution agrees very well, except for an interruption of the small stretch of instability near the intersection point of the frequencies line $|\Omega_1 - \Omega_2|$ and the frequency line $2\Omega_1$. At this special point the relation $\Omega_1/\Omega_2 = 1/3$ holds, that corresponds to a 1:3 resonance of the system. The primary resonance frequency of third order, $2\Omega_1/3$, and the combination resonance frequency of second order, $(\Omega_1 + \Omega_2)/2$ also pass through this point. Although many frequencies superimpose at this intersection point, the stabilizing anti-resonance frequency still dominates in a narrow region. The frequency $(\Omega_1 + \Omega_2)/2$ corresponds to an unstable region that causes a small dent in the main stability boundary near $\eta \approx 1.5$. Nevertheless, this region of stability loss is separated from the region of stability gain near $|\Omega_1 - \Omega_2|$ by the combination resonance of the frequency $2\Omega_1/3$. This can be confirmed by an examination of the eigenvalues of the monodromy matrix. Finally, another extremely narrow stretch of instability crossing the stable region is found for the primary resonance of second order $2\Omega_1/2 = \Omega_1$.

Although the stability bodies in Fig. 12.1 look very different, similar features can be found in both of them. Figure 12.3a shows a slice of body Fig. 12.1b at $\varepsilon = 0.3$ and Fig. 12.3b shows the corresponding analytical result. Finally a detailed comparison of the numerically and analytically derived stability boundaries is presented in Fig. 12.3c. Again an additional region of stability is achieved by a parametric excitation at a frequency at or near $\eta_0 = |\Omega_1 - \Omega_2|$. The stability boundaries according to (12.13) correspond to the thin solid lines that follow the frequency line $|\Omega_1 - \Omega_2|$. The detuning factor σ_1 is infinite at $Q = Q_{crit} = 1.73$, because here one of the modal damping parameters Θ_{ii} vanishes, as can be seen in (12.3). An additional but small stability region is created at the anti-resonance frequency of the second order $|\Omega_1 - \Omega_2|/2$. On the other hand, a significant loss of stability occurs near the combination resonance frequency $\Omega_1 + \Omega_2$. The primary resonance frequencies $2\Omega_1$ and Ω_1 have minor but still noticeable influence on the stability area and the corresponding boundaries of these unstable regions are plotted as thin solid lines.

Knowing the system parameters as in (12.6) the analytical formulae in Tab. 3.1 on page 36 allow the determination of stability borders in advance. Not only the intervals of stability, but also the troublesome and limiting intervals of instability are presented. Furthermore, the critical

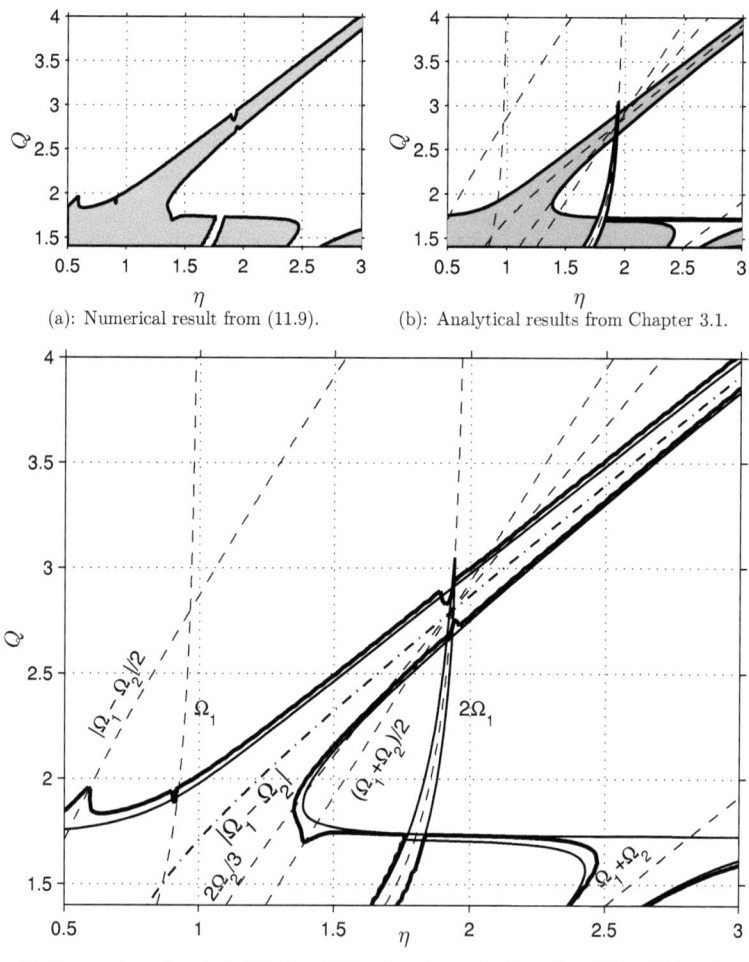

(a): Numerical result from (11.9).   (b): Analytical results from Chapter 3.1.

(c): Comparison of analytical (thin solid lines) and numerical results (thick solid lines).

Figure 12.3.: Two dimensional slice of the stability body from Fig. 12.1b for the control parameter value $\varepsilon = 0.3$ and the parameter set as defined in (12.6).

12 Harmonic stiffness variation

system parameters are calculable. The analytical results agree very well with the numerical ones. It is worth to mention that with the presented open-loop control it is not only possible to stabilize an otherwise unstable system but it can be kept completely at rest.

For the parameter set in (12.6) and an amplification factor of $\varepsilon = 0.3$, as used in Figs. 12.2 and 12.3, the following relation holds

$$\frac{\varepsilon^2 Q_{12} Q_{21}}{\Omega_1 \Omega_2 (\Theta_{11} - \Theta_{22})^2} = 1.3 \geq 1. \qquad (12.14)$$

Respecting the rescaling of the damping parameters in (12.2), the condition in (8.11) is satisfied. The fact that evaluating (12.14) for the system parameters leads to a value of 1.3, which exceeds 1.0, indicates that the maximum equivalent damping possible is achieved. Hence, the resulting time series are expected to be similar to the one presented in Fig. 8.2b, at least on the time scale $1/\varepsilon$. In general, a value larger than one in (12.14) is always advisable. On one hand, this does not enhance the equivalent damping that can be achieved, but, on the other hand, a larger value increases the detuning factor σ_1 in (12.13). Hence, the equivalent damping may not be increased, but the additionally created stability region very well. Furthermore, note that the maximum equivalent damping is only achieved near the frequency line $\eta_0 = |\Omega_1 - \Omega_2|$, while near the stability border the size of equivalent damping that is additionally introduced by parametric excitation is rather small. The maximal value is reached near the mean value, $\sigma = 0$, which corresponds to the skeleton line η_0. In the case where the expression on the left hand side in (12.14) is larger than one, the maximum equivalent damping can be achieved in a wider region near this frequency. Further details about equivalent damping can be found in Chapter 8.

In analogy to Fig. 12.1, Figure 12.4 shows three-dimensional stability bodies for the system parameters in (12.6) in dependency of the control parameter ε and η and a non-dimensional damping parameter κ_i. The same effects of stability and instability occur as described in the previous figures. Again, the stability border of the conventional system, $\varepsilon = 0$, is a straight line. In analogy to (12.3) we can calculate the critical damping values $\kappa_{1,crit}$ or $\kappa_{2,crit}$ by keeping the remaining parameters fixed,

$$\varepsilon = 0: \quad \Theta_{11} = 0 \quad \text{or} \quad \Theta_{22} = 0 \quad \rightarrow \quad \begin{cases} \kappa_{1,crit} = -0.005 & \text{for fixed } \kappa_2 \\ \kappa_{2,crit} = 0.27 & \text{for fixed } \kappa_1. \end{cases} \qquad (12.15)$$

These values border the black-colored stability body at $\varepsilon = 0$, see Figs. 12.4a and b.

Activating a parametric excitation, $\varepsilon \neq 0$, the stability boundary is deformed. The stability boundary is raised towards smaller values compared to $\kappa_{i,crit}$ near the main anti-resonance frequency $|\Omega_1 - \Omega_2| = 1.23$, as shown in (12.9), and the anti-resonance frequency of the second order $|\Omega_1 - \Omega_2|/2 = 0.61$. Three parametric resonances occur in the frequency range displayed: a large primary resonance at $2\Omega_1 = 1.86$ and a narrow one at $\Omega_1 = 0.93$, and a small combination resonance at $(\Omega_1 + \Omega_2)/2 = 1.54$. The decay of the shaded surface near $\eta \approx 2.5$ in Fig. 12.4b is caused by the destabilizing combination resonance frequency $\Omega_1 + \Omega_2$.

The main difference between the stability bodies in Fig. 12.1 and Fig. 12.4 is, that now the parametric frequencies remain unchanged for different values of the system parameters, because the natural frequencies Ω_1 and Ω_2 are independent of the damping parameter κ_i, as can be verified with (2.27). Hence, other than in the stability charts in Figs. 12.2c and 12.3c, now the frequency lines in the presented stability charts in Fig. 12.5 are straight lines.

Two interesting cuts of the stability body in Fig. 12.4b at a small and a high level of ε are presented in Fig. 12.5 and compared to analytically derived stability curves according to (12.13). As mentioned, the frequency lines are now straight lines. The thick solid boundary lines that lie symmetrical to the anti-resonance frequency $|\Omega_1 - \Omega_2|$ enclose a stable parameter domain while the thick solid lines following the primary resonance $2\Omega_1$ enclose unstable parameter values.

For a small value of the amplification factor, $\varepsilon = 0.08$, an interesting isolated region occurs, as it is shown in Fig. 12.5a. This small inclusion may not be of practical interest, but the analytical formulae are capable of predicting its border amazingly well. We need to take a closer look at the detuning factor σ_1 as defined in (12.13) at the anti-resonance frequency $\eta_0 = |\Omega_1 - \Omega_2|$ in order to understand better how such a phenomenon can happen.

Starting with high damping values κ_2, both modal damping parameters are positive, $\Theta_{11} > 0$

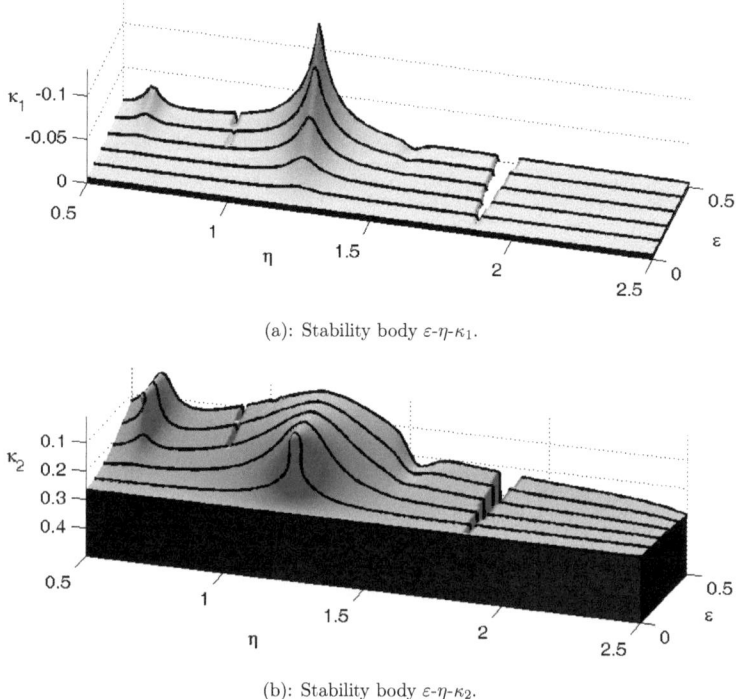

(a): Stability body ε-η-κ_1.

(b): Stability body ε-η-κ_2.

Figure 12.4.: Stability bodies of System 1 for damping coefficients κ_1 and κ_2.

12 Harmonic stiffness variation

and $\Theta_{22} > 0$, and the detuning factor σ_1 is purely imaginary. Hence, the system is stable in the entire frequency range, see Fig. 12.5a. Decreasing κ_1 down to its critical value $\kappa_{2,crit} = 0.27$ as listed in (12.15), one of the modal damping parameters Θ_{ii} vanishes and σ_1 becomes real-valued, but infinite. When further decreasing κ_2, a modal damping parameter becomes negative, while the other remains positive. In addition, $\Theta_{11} + \Theta_{22} > 0$ is satisfied and the real-valued σ_1 decreases until $\kappa_2 \approx 0.2$ is reached, where the relation in (12.5) holds, and therefore σ_1 vanishes. In the value range $\kappa_2 \approx 0.2$ down to ≈ 0.07 the detuning factor σ_1 is purely imaginary, as can be concluded from Tab. 3.1 on page 36. At $\kappa_2 \approx 0.07$ we arrive at the inclusion. At this point, again the relation in (12.5) is satisfied and the former purely imaginary σ_1 becomes real-valued, $\sigma_1 = 0$. By further decreasing, we pass the inclusion and reach the value $\kappa_2 \approx 0.02$. Lower values of κ_2 lead always to an unstable system, for which σ_1 is purely imaginary.

The stability chart for a high value of ε is shown in Fig. 12.5b. Again, the frequency line $2\Omega_1/3$ separates the stable region at the anti-resonance frequency $|\Omega_1 - \Omega_2|$ from the unstable region at $(\Omega_1 + \Omega_2)/2$.

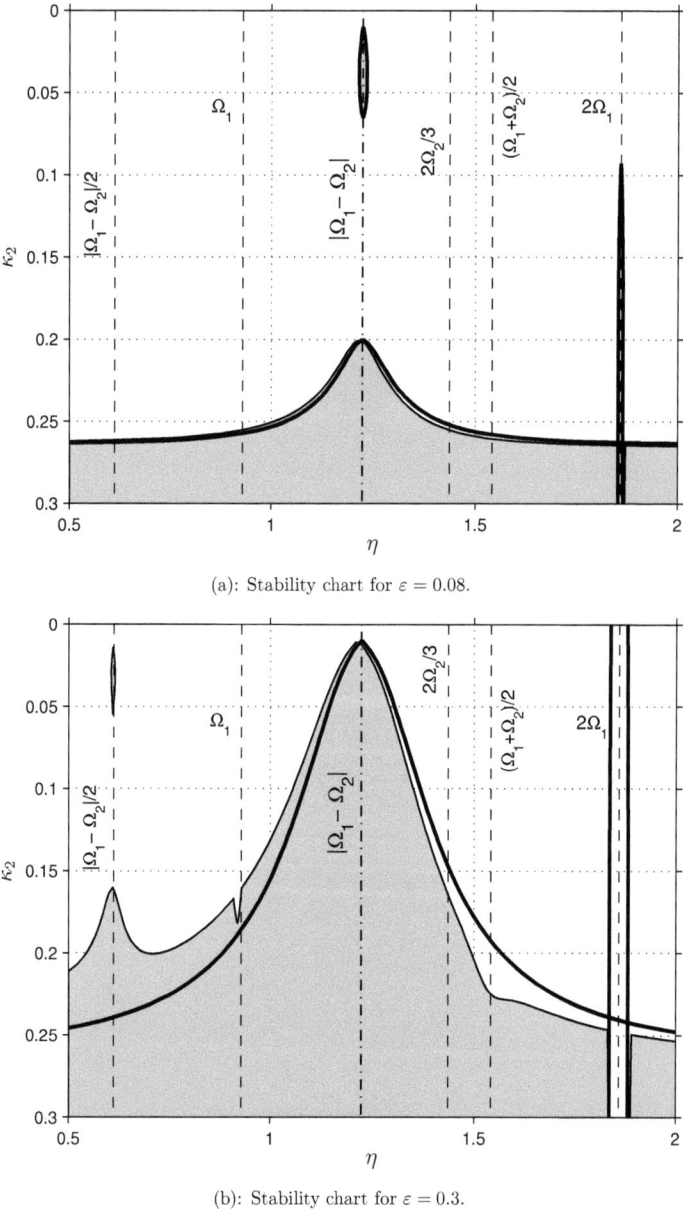

(a): Stability chart for $\varepsilon = 0.08$.

(b): Stability chart for $\varepsilon = 0.3$.

Figure 12.5.: Stability charts for the damping coefficient κ_2 at small and large value of the amplification factor ε, resulting from Fig. 12.4b. Comparison of analytical (thick solid lines) and numerical results (shaded area).

12 Harmonic stiffness variation

Finally, three-dimensional stability bodies are presented in Fig. 12.6 showing the dependency on the mass ratio M, the frequency ratio Q and the parametric excitation frequency η. The stability body for the conventional system without parametric excitation, $\varepsilon = 0$, is shown in Fig. 12.6a. In this case, of course, the frequency η has no influence on the stability boundary surface. The stability body of a system with parametric excitation at a fixed amplification factor $\varepsilon = 0.3$ can be found in Fig. 12.6b. The black-colored region at $Q = 2$ is equal to the stability chart described in Fig. 12.2a. Hence, introducing a parametric excitation deforms the stability body in Fig. 12.6a into the stability body in Fig. 12.6b. By tuning the parametric excitation frequency η, this deformation allows to operate a system for combinations of the parameters Q and M for which the conventional system would be dynamically unstable.

For multi-dimensional parameter studies the numerical Floquet method has proven to be computationally efficient and reliable. By representing the results as three-dimensional stability bodies the complex behavior of the parametrically excited system can be visualized and advantages and disadvantages of this open-loop control method become evident. The results show that parametric excitation at an appropriate frequency can be employed to extend significantly the area of stability of systems with negative damping, as for the example in (12.6). Analytical solutions for regimes of stability and of instability caused by parametric excitation were obtained in Chapter 3.1. For the investigated System 1 these results agree quite well with numerical results and are very valuable for obtaining a quick overview on the effectiveness of the proposed method.

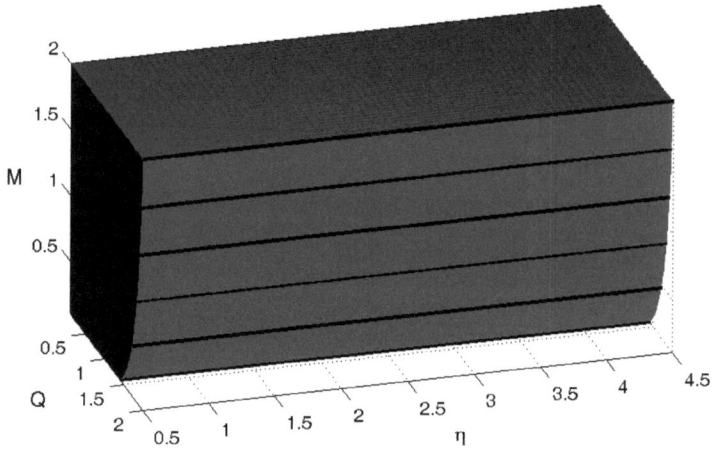

(a): System without parametric excitation $\varepsilon = 0$.

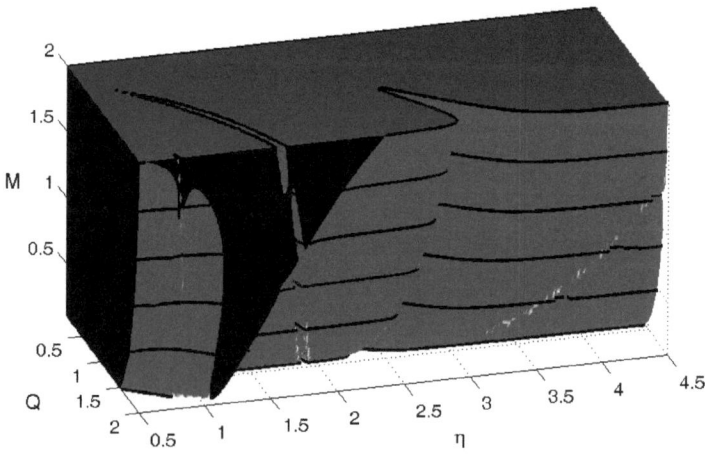

(b): System with parametric excitation $\varepsilon = 0.3$.

Figure 12.6.: Stability body M-Q-η of System 1 with and without self-excitation.

12.1.2. System 2

In this section we analyze the method of vibration suppression for System 2 as introduced in Section 2.4.2. The equations of motion in normal form representation are given by (12.1), with system matrices and corresponding coefficients defined in (2.38), (2.39) and (2.40).

The following non-dimensional characteristic system parameters, as defined in (2.34), are used as default values

$$M = 0.1, \quad Q = 1.15, \quad \kappa_1 = 0.05, \quad \kappa_2 = -0.05. \tag{12.16}$$

The damping coefficients κ_i satisfy the main stability condition (12.4),

$$\Theta_{11} + \Theta_{22} = (1 + M)\kappa_1 + \kappa_2 > 0, \tag{12.17}$$

according to (2.39). The system parameters in (12.16) lead to the natural frequencies $\Omega_1 = 0.93$ and $\Omega_2 = 1.62$ according to (2.37). From the analytical analysis as discussed in Chapter 3.1 the main parametric anti-resonance frequency is expected at

$$\eta_0 = |\Omega_1 - \Omega_2| = 0.69. \tag{12.18}$$

The critical system parameters M_{crit} and Q_{crit} are determined, in analogy to System 1, by inserting (12.6) into (12.3). In opposite to System 1, this System 2 possesses two critical values within the parameter range displayed in Fig. 12.7, which are

$$Q_{crit,1} = 0.934 \quad \text{and} \quad Q_{crit,2} = 0.964. \tag{12.19}$$

The negative damping coefficient κ_2 causes the conventional system without open-loop control ($\varepsilon = 0$) to be unstable if the frequency ratio Q is chosen outside the value range $[0.934, 0.964]$, as is the case for the parameter set chosen in (12.16).

A stability body for the set of parameters (12.16) in dependency of the frequency ratio Q and the control parameter ε and η is shown in Fig. 12.7. The critical parameter values of Q, see (12.19), represent the stability border for the conventional system without open-loop control, which can be found at the bottom of the stability body. With parametric excitation activated ($\varepsilon \neq 0$) it is possible to stabilize the system within a certain range of parameters, which lie beyond the critical system parameters $Q_{crit,1}$ and $Q_{crit,2}$ in (12.19). The original stability boundary is deformed, which creates additional stability regions. Contrary to the stability body obtained for System 1 in Fig. 12.1b, where only one critical frequency ratio occurred, the stability body of System 2 in Fig. 12.7 possesses two critical values. For System 2 two main stability regions are created, that are located to the left and to the right of the originally ($\varepsilon = 0$) stable domain. Note that the stalactite-like structure of the stability body in these regions of tapering boundary surfaces result from a too low numerical resolution in these regions. The real stability body possesses smooth boundary surfaces here.

Stability charts for small and high amplification factors, respectively, are presented in Figs. 12.8 and 12.9. The stability charts in Fig. 12.8 are typical for amplification factors ε up to the value of 0.2. For such small values of ε the stability boundaries obtained from the numerical and the analytical analysis agree quite well. At $\varepsilon \approx 0.2$ the shaded surface in Fig. 12.7 is interrupted, until further increasing of ε finally breaks the smooth stability boundary into individual parts, a boundary splitting occurs. Stability charts for such high values of ε are shown in Fig. 12.9. In these cases the analytical approximations derived in this work cannot describe the stability boundary correctly anymore. Nevertheless, a first order approximation of the stability boundary curves describes appropriately the numerical results in the vicinity of $\sigma_1 \approx 0$, as can be seen for instance in Fig. 12.8b for $Q \approx 0.72$ and $Q \approx 1.29$. Hence, the stability boundary resulting from the analytical formulae allows to predict the value range of the system parameter Q within which vibration suppression may occur.

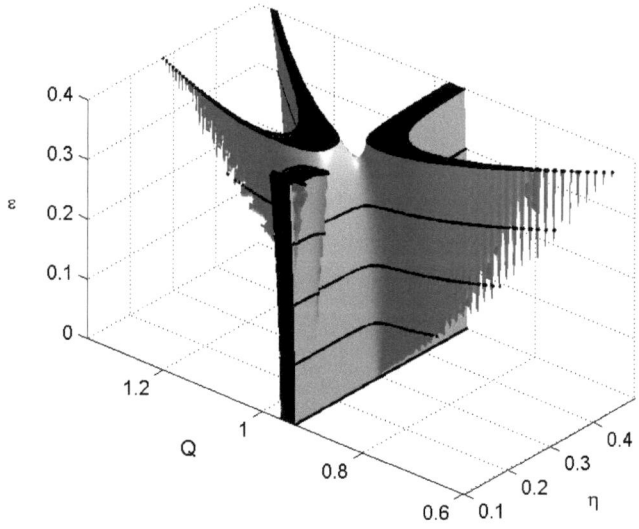

Figure 12.7.: Stability body Q-ε-η of System 2.

The analytical analysis performed in Chapter 3.1 is based on a perturbation of first order with respect to a small amplitude amplification factor $\varepsilon \ll 1$. This analysis is not able to describe split stability boundaries as it is shown in Fig. 12.8b. If one wants to capture the effects of boundary splitting of this special system appropriately in order to describe the stability border in the entire parameter domain, then an approximation by a perturbation of second order of ε, as described in [61], would be necessary.

12 Harmonic stiffness variation

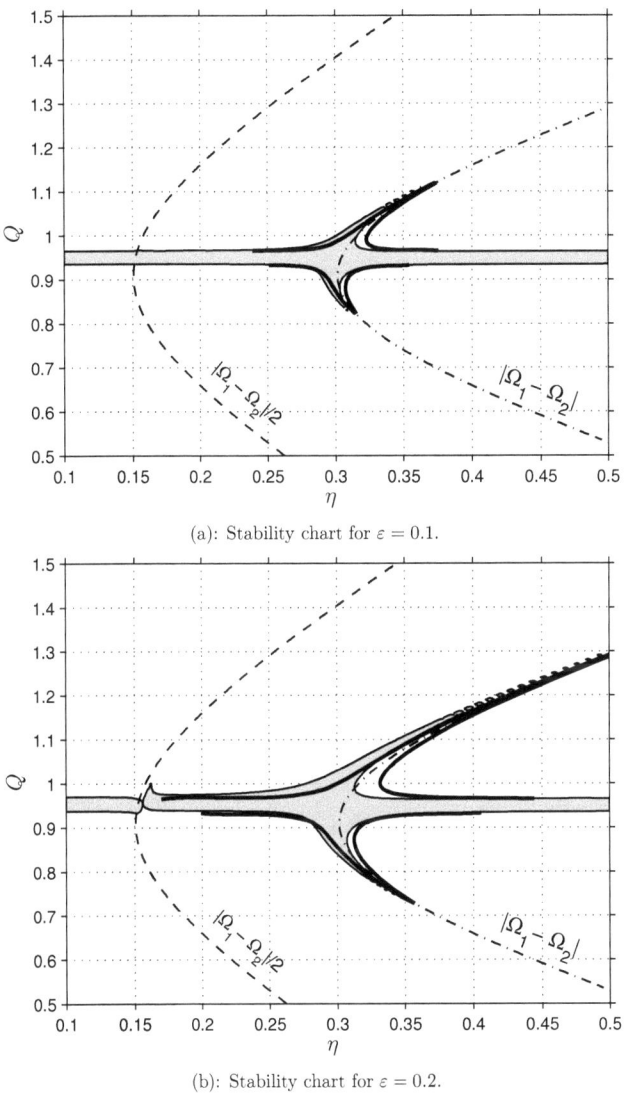

(a): Stability chart for $\varepsilon = 0.1$.

(b): Stability chart for $\varepsilon = 0.2$.

Figure 12.8.: Stability charts for frequency ratio Q at small values of the amplification factor ε, resulting from Fig. 12.7. Comparison of analytical (thick solid lines) and numerical results (shaded area).

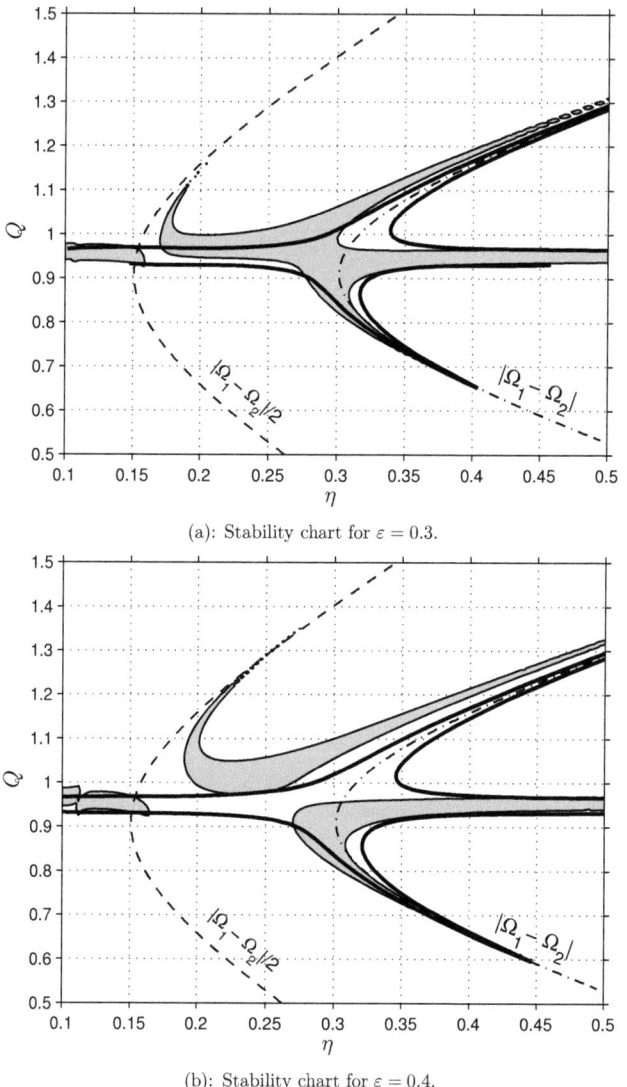

(a): Stability chart for $\varepsilon = 0.3$.

(b): Stability chart for $\varepsilon = 0.4$.

Figure 12.9.: Stability charts for frequency ratio Q at large values of the amplification factor ε, resulting from Fig. 12.7. Comparison of analytical (thick solid lines) and numerical results (shaded area).

12.1.3. Coupled pendulum system

In this section we analyze the method of vibration suppression for a coupled pendulum system, as presented in Fig. 12.10. The system consists of two pendulums with lumped masses m_1, m_2 attached at distances l_1, l_2. The angular displacement with respect to the vertical axis is denoted by φ_1, φ_2. A gravitational field g is acting on both masses in vertical direction. Viscous damping forces are assumed to act in the joints of the pendulums, described by constant damping coefficients c_1, c_2. One pendulum is self-excited, characterized by a negative damping coefficient $c_1 < 1$, while the other pendulum is parametrically excited due to a harmonic vertical motion with the amplitude ε and the frequency η of its pivot point. Both pendulums are coupled by a linear torsional stiffness k_{12}. This system was investigated in [26] and [17].

Since we are interested in the stability of the trivial solution $\varphi_1, \varphi_2 = 0$, we may approximate $\sin \varphi_i$ by its first order approximation φ_i. The normal form of the linearized equations of motion possesses the following structure (see [26] for more details)

$$\ddot{\Phi} + \Theta\dot{\Phi} + \left(\Omega^2 + \varepsilon\eta^2 \mathbf{Q}\cos(\eta\tau)\right)\Phi = 0, \qquad (12.20)$$

with the vector of angular displacements $\Phi = (\varphi_1, \varphi_2)^T$ and the symmetric parametric excitation matrix $\mathbf{Q}$. By comparing the equations (12.20) and (12.1), the difference between the pendulum system and System 1 and System 2, as analyzed in the previous sections, is revealed. The parametric excitation matrix in (12.20) is multiplied by the factor ε like in (12.1). However, in case of the pendulum system, this matrix is multiplied additionally by the expression η^2, which leads to a distortion of the resulting stability boundary curves.

In analogy to (2.34) for System 2, we introduce the characteristic parameter $Q = \omega_1/\omega_2 = \sqrt{l_2/l_1}$, with the natural frequencies of the subsystems $\omega_i = \sqrt{g/l_i}$. A stability body of system (12.20) in dependency of the parameter Q and the control parameters ε and η is presented in Fig. 12.11a. Similar to Fig. 12.8b, a stability chart for the small amplitude of $\varepsilon = 0.2$ is shown in Fig. 12.11b. Due to the fact that the parametric excitation matrix $\mathbf{Q}$ in (12.20) is symmetric, the anti-resonance frequency is located again at $|\Omega_1 - \Omega_2|$. However, for the pendulum system in Fig. 12.11b the prefactor η^2 leads to divergent boundary curves for small values of Q and a negligible spike for high values of Q, while for System 2 in Fig. 12.8b the boundary curves are of similar size for small and large values of Q, respectively.

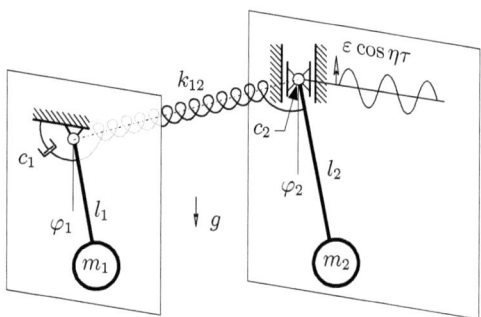

Figure 12.10.: Coupled pendulum system.

12 Harmonic stiffness variation

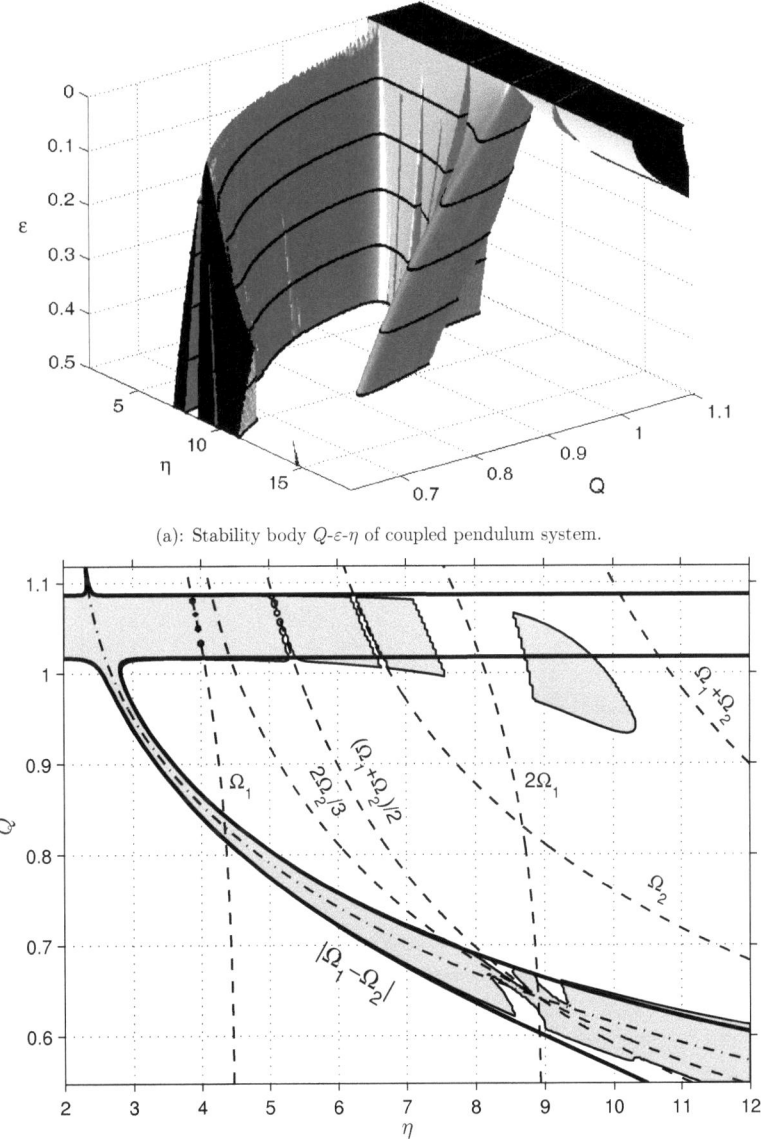

(a): Stability body Q-ε-η of coupled pendulum system.

(b): Stability chart for the control parameter $\varepsilon = 0.2$, resulting from Fig. 12.11a. Comparison of analytical (thick solid lines) and numerical results (shaded area).

Figure 12.11.: Stability analysis of coupled pendulum system for system parameter Q and control parameters ε and η.

12.2. Optimal shape of a stiffness variation

In Chapter 3.3 we performed the stability analysis of a system with a general time-periodic stiffness variation that can be represented by Fourier series, as introduced in (3.53). Applying a first order perturbation, we derived analytical stability conditions that depend on the Fourier coefficients. Based on this result, an optimization was carried out in Section 10.3 for a time-periodic stiffness variation that can be represent by Fourier cosine series, with the aim to maximize the stable frequency interval near a parametric anti-resonance frequency.

By means of these results we can optimize the parametric excitation of System 1 as defined in Section 2.4.1. We compare the effects of different parametric excitation on System 1. Time series and the corresponding Fourier coefficients of different periodic functions of the parametric excitation can be found in Tab. 10.1 on page 168. Two different shapes are considered here: the cosine function, as originally introduced in the equations of motion (12.1) of System 1, and the optimal rectangular function. The values chosen for the system parameters are found in (12.6).

By rescaling the damping parameters as introduced in (12.2), the Routh-Hurwitz conditions from (10.64), which correspond to the main parametric resonance $|\Omega_1 - \Omega_2|$ ($n = 1$), are adapted to

$$\Theta_{11} + \Theta_{22} > 0,$$
$$\Delta^{(1)} = \Theta_{11}\Theta_{22} + \frac{(\varepsilon a_1)^2}{4\Omega_1\Omega_2} Q_{12}Q_{21} > 0, \qquad (12.21)$$

where a_1 represents the leading Fourier coefficient for System 1 as defined in (3.38). The detuning factor derived in (10.63) at $|\Omega_1 - \Omega_2|$ results in

$$\varepsilon\sigma_1^{(1)} = \frac{\Theta_{11} + \Theta_{22}}{2}\sqrt{-\frac{\Delta^{(1)}}{\Theta_{11}\Theta_{22}}}. \qquad (12.22)$$

The conditions in (12.21) decide whether system is stabilized by parametric excitation at the main anti-resonance frequency or not. Condition (12.22) specifies the width of the frequency interval around this anti-resonance frequency, according to (12.13). for which vibration suppression occurs, too. By introducing a time-periodic parametric excitation, instead of a harmonic excitation, the prefactor of the parametric excitation term $Q_{12}Q_{21}$ in (12.21) changes and we can define the following equivalent amplification factor

$$\varepsilon_{eq} \equiv \varepsilon a_1. \qquad (12.23)$$

Hence, the equivalent amplification factor ε_{eq} is enhanced for $a_1 > 1$ and diminished for $a_1 < 1$.

For a time-harmonic parametric excitation $a_1 = 1$ holds and the equivalent amplification factor ε_{eq} is equal to the original factor ε. The factor ε_{eq} is larger than ε in case of the optimum parametric excitation of rectangular shape, where $a_1 = 1.27 > 1$ holds. Time series of these two time-periodic stiffness functions are plotted in Fig. 12.12, similar to Fig. 10.12, where k_m

12 Optimal shape of a stiffness variation

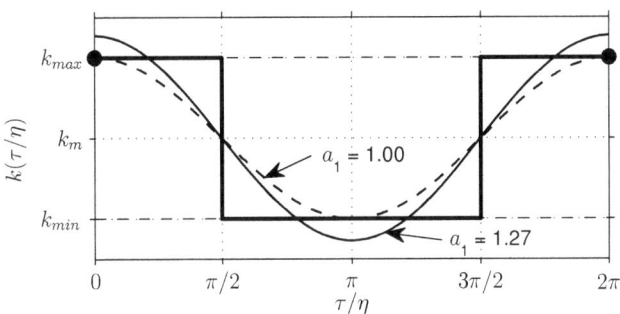

Figure 12.12.: Leading Fourier component a_1 of a harmonic and a rectangular parametric excitation.

represents the constant mean value of the stiffness parameter k_{02}, and the extremum stiffness values k_{min} and k_{max} correspond to $k_m(1 \mp \varepsilon)$. The dashed function represents the time period of a harmonic stiffness variation. The thick-lined function is the optimum rectangular function and the thin-lined function corresponds to its leading Fourier component ($n = 1$). From the point of view of the stability conditions in (12.21) and (12.22), the rectangular function simulates a higher effective amplification factor ε_{eq}, although the maximum and minimum values of both function shapes are the same. Thus, the parametric excitation term $Q_{12}Q_{21}$, which would arise for a cosine function, is adapted to $a_1^2 Q_{12}Q_{21}$. Due to the stability condition (12.21), a higher equivalent amplification factor ε_{eq} enhances the prefactor of the parametric excitation term and allows higher negative values of the modal damping term $\Theta_{11}\Theta_{22}$.

Time histories of the displacement x_1 of System 1 for a harmonic stiffness variation and for the optimal rectangular stiffness variation are illustrated in Fig. 12.13, for an amplification factor $\varepsilon = 0.2$ and a parametric excitation frequency $\eta = |\Omega_1 - \Omega_2|$. Hence, the amplification factor is higher than the critical value obtained in (12.5) resulting from (12.10), $\varepsilon > \varepsilon_{crit}$, and vibration suppression is possible. Note that the vibration amplitudes of x_2 look similar and are therefore omitted. The comparison of these time series shows that the vibration is suppressed faster using a rectangular shaped stiffness variation with an equivalent amplification factor of $\varepsilon_{eq} = 0.254$. Note that according to the condition (12.14) in Chapter 8,

$$\frac{\varepsilon^2 Q_{12}Q_{21}}{\Omega_1 \Omega_2 (\Theta_{11} - \Theta_{22})^2} = 0.58 < 1, \qquad (12.24)$$

the maximum equivalent damping is not achieved by a single harmonic stiffness variation with $\varepsilon = 0.2$. The change to a rectangular shaped stiffness variation results in a value of 0.94, which indicates that the maximum equivalent damping is achieved, although the original amplification factor ε remains constant.

Stability bodies for System 1 with a single harmonic parametric excitation were derived in Fig. 12.1. The result for an amplification factor of $\varepsilon = 0.2$ corresponds to a slice of each stability

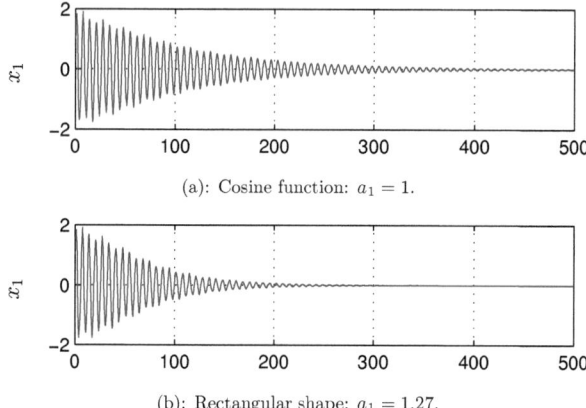

(a): Cosine function: $a_1 = 1$.

(b): Rectangular shape: $a_1 = 1.27$.

Figure 12.13.: Comparing time series for system response for different time-periodic functions at $\varepsilon = 0.2$.

body. These slices result in stability charts similar to the ones in Figs. 12.2 or 12.3 that were obtained for $\varepsilon = 0.3$, but with smaller stability regions. Using a rectangular shaped parametric excitation leads to a different stability chart, that corresponds to slices of the stability bodies at an equivalent amplification factor $\varepsilon_{eq} = 0.254$, although the actual amplification factor remains at $\varepsilon = 0.2$. This enhancement of ε_{eq} leads to a wider stability region near the anti-resonance frequency because

$$\frac{\varepsilon \sigma_1^{(n)}}{\varepsilon \sigma_1} > 1 \quad \text{for} \quad a_1 > 1,$$

according to (12.22). Moreover, a longer section of the frequency line $\eta_0 = |\Omega_1 - \Omega_2|$ lies in the stability region, see Fig. 12.1. However, this positive effect on the transient behavior near the anti-resonance frequency, reverses outside the stability region and leads to wider regions of instability near parametric resonance frequencies.

12.3. Synchronous stiffness and damping variation

In the previous sections we gained deeper insight in stabilizing dynamically unstable systems by harmonic stiffness variation. We showed that the method of damping by parametric excitation is capable of amplifying the existent damping in the system. In the following we examine System 3, as introduced in Section 2.4.3, in which two types of parametric excitation are applied, a simultaneous and synchronous stiffness and damping variation. Synchronous means that the time-harmonic variations of the damping and stiffness coefficient operate with the same phase. In practical applications this type of variation appears naturally, since the realization of a

12 Synchronous stiffness and damping variation

mechanical device for stiffness variation almost certainly induces also a change of its damping with the same phase, and therefore synchronous with its stiffness.

The normal form of the equations of motion of System 3, a system which is under the influence of this special parametric excitation, can be written as

$$\ddot{\mathbf{z}} + (\boldsymbol{\Theta} + \varepsilon \mathbf{R} \cos \eta \tau) \dot{\mathbf{z}} + \left(\boldsymbol{\Omega}^2 + \varepsilon \mathbf{Q} \cos \eta \tau \right) \mathbf{z} = \mathbf{0}, \quad (12.25)$$

with the position vector $\mathbf{z} = (z_1, z_2)^T$, the parametric excitation matrices $\mathbf{Q}$, $\mathbf{R}$ and a small amplification factor ε. Note that the modal damping matrix $\boldsymbol{\Theta}$ is rescaled in comparison to the equations introduced in Chapter 2.3 or 5, as it was performed for System 1 in (12.1). The system under consideration possesses symmetric parametric excitation matrices only. For vanishing parametric excitation matrix $\mathbf{R}$, the equations of motion (12.25) of System 3 are equal to the equations of motion (12.1) of Systems 1. In this case the parametric excitation introduced is capable of suppressing the system vibrations, as it was outlined in the last sections. On the other hand, damping by parametric excitation for a system with a vanishing parametric stiffness excitation matrix $\mathbf{Q}$, i.e. a time-periodic damping variation only, is not possible, as shown in Chapter 4. Such a system would possess only resonances, both at combination frequencies of the summation type, $(\Omega_1 + \Omega_2)/n$, and the difference type, $|\Omega_1 - \Omega_2|/n$ with $n \in \mathbb{N}$. In the following, we will investigate how a stiffness variation interacts with a synchronous variation of the damping coefficient when both coefficients are varied with the same frequency $|\Omega_1 - \Omega_2|$. The question is, whether the anti-resonance will dominate due to the stabilizing stiffness variation or the resonance due to the destabilizing damping variation.

The considered System 3 represents an extension of System 1 by an additional time-periodic damping coefficient. Therefore, the modal damping parameters Θ_{ii} and the coefficients of the stiffness excitation Q_{ij} are equal to the coefficients of System 1 as defined in (2.29) (2.30). The coefficients of the damping excitation R_{ij} can be found in (2.45). The same non-dimensional parameter set is used here as for System 1 in (12.6), which leads to the same main anti-resonance frequency $|\Omega_1 - \Omega_2| = 1.23$ as in (12.9).

Extended stability conditions (5.7) and (5.8) were obtained in Section 5.2 for this type of system. For a pure harmonic stiffness excitation, $\mathbf{R} = \mathbf{0}$, these conditions simplify to the equivalent conditions derived in Chapter 3.1. For a pure harmonic damping excitation, $\mathbf{Q} = \mathbf{0}$, these conditions simplify to the equivalent conditions derived in Chapter 4. The frequency interval of vibration suppression for System 3 follows from (5.8) and (3.4) to

$$|\Omega_1 - \Omega_2| + \varepsilon \sigma_s - \varepsilon \tilde{\sigma}_{k+c} \leq \eta \leq |\Omega_1 - \Omega_2| + \varepsilon \sigma_s + \varepsilon \tilde{\sigma}_{k+c}, \quad (12.26)$$

where $2\varepsilon \tilde{\sigma}_{k+c}$ determines the width of the frequency interval. The expression $\varepsilon \sigma_s$ represents a shift of the anti-resonance frequency $|\Omega_1 - \Omega_2|$ as it was shown in Fig. 5.1 on page 81. Note that the resultant stability width $\tilde{\sigma}_{k+c}$ of the synchronous stiffness and damping variation is not equivalent to the sum of the stability width due to the stiffness variation $\tilde{\sigma}_k$ and the (in)stability width due to the damping variation $\tilde{\sigma}_c$, see (5.8), (3.24) and (4.14),

$$\tilde{\sigma}_{k+c} \neq \tilde{\sigma}_k + \tilde{\sigma}_c. \qquad (12.27)$$

In case of a pure stiffness variation as in (12.1), or $\mathbf{R} = \mathbf{0}$ in (12.25), as well as in case of a pure damping variation, $\mathbf{Q} = \mathbf{0}$ in (12.25), the displacement $\varepsilon\sigma_s$ of the skeleton line of the stability boundary curve does not occur. In both cases the skeleton line is defined by the anti-resonance frequency $|\Omega_1 - \Omega_2|$.

In order to examine the difference between System 1 and System 3, we compare the analytical results obtained for vibration suppression. Figure 12.14 shows stability charts of System 1 and System 3, respectively, in dependency of the non-dimensional system parameter Q and the non-dimensional parametric excitation frequency η for an amplitude amplification factor $\varepsilon = 0.3$. The stability charts in Fig. 12.14a and Fig. 12.3c are equivalent. The shaded region indicates the numerically calculated stability domain. The thick-lined curve shows the analytical stability border in Fig. 12.14a calculated from (3.24) and in Fig. 12.14b from (5.8). The parametric resonance frequencies of the first kind, $2\Omega_1/n$ and $2\Omega_2/n$, and those of the second kind $(\Omega_1 \mp \Omega_2)/n$, with $n = 1, 2$ are plotted as dashed lines. The thin dash-dotted line in Fig. 12.14a indicates the frequency line of the anti-resonance frequency $|\Omega_1 - \Omega_2|$, which represents already the skeleton line of the corresponding thick-lined stability boundary curve for System 1, according to (12.13). This frequency line is plotted for System 3 in Fig. 12.14b as a dashed line. According to (12.26) an additional shift $\varepsilon\sigma_s$ from this anti-resonance frequency occurs, which is indicated by a dash-dotted line. This shift becomes infinite near the stability boundary of the conventional system with $\varepsilon = 0$, where $\Theta_{11} = 0$ or $\Theta_{22} = 0$ is satisfied, analogous to the width $\tilde{\sigma}_{k+c}$ of the stability interval. The solid line indicates the final stability boundary curve. The analytical stability boundaries corresponding to the resonance frequencies $2\Omega_1$, $2\Omega_2$ and $\Omega_1 + \Omega_2$ are not drawn here, to make the plots easier to read.

Comparing the stability charts in Fig. 12.14a and b reveals the effect of the additional shift as predicted by (12.26). The dash-dotted skeleton line for pure stiffness variation in Fig. 12.14a is moved to the dash-dotted skeleton line for a combined synchronous stiffness and damping variation in Fig. 12.14b. Compared to the stability boundary for System 1, the stability boundary for System 3 is deformed due to the additional displacement $\varepsilon\sigma_s$. Note that although a pure damping variation is always destabilizing, in combination with a synchronous stiffness variation even a small improvement near $\eta = 1.5$ is obtained. On the other hand, the deformation of the skeleton line near $\eta = 0.7$ corresponds to a loss of stability. Nevertheless, the stability domain near the anti-resonance frequency $|\Omega_1 - \Omega_2|$ is only disturbed, but not destroyed by the interaction with the destabilizing damping variation. Consequently, the stability region gained and created by the stiffness variation is only disturbed, but not destroyed by the interaction with the destabilizing damping variation. The main stable domain is conserved, in some regions even enlarged. The analytical analysis predicts an additional shift of the skeleton line of the

12 Synchronous stiffness and damping variation

stability boundary due to the combination of stiffness and damping variation that explains this small loss and the small enlargement of the stable domain very well.

Finally, a three-dimensional stability body for a fixed amplification factor of $\varepsilon = 0.3$ in dependency of the mass ratio M, the frequency ratio Q and the parametric excitation frequency η is presented in Fig 12.15. A slice of this body at $M = 0.5$ is depicted by the stability chart in Fig. 12.14b. Hence, introducing a parametric excitation in System 3 deforms the stability body of the conventional system without parametric excitation in Fig. 12.6a, or in Fig. 12.15a for an extended value range, to the stability body in Fig. 12.15b. Again as for System 1, choosing the proper parametric excitation frequency η allows to use combinations of M and Q that would otherwise destabilize the conventional system, $\varepsilon = 0$.

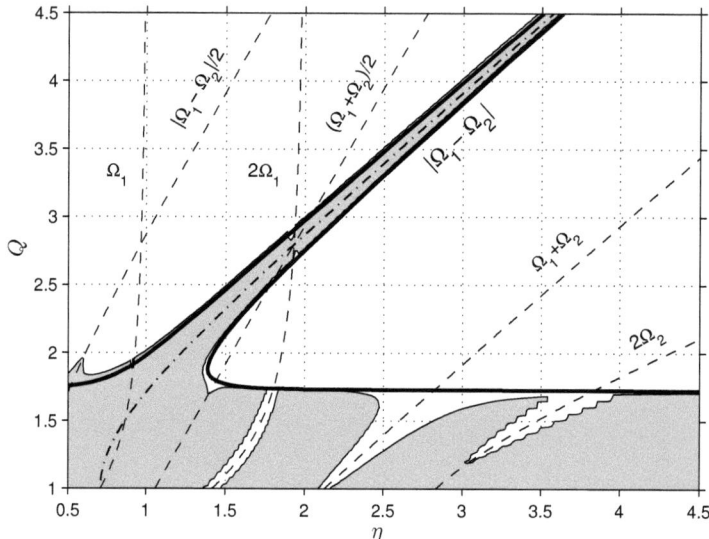

(a): Comparison of analytical (thick solid line) and numerical analysis (shaded area) for pure stiffness variation in System 1 (see Fig. 12.3c).

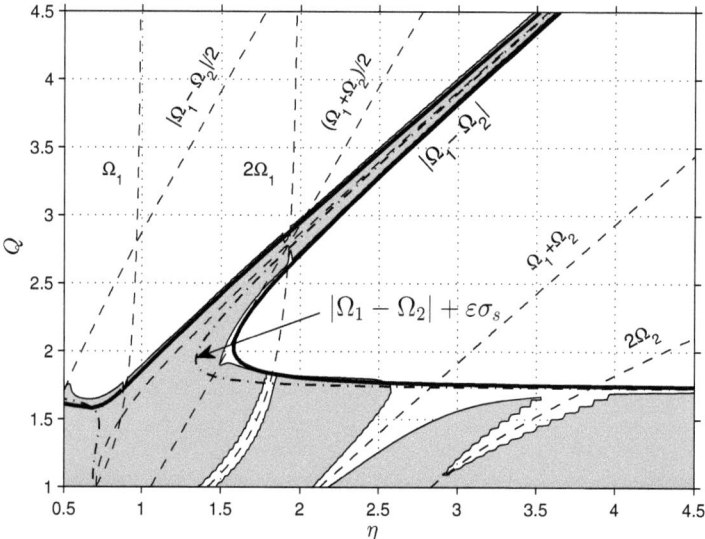

(b): Comparison of analytical (thick solid line) and numerical analysis (shaded area) for synchronous stiffness and damping variation in System 3.

Figure 12.14.: Stability charts Q-η of System 1 and System 3 for the control parameter $\varepsilon = 0.3$.

12 Synchronous stiffness and damping variation

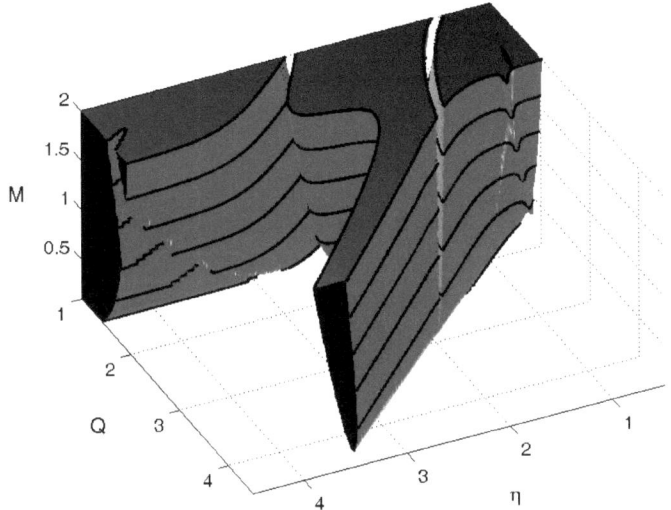

(a): Pure stiffness variation in System 1 (see Fig. 12.6b).

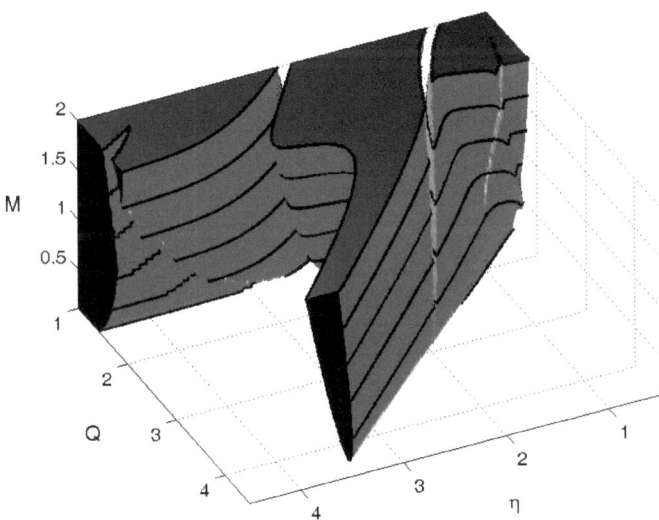

(b): Synchronous stiffness and damping variation in System 3.

Figure 12.15.: Comparison of stability bodies M-Q-η of System 1 and System 3 for the control parameter $\varepsilon = 0.3$.

12.4. Inertia variation

The investigations of System 1, System 2 and System 3 verify that the vibrations of a dynamically unstable system can be fully suppressed by introducing a parametric excitation with a frequency η near to the anti-resonance frequency $|\Omega_1 - \Omega_2|$. In this section we analyze whether a harmonic inertia variation, as defined in System 4 in Section 2.4.4, is also capable of stabilizing an unstable system. For the exemplary lumped mass system in Fig. 2.1 on page 11, it is hard to imagine how to realize a time-periodic variation of one or more mass coefficients. On the other hand, if we think of a rotary system, where the inertia coefficients correspond to inertia moments, instead of masses, a harmonic inertia variation can be realized simply by variation of the radius of gyration. In this case an inertia variation degenerates to a variation of a length, which is an easily tunable geometric size. For rotary systems the type of the degrees of freedom change from displacements to rotary angles.

The normal form of the equations of motion of System 4 results from the first order approximation in (2.52) or (6.2) and is given by

$$\ddot{z} + \left(\Theta + \varepsilon\eta \mathbf{S} \sin \eta\tau\right)\dot{z} + \left(\Omega^2 + \varepsilon \mathbf{Q} \cos \eta\tau\right)\mathbf{z} = \mathbf{0}, \tag{12.28}$$

with the position vector $\mathbf{z} = (z_1, z_2)^T$, the parametric excitation matrices $\mathbf{Q}$, $\mathbf{S}$ and a small amplification factor ε. Note that the modal damping matrix Θ is rescaled in comparison to the equations introduced in Chapter 2.3 or 6, as it was performed for System 1 in (12.1). Hence, in a first order approximation a harmonic inertia variation is equivalent to an simultaneous stiffness and damping variation with a phase shift of $\pi/2$, an asynchronous stiffness and damping variation.

The non-dimensional characteristic system parameters used are found in (12.6), except for a modification of the mass ratio, $M = 4.5$. Due to the fact that the original system matrices in (2.52) possess symmetric and skew-symmetric parts, we cannot conclude immediately whether the anti-resonance will occur at the combination frequency $|\Omega_1 - \Omega_2|$ or at $(\Omega_1 + \Omega_2)$. Hence, we have to insert the values chosen for the system parameters chosen into condition (6.11) and (6.15), respectively. For a parametric excitation at frequency $\eta_0 = \Omega_1 - \Omega_2$, with $\Omega_1 > \Omega_2$, in (6.11) as well as at frequency $\eta_0 = \Omega_1 + \Omega_2$ in (6.15) the parametric excitation terms result in

$$(Q_{12} + \eta_0 \Omega_2 S_{12})(Q_{21} - \eta_0 \Omega_1 S_{21}) = 0.24 = (Q_{12} - \eta_0 \Omega_2 S_{12})(Q_{21} - \eta_0 \Omega_1 S_{21}). \tag{12.29}$$

Respecting that for the parameter values chosen System 4 is dynamically unstable and the relation $\Theta_{11}\Theta_{22} < 0$ holds, the conditions (6.11) and (6.15) reveal that the combination frequency $\eta_0 = \Omega_1 - \Omega_2$ corresponds to an anti-resonance frequency and the frequency $\Omega_1 + \Omega_2$ to a resonance frequency. For stable system behavior the parametric excitation frequency η has to lie inside the interval

$$\Omega_1 - \Omega_2 - \varepsilon\sigma_1 \leq \eta \leq \Omega_1 - \Omega_2 + \varepsilon\sigma_1, \tag{12.30}$$

12 Inertia variation

as defined in (12.13). The detuning factor σ_1 can be calculated from (6.12).

A comparison between analytical predictions and numerical calculations for an amplitude amplification factor $\varepsilon = 0.3$ is shown in Fig. 12.16, where the shaded area indicates parameter combinations of Q and η that lead to a stable system behavior, and therefore damping by parametric excitation can be achieved. The thin dash-dotted line corresponds to the main anti-resonance frequency $\Omega_1 - \Omega_2$, that represents the skeleton line of the corresponding thick-lined stability region according to (12.30). Instability boundaries corresponding to the resonance frequencies $2\Omega_1$ and $\Omega_1 + \Omega_2$ are determined from the relations in (6.8) and (6.16). The combination frequency $(\Omega_1 - \Omega_2)/2$ leads to a very small stability gain near $Q \approx 2$, while the combination frequency $(\Omega_1 + \Omega_2)/2$ yields a small stability loss. Finally, as discussed in Fig. 12.2c for System 1, the frequency line $2\Omega_2/3$ separates the thick-lined stable region that corresponds to $|\Omega_1 - \Omega_2|$ from the unstable region corresponding to $(\Omega_1 + \Omega_2)/2$.

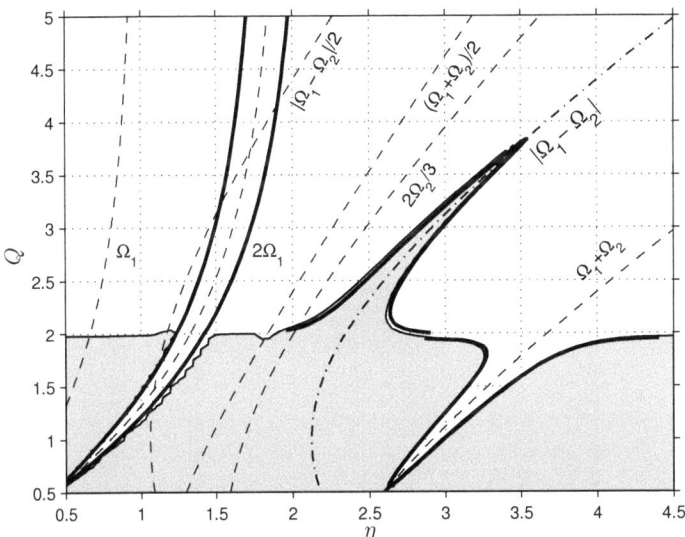

Figure 12.16.: Stability chart Q-η for the control parameter $\varepsilon = 0.3$. Comparison of analytical (thick solid line) and numerical results (shaded area).

A three-dimensional stability body, similar to Fig. 12.15, is presented in Fig. 12.17 for a fixed amplitude amplification factor $\varepsilon = 0.2$. The surface at the top of this body corresponds to the stability chart in Fig. 12.16.

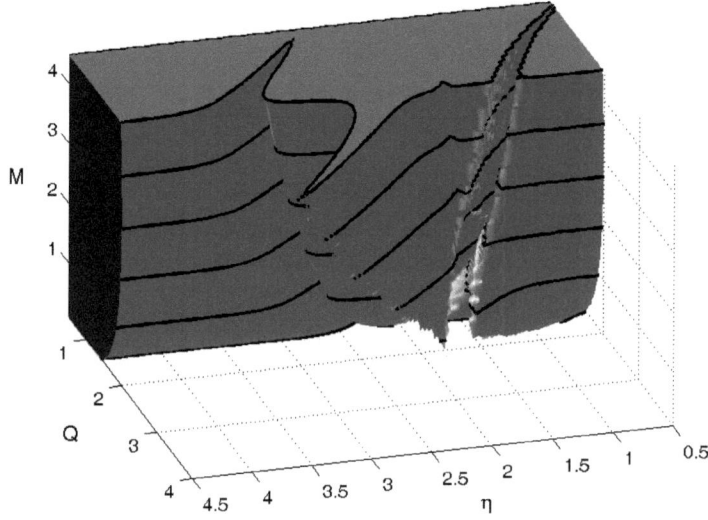

Figure 12.17.: Stability body M-Q-η of System 4 for the control parameter $\varepsilon = 0.3$.

13. Conclusions and Outlook

This work deals with the investigation of mechanical systems with linear equations of motion with time-periodic coefficient matrices. In Chapter 2 these equations of motion are transformed to their quasi-normal forms. Four exemplary mechanical systems with time-periodic coefficients are presented. These systems possess just two degrees of freedom and represent examples for minimum configurations that are capable of achieving damping by parametric excitation.

The further investigations are subdivided into three parts. In the first part a thorough stability analysis for the proposed method of damping by parametric excitation is carried out for systems with two degrees of freedom. Such systems may be lumped mass systems or may represent two modes of a continuous structure. This part is rather generally formulated and can be used as a guide for designing a device for vibration suppression by parametric excitation. From previous studies it is known that time-periodic stiffness coefficients may introduce additional damping into a system if the parametric excitation frequency is near to a parametric combination frequency of the system. In this work it is shown that time-periodic coefficients cannot create damping in a system, but amplify the already existent damping of the system. In particular, it is revealed that damping by parametric excitation can even be applied to systems without self-excitation. Special attention is paid to symmetry properties of the parametric excitation matrices and their influence on the location of the parametric anti-resonance frequencies. If the parametric excitation matrices are symmetric, the anti-resonance frequency is located near to a parametric combination frequency of the difference type, $|\Omega_i - \Omega_j|/n$, $n \in \mathbb{N}$ and for skew-symmetric parametric excitation matrices near a parametric combination frequency of the summation type, $(\Omega_i + \Omega_j)/n$. It is demonstrated that vibration suppression by time-periodic damping coefficients is possible only in case of a skew-symmetric parametric excitation matrix. In opposite to time-periodic stiffness coefficients, in this case vibration suppression can be achieved at the parametric combination frequency of the difference as well as the summation type. Moreover, the interaction between time-periodic stiffness and damping coefficients is considered. Such an interaction is capable of suppressing vibrations as well, but leads to an additional shift of the parametric anti-resonance frequency from the corresponding parametric combination frequency. It is demonstrated that in a first order approximation a time-harmonic inertia variation is equivalent to a simultaneous stiffness and damping variation with a phase shift of $\pi/2$ and that it is capable of being used as a tool for damping by parametric excitation as well. Finally, conditions for vibration suppression and the influence of symmetry of the parametric excitation matrices for the most general linear system with time-harmonic coefficients are presented.

In Part II, the great benefit of the analytically approximated stability boundary curves becomes evident. The investigations allow a deeper insight into the proposed method of vibration suppression. Using the formulae from Part I the maximum eigenvalue of the dominant motion, which is related to the equivalent damping, is estimated. It is revealed that a parametric excitation leads to a coupling of eigenvalues of the homogenous equations of motion. Moreover, the maximum equivalent damping achievable by parametric excitation is derived. These studies are the basis for an extension to systems with more degrees of freedom, which shows that under certain conditions a parametric excitation at an anti-resonance frequency $|\Omega_i - \Omega_j|/n$ or $(\Omega_i + \Omega_j)/n$ leads to a coupling of the ith and jth mode, while the remaining modes are not affected. The important consequence is that a parametric excitation with a single frequency can only stabilize one unstable mode. Moreover, it is shown that by activating a parametric excitation energy is transferred into the system, which leads to a more effective energy dissipation than in the conventional system. The existent system damping is amplified and vibrations are suppressed successfully.

Finally, an optimization for a system with multiple time-harmonic stiffness coefficients is performed for arbitrary amplitude amplification factors and phases between these variations. For a restricted overall parametric excitation, the optimum is found to be a configuration with just one maximum amplification factor, while the remaining factors vanish. It depends on the system parameters which one of these factors should be maximized. The effect of additional constraints on the design of the amplitude amplification factors is investigated, too, and leads to an adaptation to the result above. Additionally, for a system with a single time-periodic stiffness coefficient the optimum function shape in case of restricted amplitude amplification factors is found to be the simple rectangular shape, which corresponds to a bang-bang open-loop control. It is shown that only by adjusting the shape of the parametric excitation the effective amplitude amplification factor is enlarged by 27% and leads to higher equivalent damping values.

In Part III, the predictions of stability and instability boundary curves derived analytically in Part I are compared to results obtained numerically by using the computationally efficient Floquet theory. By representing the numerical results as three-dimensional stability bodies the complex behavior of the parametrically excited system can be visualized and advantages as well as disadvantages become evident. Systems with time-harmonic stiffness, damping and inertia coefficients are considered. Without loss of generality, only systems with symmetric parametric excitation matrices and self-excitation are investigated. For the systems investigated analytical predictions agree quite well with numerical results and are very valuable for obtaining a quick overview on the effectiveness of the proposed method, provided that the amplitude amplification factor is small. From a mathematical point of view, first order perturbation, as performed in Part I, is not allowed for large values of this factor, but may be a good choice for obtaining the range of system parameter values within which damping by parametric excitation occurs. The analytically derived formulae represent a reliable tool to describe the additional stability domain created by parametric excitation near an anti-resonance frequency as well as the instability domain caused by parametric excitation near resonance frequencies. Even locally isolated stability domains are described appropriately by the analytical approximation. Moreover, it is shown that in case of simultaneous appearance of time-harmonic stiffness and damping coefficients, the additional stability domain created by the stiffness variation is only disturbed, but not destroyed, by the interaction with the destabilizing damping variation. The main stable domain is conserved, in some regions even enlarged.

The results demonstrate that parametric excitation at an appropriate frequency can be employed to extend significantly the area of stability of systems with or without self-excitation. The proposed method shows great potential in practical applications when a destabilization due to self-excitation occurs or when the damping of weakly damped systems shall be enhanced. This method can be even applied to shorten the period of the transition behavior of weakly damped systems with impulse-like external force acting on them. In this case, the damping during the transient behavior is amplified and the transition is accelerated by parametric excitation at an anti-resonance frequency, while the reaction due to the external excitation remains unchanged – a key for the design of sensors with short response time, but high sensitivity.

There are many directions of further research in which the present work can be extended. Since this work deals exclusively with single-frequent parametric excitation, one important topic is to analyze multi-frequency parametric excitation, especially, in the context of systems with multiple degrees of freedom. The short analysis in Section 8.4 reveals that a single frequency parametric excitation is capable of stabilizing only one unstable mode, that motivates the conclusion that multi-frequency parametric excitations with different anti-resonance frequencies are capable of stabilizing several unstable modes. However, an increase in the number of parametric excitation frequencies, increases the number of possible parametric resonance frequencies as well.

Another task would be to perform higher order approximations of the dominant motion for, at least, two reasons. On one hand, to be able to obtain analytical formulae that describe the stability boundary curves of systems similar to System 2 (see results in Fig. 12.8 on page 193). These formulae enable even closer approximations of boundary curves of systems similar to System 1, see results in Fig. 12.3. On the other hand, such an approximation allows to derive an explicit expression for distinguishing the completely different shapes of the stability boundary curves of System 1 and System 2 – a switching condition that decides whether an approximation of first order or of second order should be applied.

Since in a real system parameters may fluctuate or may not always hit the predetermined design values, a further task would be to investigate the influence of such disturbances on the effectiveness of damping by parametric excitation. By using the analytical formulae presented in Part I it is an easy task to study the sensitivity and robustness of the proposed method of vibration suppression. Another important task is to examine how robust the natural frequencies can be identified, even online, for a specific system.

This work was mainly motivated by the pioneering work of A. Tondl [53]. Since the completion of the present work as [10] several analytical and numerical investigations on systems with variable stiffness [14, 15, 16], damping [13] and inertia-like [23] parameters were performed and application to new systems examined [18, 20, 21, 23]. Especially important is the work [24] in which the stability analysis based on a first order perturbation is compared to an analysis of second order perturbation. The aforementioned switching condition to decide which approximation order is sufficient could not be found so far.

From the viewpoint of a practical application, a very important task not outlined in this work is to verify the method of vibration suppression experimentally. In the meantime, the concept was proven experimentally on laboratory systems. Two realizations were designed and finished by the author and collaborators. For the first proof-of-concept experiment a tension-controlled elastic structure was attached to elastically suspended cars on an air track [22]. The next step towards real systems was the study [20] in which a mechatronic device consisting of current-controlled coils in interaction with permanent magnets controlled the stiffness of a flexible cantilever system. Real-life experiments are still missing but the preliminary results are promising. Hopefully this work serves as a motivation for further activities in the field of parametric excitation.

Bibliography

[1] Abadi. *Nonlinear dynamics of self-excitation in autoparametric systems*. PhD thesis, Utrecht University, Netherlands, 2003.

[2] H. Bartsch. *Taschenbuch Mathematischer Formeln*. Fachbuchverlag Leipzig, 19 edition, 1999.

[3] H. Billharz. Bemerkung zu einem Satze von Hurwitz. *Zeitschrift für angewandte Mathematik und Mechanik (ZAMM), Applied Mathematics and Mechanics*, 24(2):77–82, 1944.

[4] R. Bishop. *Vibration*. Cambridge University Press, 1979.

[5] I. Blekhman. *Selected topics in vibrational mechanics*, volume 11 of *Stability, vibration and control of systems*. World Scientific, 2nd edition, 2004.

[6] V. Bolotin. *Kinematische Stabilität elastischer Systeme*. VEB Deutscher Verlag der Wissenschaften, Berlin, 1961.

[7] M. Cartmell. *Introduction to linear, parametric and nonlinear vibrations*. Chapman and Hall, 1990.

[8] F. Dohnal. Application of the averaging method on parametrically excited 2dof-systems. Technical report, Institute for Machine Dynamics and Measurement, Vienna University of Technology, 2003.

[9] F. Dohnal. Suppression of self-excited vibrations by non-smooth parametric excitation. In *Proceedings of Dynamics of Machines 2004*, pages 27–34, Prague, Czech Republic, Feb 2004. Institute of Thermomechanics, Academy of Sciences of the Czech Republic.

[10] F. Dohnal. *Damping of mechanical vibrations by parametric excitation*. PhD thesis, Vienna University of Technology, Austria, 2005.

[11] F. Dohnal. Suppression of self-excited vibrations by synchronous stiffness and damping variation. In *Proceedings of Dynamics of Machines 2005*, pages 23–30, Prague, Czech Republic, Feb 2005. Institute of Thermomechanics, Academy of Sciences of the Czech Republic.

[12] F. Dohnal. Vibration suppression of self-excited oscillations by parametric inertia excitation. In *Proceedings of the 76th Int. Conf. of Gesellschaft für Angewandte Mathematik und Mechanik (GAMM)*, Luxembourg, 2005.

[13] F. Dohnal. Suppressing self-excited vibrations by synchronous and time-periodic stiffness and damping variation. *Journal of Sound and Vibration*, (306):136–152, 2007.

[14] F. Dohnal. Damping by parametric stiffness excitation – resonance and anti-resonance. *Journal of Vibration and Control*, (14):669–688, 2008.

[15] F. Dohnal. General parametric stiffness excitation – anti-resonance frequency and symmetry. *Acta Mechanica*, (196):15–32, 2008.

[16] F. Dohnal. Optimal dynamic stabilisation of a linear system by periodic stiffness excitation. *Journal of Sound and Vibration*, (320):777–792, 2009.

[17] F. Dohnal and H. Ecker. Suppressing self-excited vibrations in a coupled pendulum system. In *CD-Rom Proceedings of the 21st International Congress of Theoretical and Applied Mechanics (ICTAM)*, Warsaw, Poland, August 2004.

[18] F. Dohnal, H. Ecker, and H. Springer. Enhanced damping of a cantilever beam by a periodic axial force. *Archive of Applied Mechanics*, (78):935–947, 2008.

[19] F. Dohnal, H. Ecker, and A. Tondl. Vibration control of self-excited oscillations by parametric stiffness excitation. In *Proceedings of the 11th International Congress of Sound and Vibration (ICSV11)*, pages 339–346, St. Petersburg, Russia, July 2004.

[20] F. Dohnal and B.R. Mace. Amplification of damping of a cantilever beam by parametric excitation. In *Proceedings of the Ninth International Conference on Motion and Vibration Control (MOVIC)*, page Paper ID 1248, Munich, Germany, September 2008.

[21] F. Dohnal and B.R. Mace. Damping of a flexible rotor by time-periodic stiffness and damping variation. In *Proceedings of the Ninth International Conference on Vibrations in Rotating Machinery (VIRM)*, pages 775–786, Exeter, UK, September 2008.

[22] F. Dohnal, W. Paradeiser, and H. Ecker. Experimental study on cancelling self-excited vibrations by parametric excitation. In *Proceedings of ASME 2006 International Mechanical Engineering Congress and Exposition (IMECE)*, page 10 pages, Chicago, IL, USA, November 2006.

[23] F. Dohnal and A. Tondl. Suppressing flutter vibrations by parametric inertia excitation. *Journal of Applied Mechanics*, (76):8 pages, 2009.

[24] F. Dohnal and F. Verhulst. Averaging in quenching by parametric stiffness excitation. *Nonlinear Dynamics*, (54):231–248, 2008.

[25] H. Ecker. Optimal phase relationship for a system with multi-location parametric excitation. In *Proceedings of Dynamics of Machines 2005*, pages 47–54, Prague, Czech Republic, Feb 2005. Institute of Thermomechanics, Academy of Sciences of the Czech Republic.

[26] H. Ecker and F. Dohnal. Dynamic stabilization of a self-excited coupled pendulum system. In *Proceedings of the 7th International Conference on Motion and Vibration Control (MOVIC)*, St. Louis, USA, August 2004.

[27] H. Ecker and A. Tondl. Supression of flow-induced vibrations by a dynamic absorber with parametric excitation. In *Proceedings of the 7th Int. Conf. on Flow-Induced Vibrations (FIV00)*, 2000.

[28] N. Eicher. *Einführung in die Berechnung parametererregter Schwingungen*. Technische Universität Berlin, TUB-Dokumentation Weiterbildung, 1981.

[29] M. Farkas. *Periodic motions*. Springer, Berlin, 1994.

[30] S. Fatimah. *Bifurctations in dynamical systems with parametric excitation*. PhD thesis, Utrecht University, Netherlands, 2002.

[31] S. Fatimah and F. Verhulst. Suppressing flow-induced vibrations by parametric excitation. *Nonlinear Dynamics*, (23):275–297, 2003.

[32] F. Gantmacher. *Matrizenrechnung Teil II, spezielle Fragen und Anwendungen*. Hochschulbücher für Mathematik. VEB Deutscher Verlag der Wissenschaften, Berlin, 1966.

[33] Y. Kuznetsov and V. Levintin. CONTENT: Integrated environment for the analysis of dynamical systems. Technical report, Centrum voor Wiskunde en Informatica, Amsterdam, Netherlands, 1997.

[34] K. Magnus and K. Popp. *Schwingungen*. Teubner Studienbücher, Mechanik, 1997.

[35] W. Magnus and S. Winkler. *Hill's equation*. John Wiley, New York, 1966.

[36] N. Maia and J. Silva. *Theoretical and experimental modal analysis*. Research studies press ltd., 1992.

[37] A.A. Mailybaev. On stability domains of non-conservative systems under small parametric excitation. *Acta Mechanica*, 154(1-4):11–33, 2002.

[38] K. Makihara, H. Ecker, and F. Dohnal. Stability analysis of open-loop stiffness control to suppress self-excited vibrations. *Journal of Vibration and Control*, 11:643–669, 2005.

[39] M. Marden. *Geometry of polynomials*. Mathematical Surveys No.3, American Mathematical Society, Providence, Rhode Island, 1966.

[40] R. Mickens. *An introduction to nonlinear oscillations*. Cambridge Press, 1981.

[41] P. Müller and W. Schiehlen. *Linear vibrations: a theoretical treatment of multi-degree-of-freedom vibrating systems*. Martinus Nijhoff Publishers, Kluwer Academic, 1985.

[42] A. Nayfeh. *Method of normal forms*. John Wiley & Sons Inc., 1992.

[43] A. Nayfeh and D. Mook. *Nonlinear Oscillations*. John Wiley, New York, 1979.

[44] N.N. MATLAB 6.5. Copyright © The Mathworks, Natick, USA, 2002.

[45] N.N. ACSL 11.8 – Advanced Continuous Simulation Language. Copyright © AEgis Technology Group, Huntsville, AL, USA, 2003.

[46] N.N. MAPLE 9.0. Copyright © Waterloo Maple Inc, 2003.

[47] H. Nowotny. Elektrodynamik und Relativitätstheorie. Skriptum zur Vorlesung, Technische Universität Wien, 2003.

[48] S. Rao. *Mechanical vibrations*. Pearson Prentice Hall, 4th edition, 2004.

[49] G. Schmidt. *Parametererregte Schwingungen*. VEB Deutscher Verlag der Wisschenschaften, Berlin, 1975.

[50] H. Schmieg. *Kombinationsresonanz bei Systemen mit allgemeiner harmonischer Erregermatrix*. PhD thesis, University Karlsruhe, Germany, 1976.

[51] A. Seyranian and A. Mailybaev. *Multiparameter stability theory with mechanical applications*, volume 13 of *A, Stability, vibration and control of systems*. A. Guran, 2003.

[52] J. Thomsen. *Vibrations and stability – Advanced theory, analysis and tools*. Springer Verlag Berlin Heidelberg New York, 2nd edition, 2003.

[53] A. Tondl. *On the interaction between self-excited and parametric vibrations*. Number 25 in monographs and memoranda. National Research Institute for Machine Design, Prague, 1965.

[54] A. Tondl. *Some problems of rotordynamics*. Chapman and Hall, London, 1965.

[55] A. Tondl. *Quenching of self-excited vibrations*. Academy of sciences Czech Republic, 1991.

[56] A. Tondl. To the problem of quenching self-excited vibrations. *Acta Technica CSAV*, (43):109–116, 1998.

[57] A. Tondl. Combination resonances and anti-resonances in systems parametrically excited by harmonic variation of linear damping coefficients. *Acta Technica CSAV*, (3):239–248, 2003.

[58] A. Tondl and H. Ecker. Cancelling of self-excited vibrations by means of parametric excitation. In *Proceeding of the 1999 ASME Design Engineering Technical Conference (DETC99)*, 1999.

[59] H. Troger and A. Steindl. *Nonlinear stability and bifurcation theory: An introduction for engineers and applied scientists*. Springer Verlag, 1991.

[60] F. Verhulst. *Nonlinear differential equations and dynamical systems*, volume 50 of *Texts in applied mathematics*. Springer-Verlag Berlin Heidelberg New York, 2000.

[61] F. Verhulst. *Methods and applications of singular perturbations*. Springer-Verlag Berlin Heidelberg New York, 2005.

[62] J. Wittenburg. *Schwingungslehre: lineare Schwingung, Theorie und Anwendung*. Springer Verlag, 1996.

[63] T. Yamamoto and A. Sato. *Memoirs of the Faculty of Engineering*, chapter On the vibrations of 'summed and differential types' under parametric excitation. Nagoya University, Japan, 1970.

A. Routh-Hurwitz theorems

The classical stability theorem of Routh-Hurwitz for real-valued polynomials is found for instance in [32]:

Theorem A.1. *Consider the characteristic equation*

$$|\lambda \mathbf{I}_n - \mathbf{A}| = \lambda^n + a_1 \lambda^{n-1} + \ldots + a_{n-1}\lambda + a_n = 0$$

determining the n eigenvalues λ of a real $n \times n$ square matrix $\mathbf{A}$, where $\mathbf{I}$ is the identity matrix of dimension n and $a_i \in \mathbb{R}$. Then the eigenvalues all have negative real parts if and only if the corresponding Routh-Hurwitz determinants satisfy

$$\Delta_1 > 0, \ \Delta_2 > 0, \ldots, \ \Delta_n > 0,$$

where

$$\Delta_k = \begin{vmatrix} a_1 & 1 & 0 & 0 & 0 & 0 & \cdots & 0 \\ a_3 & a_2 & a_1 & 1 & 0 & 0 & \cdots & 0 \\ a_5 & a_4 & a_3 & a_2 & a_1 & 1 & \cdots & 0 \\ \vdots & \vdots & \vdots & \vdots & \vdots & \vdots & \ddots & \vdots \\ a_{2k-1} & a_{2k-2} & a_{2k-3} & a_{2k-4} & a_{2k-5} & a_{2k-6} & \cdots & a_k \end{vmatrix}.$$

Especially for a polynomial of order four

$$\lambda^4 + a_1 \lambda^3 + a_2 \lambda^2 + a_3 \lambda + a_4 = 0$$

these conditions demand that

$$\Delta_1 = a_1 > 0,$$

$$\Delta_2 = \begin{vmatrix} a_1 & 1 \\ a_3 & a_2 \end{vmatrix} = a_1 a_2 - a_3 > 0,$$

$$\Delta_3 = \begin{vmatrix} a_1 & 1 & 0 \\ a_3 & a_2 & a_1 \\ 0 & a_4 & a_3 \end{vmatrix} = a_1 a_2 a_3 - a_3^2 - a_1^2 a_4 > 0, \qquad (A.1)$$

$$\Delta_4 = \begin{vmatrix} a_1 & 1 & 0 & 0 \\ a_3 & a_2 & a_1 & 1 \\ 0 & a_4 & a_3 & a_2 \\ 0 & 0 & 0 & a_4 \end{vmatrix} = a_4 \Delta_3 > 0 \Rightarrow a_4 > 0.$$

The Δ_1- and Δ_3-conditions are the most restrictive ones. But of course the Δ_2- and Δ_4-condition have to be fulfilled, too.

The Routh-Hurwitz theorem as stated above is only applicable if the coefficients of the examined polynomial are all real-valued. Sometimes it might be necessary to determine the stability of a polynomial with complex coefficients. One of the first works concerning the number of zeros with negative real parts is [3]. Later this results were presented in [39, pp179] and [32, pp220] as the *extended* Routh-Hurwitz theorem:

Theorem A.2. *Given the polynomial having no pure imaginary zeros*

$$F(z) = z^n + (a_1 + jb_1) z^{n-1} + \cdots + (a_n + jb_n),$$

where $a_j, b_j \in \mathbb{R}$ *and* $z \in \mathbb{C}$, *let us form the determinants* $\Delta_1 = A_1$ *and*

$$\Delta_k = \begin{vmatrix} a_1 & a_3 & a_5 & \cdots & a_{2k-1} & -b_2 & -b_4 & \cdots & -b_{2k-2} \\ 1 & a_2 & a_4 & \cdots & a_{2k-2} & -b_1 & -b_3 & \cdots & -b_{2k-3} \\ \vdots & \vdots & \vdots & \cdots & \vdots & \vdots & \vdots & \cdots & \vdots \\ 0 & 0 & 0 & \cdots & a_k & 0 & 0 & \cdots & -b_{k-1} \\ 0 & b_2 & b_4 & \cdots & b_{2k-2} & a_1 & a_3 & \cdots & a_{2k-3} \\ 0 & b_1 & b_3 & \cdots & b_{2k-3} & 1 & a_2 & \cdots & a_{2k-4} \\ \vdots & \vdots & \vdots & \cdots & \vdots & \vdots & \vdots & \cdots & \vdots \\ 0 & 0 & 0 & \cdots & b_k & 0 & 0 & \cdots & a_{k-1} \end{vmatrix}$$

for $k = 2, 3, \ldots, n$, *with* $a_j = b_j = 0$ *for* $j > n$. *Let us denote by* p *and* q *the number of zeros of* $F(z)$ *in the half-planes* $\mathcal{R}\{z\} > 0$ *and* $\mathcal{R}\{z\} < 0$ *respectively. If* $\Delta_k \neq 0$ *for* $k = 1, 2, \ldots, n$, *then*

$$p = \mathcal{V}(1, \Delta_1, \Delta_2, \ldots, \Delta_n),$$

and

$$q = \mathcal{V}(1, -\Delta_1, \Delta_2, \ldots, (-1)^n \Delta_n).$$

Hence, a polynomial has only roots with negative real parts if all determinants are positive, $q = 0$. Especially for a complex polynomial of order two

$$\lambda^2 + (a_1 + jb_1)\lambda + (a_0 + jb_0) = 0$$

the necessary and sufficient conditions for roots with negative real parts are

$$\Delta_1 = a_1 > 0$$

$$\Delta_2 = \begin{vmatrix} a_1 & 0 & -b_0 \\ 1 & a_0 & -b_1 \\ 0 & b_0 & a_1 \end{vmatrix} = a_1^2 a_0 - b_0^2 + a_1 b_1 b_0 > 0 \quad\quad (A.2)$$

Especially if a real polynomial of order four can be transformed into a complex polynomial of order two, the number of the necessary conditions to be considered is reduced as one can see from comparing (A.1) with (A.2).

B. Trigonometric decomposition theorems

With the help of trigonometric addition formulas, see [2, pp.351], the arising products of the harmonic terms on the right hand sides of (3.8, 3.32, 3.54, 4.3, 5.3, 6.4) can be rearranged as a sum of basic trigonometric terms:

$\Theta_{ii}, \sigma:\qquad s_i s_i = \frac{1}{2}\left[1 - \cos(2\varpi_i t)\right],$

$\qquad\qquad c_i s_i = \frac{1}{2}\sin(2\varpi_i t),$

$\qquad\qquad c_i c_i = \frac{1}{2}\left[1 + \cos(2\varpi_i t)\right],$

$\Theta_{ij}:\qquad s_i s_j = \frac{1}{2}\left[\cos(\varpi_i - \varpi_j)t - \cos(\varpi_i + \varpi_j)t\right],$

$\qquad\qquad c_i s_j = \frac{1}{2}\left[\sin(\varpi_j - \varpi_i)t + \sin(\varpi_i + \varpi_j)t\right],$

$\qquad\qquad c_i c_j = \frac{1}{2}\left[\cos(\varpi_i - \varpi_j)t + \cos(\varpi_i + \varpi_j)t\right],$

$Q_{iin}:\qquad c_i \cos nt\, c_i = \frac{1}{4}\left[\,\cos(2\varpi_i - n)t + 2\cos nt + \cos(2\varpi_i + n)t\right],$

$\qquad\qquad c_i \cos nt\, s_i = \frac{1}{4}\left[\sin(2\varpi_i - n)t + \sin(2\varpi_i + n)t\right],$

$\qquad\qquad s_i \cos nt\, s_i = \frac{1}{4}\left[-\cos(2\varpi_i - n)t + 2\cos nt - \cos(2\varpi_i + n)t\right],$

$Q_{ijn}:\qquad c_i \cos nt\, c_j = \frac{1}{4}\left[\,\cos(\varpi_i + \varpi_j - n)t + \cos(\varpi_j - \varpi_i + n)t\right.$
$\qquad\qquad\qquad\qquad\left. + \cos(\varpi_j - \varpi_i - n)t + \cos(\varpi_i + \varpi_j + n)t\right],$

$\qquad\qquad c_i \cos nt\, s_j = \frac{1}{4}\left[\,\sin(\varpi_i + \varpi_j - n)t + \sin(\varpi_j - \varpi_i + n)t\right.$
$\qquad\qquad\qquad\qquad\left. + \sin(\varpi_j - \varpi_i - n)t + \sin(\varpi_i + \varpi_j + n)t\right],$

$\qquad\qquad s_i \cos nt\, s_j = \frac{1}{4}\left[-\cos(\varpi_i + \varpi_j - n)t + \cos(\varpi_j - \varpi_i + n)t\right.$
$\qquad\qquad\qquad\qquad\left. + \cos(\varpi_j - \varpi_i - n)t - \cos(\varpi_i + \varpi_j + n)t\right],$

$P_{iin}\qquad c_i \sin nt\, c_i = \frac{1}{4}\left[-\sin(2\varpi_i - n)t + 2\sin nt + \sin(2\varpi_i + n)t\right],$

$\qquad\qquad c_i \sin nt\, s_i = \frac{1}{4}\left[\cos(2\varpi_i - n)t - \cos(2\varpi_i + n)t\right],$

$\qquad\qquad s_i \sin nt\, s_i = \frac{1}{4}\left[\,\sin(2\varpi_i - n)t + 2\sin nt - \sin(2\varpi_i + n)t\right],$

$$P_{ijn}: \quad c_i \sin nt \, c_j = \tfrac{1}{4}\left[-\sin(\varpi_i + \varpi_j - n)t + \sin(\varpi_j - \varpi_i + n)t\right.$$
$$\left. - \sin(\varpi_j - \varpi_i - n)t + \sin(\varpi_i + \varpi_j + n)t\right],$$

$$c_i \sin nt \, s_j = \tfrac{1}{4}\left[\ \cos(\varpi_i + \varpi_j - n)t - \cos(\varpi_j - \varpi_i + n)t\right.$$
$$\left. + \cos(\varpi_j - \varpi_i - n)t - \cos(\varpi_i + \varpi_j + n)t\right],$$

$$s_i \sin nt \, s_j = \tfrac{1}{4}\left[\ \sin(\varpi_i + \varpi_j - n)t + \sin(\varpi_j - \varpi_i + n)t\right.$$
$$\left. - \sin(\varpi_j - \varpi_i - n)t - \sin(\varpi_i + \varpi_j + n)t\right],$$

with $n = 1, 2, ..., N$, and N is number of respected frequencies in the Fourier series in (3.53). For a simple system with two modes we obtain $(3 + 9 \times N)$ different periods.

Südwestdeutscher Verlag
für Hochschulschriften

Wissenschaftlicher Buchverlag bietet
kostenfreie
Publikation
von
Dissertationen und Habilitationen

Sie verfügen über eine wissenschaftliche Abschlußarbeit zu aktuellen oder zeitlosen Fragestellungen, die hohen inhaltlichen und formalen Anspruchen genügt, und haben **Interesse an einer honorarvergüteten Publikation?**

Dann senden Sie bitte erste Informationen über Ihre Arbeit per Email an:
info@svh-verlag.de.

Unser Außenlektorat meldet sich umgehend bei Ihnen.

Südwestdeutscher Verlag für Hochschulschriften
Aktiengesellschaft & Co. KG

Dudweiler Landstr. 99
D – 66123 Saarbrücken

www.svh-verlag.de

Printed by Books on Demand GmbH, Norderstedt / Germany